Classics in Environmental Studies

Series **Environmental Studies**

NICO NELISSEN,
JAN VAN DER STRAATEN &
LEON KLINKERS (EDS.)

Classics in Environmental Studies

An Overview of Classic Texts in Environmental Studies

INTERNATIONAL BOOKS, 1997

The cover photo shows the pioneer ecosystem in a part of the newly formed polder Oostvaardersplassen in the Netherlands.

© International Books, see for the copyrights of each seperate article page 10

CIP-Data Koninklijke Bibliotheek, The Hague

Classics

Classics in Environmental Studies : An Overview of Classic Texts in Environmental Studies / Nico Nelissen, Jan van der Straaten & Leon Klinkers (eds.). – Utrecht : International Books.
ISBN 90-6224-963-9 cloth
ISBN 90-6224-973-6 paperback
NUGI 661/825

Keywords: Environmental Studies, Philosophy, Environment

Cover design: Marjo Starink, Amsterdam, The Netherlands
Cover photo: V.L. Wigbels, Biofaan, Lelystad, The Netherlands

International Books, A. Numankade 17, 3572 KP Utrecht,
The Netherlands, +31-30-2731840

Foreword

One sometimes wonders whether the author of a book which we now classify as a 'Classic in Environmental Studies' fully understood the potential of his or her work at the time it was drafted. One does not sit down to write a 'classic', I guess. It is up to others -in retrospect- to define a book as such. While finalizing 'Silent Spring' in 1962, Rachel Carson wrote to a good friend:

> *'The beauty of the living world I was trying to save has always been uppermost in my mind - that, and anger at the senseless, brutish things that were being done . . . Now I can believe I have at least helped a little.'*

She could have hardly understood the enormous impact that her work was going to have.

The book that lies before you brings together many of these classics which shook us awake, made us aware of the beauty of the web of life and the way we were destroying it. Sometimes pure reports of scientific evidence, sometimes cries from the heart ... Bringing so many of these together in one volume, this book will provide students and professionals in the field of environmental studies (but hopefully also in other sciences) with a concise history of the evolution in the knowledge and understanding of environmental deterioration.

Classics in Environmental Studies is a random indication of cutting edge wisdom. As we learn to better understand the physical, societal, and political complexities of sustainable development, both domestically and internationally, there will be more 'classics' in the future. I express the hope that this volume will encourage readers to put this wisdom into practice in their daily work and to enlighten us with their own creative thinking.

Margaretha de Boer,
Minister for Housing, Spatial Planning and the Environment, the Netherlands

Table of Contents

Part V. The Current State of Affairs

Part VI. The Future

Copyrights

Preface

After many years of teaching, we have come to the conclusion that environmental studies has reached maturity. Moreover, it is becoming increasingly clear that this field of study already has a history of many decades, if not centuries. In the bachelor's and master's degree programmes, however, very little attention is given to the history of environmental sciences. Often it is said that it is a new discipline that came into existence in the 1960s. Consequently, all the attention in courses is focused on current problems in the environment, concepts which help us to interpret this situation, the causes of the environmental crisis, and strategies for a possible solution. It has, however, become apparent that it is essential that students be made aware of how the discipline has developed and what the cornerstones of the discipline are. Students should master the classical authors and their publications. It is useful for students (and, of course, practitioners as well) in the field of environmental studies to be familiar with the roots of their discipline.

One way to accomplish this would be to go to the original textbooks containing these classic studies. Reading all this literature would, however, be a task of Herculean proportions. Instead, we decided to make a selection of classical books and select those sections that provide a representative idea of the basic insights of the studies. The result is a compilation of fragments of studies that can be considered as classic books and documents in the field of environmental studies.

With this idea in mind, we needed to find existing books with a similar format. There are some books which use the same approach and have the same purpose as this one, but differ, however, in one essential property. There are, for example, books which aim to present a concise history of the development of environmental studies. They do so by presenting a selection of fragments of the classical studies through a short description of the studies. Our book, however, tries to 'let the classic studies speak for themselves', in that only a short introduction and brief comments are given on each study which is followed by an actual excerpt of the given study to show how environmental issues are analyzed in that particular piece.

In writing this book, we were assisted by several individuals. Carola Fasol served as our 'library liaison'. She collected all the classic studies and passed them on to the editors. She checked all bibliographical data and assisted the editors as a secretary. Rob van Krimpen and Marianne Sanders of the Tilburg University Language Center were helpful in correcting the English texts. ECNC (European Centre for Nature Conservation) supported the project in financial and logistical matters. The Foundation for the Protection of the Environment (Stichting Bescherming Milieu) supported the project financially. Without their help, the composition and publication of the book would not have been possible. We would like to thank them all for their assistance.

Tilburg, October 1996
NN, JvdS, LK

General Introduction

J.M. Shafritz & A.C. Hyde start their *'Classics of Public Administration'* (Pacific Grove, 1992) with the proclamation: 'Be assured—the editors are not so bold as to assert that these are 'the' Classics of Public Administration'. The same comment could be made of our 'Classics in Environmental Studies.' The field is so diverse that just one list of classics is impossible to compile. We have attempted to make a list of studies that from our point of view can be considered as classics in the field of environmental studies. We define environmental studies as the interdisciplinary field of studies concerned with problems in the relationships between man, society, and environment. The basic ideas for this discipline originated in different disciplines: biology, ecology, physics, chemistry, sociology, psychology, law, administrative and political science, etc. In focusing on environmental problems, an interdisciplinary science developed that is now recognized as a domain in academic studies. In the last decade, the discipline has matured enormously. Nearly every (if not every) university in the western world (and even in other parts of the world) has at least one course in environmental studies. Many of them have complete programmes both at the bachelor's and master's level.

What are 'environmental studies'?

There are many disciplines that are concerned with the environment, but not all can be considered environmental studies. There are numerous studies on environmental problems in which the subject matter is treated from a monodisciplinary point of view. These studies are not included in our category of environmental studies. We use the term environmental studies to refer to texts in which environmental problems are studied from a more or less interdisciplinary point of view. The dividing line between disciplines and disciplinary approaches is difficult to make and sometimes it can even be said that some studies, for example, Pigou and Mishan, are in themselves disciplinary studies of economics. We have included, however, the contributions of these authors as we are convinced that their influence on many other scholars, even outside the realm of economics, has been very substantial. Nevertheless, we believe that the whole group of selected studies can be considered as being interdisciplinary. They have all contributed conspicuously to our understanding of environmental problems. They can be considered 'classics', a term which requires further discussion.

What are 'classics' in environmental studies?

Which studies can be considered 'classics' in environmental studies? The studies themselves are not presented by the authors under the denominator 'classics'. The term 'classic' is a qualification given by the professional elite. In the network of environmental scientists, some studies are often mentioned or referred to. These studies are often cited and the modern approach is strongly influenced by the paradigms of these books. Of course, owing to different views on this field, some studies are considered to be more classic than others. But the scientific community agrees on the importance of some studies and considers these to be classic. This means that the criterion of reputation does play a very important role in determining whether a study is a classic. The selection procedure took place with this fact in mind.

The selection procedure

How does one make a selection from the body of 'classics in environmental studies'? In this case, the following method was used. Initially, each of the editors compiled a preliminary list of classics based on his own knowledge. Secondly, we compared our lists. Those publications which were considered by all of us to be classics were retained. Next, we sent this list to colleagues in the Netherlands and requested that they add classics which were, in their opinion, worthy of mention and to delete titles which they did not consider to be classic. This resulted in a new list which we used as the 'intersubjective' result of the definition of 'classics in environmental studies'. We are aware that this procedure can be criticized, but we are convinced that every other procedure would have had its own drawbacks and would have proven to be less satisfactory than this one.

The structure of the book

This book is divided into several distinct parts. The structure of the book is based on a chronological principle. This means that the classics have been ordered in categories related to their publication date. It is a historical approach, as this provides the reader with greater insight into the development of theories in the course of time as a response to societal problems. In this context, the approaches of the authors are discussed. Six different stages were distinguished.

Part I. The Beginning

In this section, studies were selected which could be seen as *the* starting points of the discipline. The studies are all very different. The first one, for example, emphasizes the role of population growth in relation to the wealth of the population. Another focuses on the evolution of species through the principle of the survival of the fittest. In this period, the study of environmental issues did not yet exist. The foundations for the discipline were being laid. These studies became a source of inspiration and a point of departure for later scientists.

Part II. The Revival

In the 1950s and 60s, with the expansion of environmental issues, some authors began to analyze environmental issues from a more systematic point of view. These authors were deeply concerned about the growth of environmental problems and the increasing decay of the natural environment. They were searching for the historical roots of the environmental crisis and emphasized the costs of economic growth. They were convinced of the necessity of radical change in the social, economic, and political order of society.

Part III. The Bestsellers

In the 1970s, environmental studies became popular, as a result of a number of publications which reached out to a readership beyond the academic group. Among them was the Report to the Club of Rome (Meadows publication) *The limits to growth*. This was the starting point of a series of studies initiated by the Club of Rome. In this period, the United Nations became aware of the impact of environmental problems on the future of the planet. The first significant international conference on the environment took place in Stockholm in 1972. This proved to be an important step in the development of national (and international) policy programmes aimed at reducing environmental problems.

Part IV. The Eighties

In the 1980s, more knowledge on environmental problems became available; in addition to this increased knowledge, however, there were also more policy instruments to deal with the problems. Of great importance was the publication of the Brundtland Report, a report by the World Commission on Environment and Development. This document, entitled *Our Common Future*, put forward the now well-known concept of 'sustainable development'. Although this concept was first introduced by the International Union for the Conservation of Nature in *The World Conservation Strategy* in 1980, it was the Brundtland Commission Report which put this issue on the international agenda. This concept became the central concept for numerous environmental programmes, at the local, regional, national, and international level. In every sector in society, the idea of sustainability was introduced, and we were confronted with sustainable agriculture, sustainable tourism, sustainable industry, sustainable traffic, etc.

Part V. The Current State of Affairs

What is the current state of affairs, the actual body of knowledge in environmental studies? Lester Brown from the World Watch Institute produces 'the state of the world' each year, with a heavy emphasis on environmental issues. Some people argue that we are now facing the first world revolution, that is to say, that radical changes are taking place on a global basis. The need for this global revolution is urgent, as we are currently acting beyond the limits. We are confronted with a global collapse; we may, however, achieve societal development in the direction of a sustainable future. On a world-wide scale, we are developing programmes to reach sustainability. This is being strived for by a number of national governments and also by supranational institutions such as the European Union.

Part VI. The Future

What will the future bring? The need for a balance has been widely formulated. National and international governments have formulated and are implementing environmental policies. Significant in this regard is the Conference of the United Nations held in Rio in 1992. It was there that an agenda for the 21st century was formulated. Public opinion on matters related to the environment is moving in the direction of environmental awareness and positive changes in environmental behaviour. It is not only the government that is active: industry is changing its course too. What was once a reactive strategy to the environment is changing into a proactive strategy. Most of these programmes are for the short and middle terms. In some countries, committees have been set up to look into long-term developments in an attempt to move social systems in the direction of a sustainable future.

Introductory remarks

Each part of the book is introduced briefly by the editors. A general description is given of the Zeitgeist as well as a summary of the contents of the studies included in each given section.

Conclusions

At the end of the book a number of conclusions are given and the perspective of the environment and of environmental studies is discussed.

Part I

THE BEGINNING

Introduction to Part I: The Beginning

The period in which the writings included in Part I were published covers almost 150 years. In this period, the social, cultural, economic, and political conditions that influenced these classics were subject to many changes.

The French and the American Revolutions

The period of the French and the American Revolutions was not without turmoil. The period before the French Revolution is often considered as a time of enlightenment, a time in which traditional hierarchical relationships were no longer seen as a sound basis for societal and individual behaviour. The structures that were then present had been developed, and were being maintained, to guarantee the vested interest of elites. The French Revolution can be seen as a crystallisation point in this process of change. In the New World, a similar process took place and culminated in the American Revolution. Rationalisation, freedom, and equality were central issues in these revolutions. These principles were apparent in the general belief in progress that emerged. These basic principles have influenced the ideology of some important schools of societal and philosophical thinking such as socialism and liberalism.

Imperialism

Both the American and French Revolutions coloured the early 1800s. Many European countries (the United Kingdom, France, and other European powers) gathered wealth in the colonies. They were able to guarantee a constant flow of cheap raw materials and resources from their colonies into their national economies. This imperialistic behaviour was, of course, a far cry from the viewpoint of enlightenment. These countries were not willing to give up their dominant global position and they used armies, slavery, trade, and exploitation to sustain these global processes. Even now we can see the impact of this period in the relationship between first and third world countries. The latter export mainly cheap primary goods to the developed countries which set the prices. Indeed, these countries are still economically dependent on the first world countries.

Industrial Revolution

Rationalisation, mechanisation, and automisation of the labour process and, consequently, the regrouping of the work-force into larger production units was made possible by industrialisation. The industrial revolution brought about some new phenomena. The degree of urbanisation increased and cities became of great importance to the economy. Centralisation and scaling-up stimulated the growth of the cities. Infrastructures (roads, railways, ports and rivers) were remodelled or built for the transport of raw materials and freight. Even new services and products came into existence. Local economies became regional, national, and sometimes global economies. The volume of freight transport boomed and the related speed of communication increased dramatically. Everything seemed to be undergoing rapid development. One should not overlook the crucial impact that the vast majority of so-called modern developments had in the second part of the nineteenth century. Steam engines, trains, combustion engines, electricity, telephones, steel mills and coal mines, and radiation changed nineteenth century society completely in a few decades and cleared the way for the industrial era.

Criticism concerning industrialisation

The rapid and massive industrialisation that took place did not benefit all parts of society. Industrial workers often lived in miserable circumstances. The new developments in economic processes were not in line with the basic principles of the Enlightenment; traditional power relations were more or less replaced by new industrial power structures. Marx and Durkheim were vocal critics of these new developments. Capitalistic exploitation, social disruption, child labour, and alienation were central topics in Marx's analysis. Furthermore, many other factors—such as the abatement of the old regimes, the forming of national states, and general social and economic turbulence—played a significant role in international relationships. Although it is difficult to explain the beginning of the First World War, all these developments had an impact. In Russia, the Russian Revolution was propagated as a reaction to the capitalistic order in which inequality and class conflicts lead to the dominance of one class over the other. The new socialist society was to rid the people of this ordeal.

World Wars

The First World War did not decrease the instability between western nations. Mass unemployment, new nationalistic ideas, in particular, in Germany, and the heavy financial burdens placed on Germany by the Allied forces in the Versailles Treaty created a climate in which military confrontation was seen as the solution to the principal problems. The complete destruction of society resulting from the Second World War required the European countries to rebuild and reconstruct. National identity and the role of Europe demanded the re-shaping of most, if not all, fundamental structures.

The Cold War

The post-war period was the period of the rebuilding of Western Europe and the need for international co-operation was met by founding new international institutions. Relations between the USSR and the US deteriorated and led to the cold war.

An overview of this period

The events of this period are so diverse that it is very hard to find a central theme and therefore an all-embracing label. To choose 'Revolution' as the central theme for this period is very tempting as it was a time of rapid and violent changes. At the start of this period, two kinds of revolution took place: the Industrial Revolution, symbolised by the invention of the steam engine, and the Political Revolution that resulted in the American and the French Revolutions. Both revolutions are still in progress in our time: everywhere in the world people are striving for industrialisation and democracy. Western science and western political ideology dominate the current era. Everything is based on the idea of progress.

The shadowside of progress

Progress, achieved by ongoing large-scale production, an increase in production volume, and an increase in transport and new production methods, has taken its toll on the natural environment. At first, local effects became tangible. The emission of particles of soot covered nature with a grey blanket. The moth population was affected by the pollution of the area: in industrial areas, there were many more black specimens than in less polluted areas. This is a form of evolutionary adjustment. Many vulnerable animal and plant species in these areas disappeared. Rivers and lakes were polluted. Forests were cut down and replanted with types that were better for industrial use. The planting of only one kind of tree reduced the diversity of original species. In fen areas, peat was cut and the mining of pit and brown coals flourished to supply industry with sufficient fuel. Vast areas were dehydrated for the use of farming or buildings. Slowly, landscapes changed and regional effects started to become evident. No global effects could be found in the 1850s because of the nature of production in this era. The natural balance had not yet been disturbed. The emissions of carbon-dioxide could still be compensated by the available oxygen-producing woods. Nature's buffer capacity, however, is not without limits, and carbon-dioxide concentrations have been increasing since the beginning of this century.

Deterioration of the quality of the environment

Slowly but relentlessly the quality of the natural environment deteriorated. The effects at the local level were soon followed by effects at the regional level. Ecosystems were thrown out of balance and endangered plant and animal species that are interdependent. Because of ongoing industrialisation and consumption, ecosystems that are of influence on a global level were eventually affected resulting in the greenhouse effect and the related rise in temperature and the sea level, the holes in the ozone layer, and the increasing danger of skin cancer. Step by step, the natural environment was being destroyed. More and more people began to feel that the negative effects on the quality of our natural environment outweighed the benefits of growing prosperity.

The first classics

The classics presented in Part I are classics in every possible way. Not only are they classics in the field of environmental studies, but they are also considered basic texts in other new disciplines. The works of Malthus: *An Essay on the Principles of Population* and *The Law of Diminishing Returns* are still works of authority within economics. Darwin's work: *On the Origin of Species by Means of Natural Selection or the Preservation of Favoured Species in the Struggle for Life* is a landmark in the field of biology and in particular the field of evolutionary biology and ecology.

These books have often been eye-openers for different groups in society because of their views on many different issues. As Pigou asks himself: Is the Gross National Product the right measure to calculate welfare and how does this measure include negative external effects such as the depletion of natural resources, the influence of pollution and contamination to the public health, etc.? Reid gives us an ecological frame of reference and shows us how everything is related in the natural environment. One could speak of a 'web of life' in which organisms, plants, food pyramids, and ecosystems are interrelated. Every link (every animal and every plant) is part of a larger whole and has an influence on that larger whole. There is, however, an organism that is inclined to think only in terms of its constituent parts and, consequently, loses its relation with the larger whole, namely, the human being. Carson speaks of one of the derailments of human beings: the use of chemicals. Many were shocked to see the persistence of pesticides such as DDT that appeared not only in the eggs of penguins in areas thousands of miles from those areas in which the pesticides were used, but also in human breast milk.

Limitations of the first classics

These classics do, however, have some limitations. The studies are mainly mono-disciplinary, are in need of a policy reference, often include no political statements, and are mainly of a scientific nature. But they are fundamental to modern environmental studies in the sense that the issues and statements they introduced are still topics of current policy. Some issues and statements are no longer being disputed; they have been accepted as starting points of the policies that are to be implemented. Some others, such as birth control, are still tricky subjects.

Malthus: An Essay on the Principle of Population

At the end of the 18th century, Malthus' *An Essay on the Principle of Population; or a View of its Past and Present Effects on Human Happiness: with an inquiry into prospects respecting the future removal or mitigation of the evils which it occasions* (1798) was published in England. It is an alarming book in the sense that it makes clear that natural resources are finite and that population growth is an important factor in the use of these resources. The book is based on the ideas of the Law of Diminishing Returns which was a generally accepted theory in England in the second part of the eighteenth century and the beginning of the nineteenth century. The Law of Diminishing Returns has its roots in the agricultural society of England of that period. The Enclosure Act facilitated the enclosure of common land by landed nobility. In the second half of the eighteenth century, England had a sharp increase in the size of its population. However, corn production did not increase in volume, leading to rising prices for corn and, consequently, starvation for many members of the growing population. As the Enclosure Movement initially used the most fertile soils and later on the less fertile ones, the marginal yields of agriculture declined. Malthus believed this agricultural development was relevant to all sectors of society.

When we take the Law of Diminishing Returns as a starting point, the conclusion with regards to the population is rather clear. If there are no checks on birthrates, the human population will increase up to, and including, the point of starvation. The following topics are relevant:
- Malthus was of the opinion that there are limited possibilities regarding the use of natural resources. Natural resources can only support a limited population.
- There is a point of morality which was given considerable attention in the ni-

neteenth century. Are human beings dependent on the conditions of nature or can they change these conditions by personal interference, for example, in the field of birth control? So far the Roman Catholic church has rejected all forms of birth control.

– Technological development did not play any role in the ideas of Malthus. Malthus assumed that technology would only have a slight or modest influence on the level of production. The Malthusian approach is a static one.

The basic shortage of resources in the economic process and the inadequacy of technological development to make a substantially higher level of resources available to society are major topics in Malthus' theories. His view is mainly based on an agricultural society in which the use of resources still had the character of a flow. The increased use of stock resources in the process of the Industrial Revolution generated a completely new starting point. This is the reason why Marshall, one of the founding fathers of neo-classical economics, could argue in 1890 that Malthus should not be blamed for his negative viewpoint on economic development since he could not foresee the new developments resulting from the use of steam engines and the availability of steel and coal. On the other hand, Malthus was not aware of 'modern' environmental problems, in particular, the relationship between ecology and economy recently discussed in The Report to the Club of Rome and in the Brundtland Report *Our Common Future*.

By placing birth control as a form of self-determination in the core of his argument, Malthus is a typical representative of the Enlightenment. By doing so, he was challenging the dominant ideas in Christian circles where every form of self-determination was excluded. This controversy still exists in underdeveloped Catholic countries.

Darwin: On the Origin of Species

Darwin's book *On the Origin of Species by Means of Natural Selection, or the Preservation of Favoured Races in the Struggle for Life*, was first published in 1859. While on board the H.M.S. Beagle, Darwin was struck by certain facts in the distribution of the organisms inhabiting South America and in the geological relations of present to extinct inhabitants of the continent and its islands. When he came home in 1837, he attempted to shed some light on the origin of species by patiently accumulating and reflecting on all sorts of facts which could possibly have a bearing on it. And as he said: 'Although much remains obscure, and will long be obscure, I can entertain no doubt, after the most deliberate study and dispassionate judgement of which I am capable, that the view of most naturalists until recently entertained, and which I formerly entertained—namely, that each species has been independently created—is erro-

neous. I am fully convinced that species are not immutable; but that those belonging to what are called the same genera are linear descendants of some
other and generally extinct species, in the same manner as the acknowledged
varieties of any one species are the descendants of that species. Furthermore, I
am convinced that natural selection has been the most important but not the
exclusive, means of modification.' The influence of the Enlightenment on Darwin's opinion brought him to new viewpoints about the Creation and the role
of the Creator. He was fully aware that his ideas were at odds with traditional
thinking and he therefore delayed publication considerably.

Darwin's study started a scientific revolution. He laid the foundation for
evolutionary biology and ecology. Darwin's influence cannot be overlooked in
the field of ecological economics which has recently emerged. This is caused by
the fact that in ecological economics, the relationship between ecology and
economy has been newly defined. He gave us insight into the time needed for a
particular species to develop, how certain plant and animal species have adjusted over time and have become dependent on one another, and how vulnerable ecosystems are. Only in the second half of this century is attention being
given to these interdependent relations and not just to its separate constituent
elements, but also to its influence on the whole ecological system.

Pigou: The Economics of Welfare

In his book *The Economics of Welfare* (1920), Pigou raises fundamental questions in the field of welfare, national dividend, and the distribution of that dividend among different groups. Special attention is given to the definition of the
national dividend, which was given the name GNP in later periods, and the
meaning of that definition for the measurement of welfare. Part I is entitled
Welfare and the National Dividend; Part II, the Size of the National Dividend
and the Distribution of Resources among Different Uses; Part III, the National
Dividend and Labour; Part IV, the Distribution and the National Dividend.

Environmental issues were embedded in this general framework. Environmental issues were seen as a problem when considering the relation between
national dividend and an optimal allocation of production factors. Environmental economists often pay attention to Pigou, which might give us the idea
that Pigou was an environmental economist 'avant la lettre'. By doing so, the
relative position of Pigou in this period is overlooked. Pigou was not dealing
with environmental issues, but with the measurement of welfare in which environmental issues could play a role. This is in line with the low profile which
was given to the environmental ideas of Pigou in economic textbooks between
1920 and 1965.

The significance of Pigou is found in his early recognition of the shortcomings of a market economy when certain issues come up for discussion. The concept of negative external effects made it possible to analyse environmental problems from an economic point of view long before they became an important societal problem.

In the Pigovian approach, negative external effects have to be defined by authorities which allows them to monetarise these effects. An economic tax should be used to shift the burden of these social costs to the polluting industry, restoring optimal allocation of production factors. In reality, this approach has some inherent problems. In many cases, the impact of polluting activities on nature and the environment are uncertain. Furthermore, authorities are not able to calculate the correct level of the environmental costs as pollution often does not have a price and therefore it is not possible to determine the correct level of the levy.

Reid: The Sociology of Nature

Reid's book, *The Sociology of Nature* (1962), is mainly a compilation of examples of how plant and animal species need each other. At the same time, it has a codificational function in the sense that it introduces the ecological frame of reference. It is one of the first books that takes an ecocentric view, rather than an anthropocentric view, of the environment in which man and nature are equal partners.

The purpose of Reid's book is to outline a complicated system and sketch, imperfectly and selectively, a huge and varied panorama. Humans have difficulty giving weight to both the constituent parts and to the whole of an ecosystem. Nevertheless, it is important that an attempt in this direction be made, for only then may we hope to grasp that attribute of fundamental unity which is the very essence of the natural world. Only then can we begin to apprehend the fact that each single phenomenon of nature has no meaning, no existence, except by virtue of its relationship with all the other phenomena; that the beauty of the panorama, its magnificance, depends on this realization.

The principle of dependence governs the lives of all creatures. Even the struggle for existence is an organised struggle, and in addition to keen and sometimes bitter competition, there is also co-operation between animals and plants, between one animal species and the other, and finally between animals of the same species. We know now, and we must studiously pay heed to the knowledge, that it is only through intimate understanding of the web that binds all life on this earth, through so ordering things that man lives in harmony with nature, that disaster can be averted and the survival of humanity assured.

Carson: Silent Spring

In *Silent Spring* (1962), Rachel Carson describes the effects of exposure resulting from the indiscriminate use of chemicals. She describes, using many examples, how pesticides and insecticides are applied almost universally to farms, forests, gardens, and homes with scant regard to the consequent contamination of our environment and the widespread destruction of wildlife. Today's world has to face not an occasional dose of poison which has accidentally got into some article of food, but a persistent and continuous poisoning of the whole human environment. The 'control of nature' is a phrase conceived in arrogance, born in the Neanderthal age of biology and philosophy, when it was supposed that nature exists for the convenience of man. The concepts and practices of applied entomology date for the most part from the Stone Age of science. It is our alarming misfortune that so primitive a science has armed itself with the most modern weapons, and that in turning them against the insects, it has also turned them against the earth.

The book was a wake-up call for many and led to the foundation of a Senate Committee for Environmental Affairs by John F. Kennedy in the United States. The excesses in the use of chemicals were evident. Everywhere, the US Government used these chemicals; evidence to this effect could be easily found. The effects did not confine themselves to the short term. The concentration of toxic substances is highest near the top of the food chain, e.g., in predatory birds and the bodies of humans. The poison had an immediate effect on the thickness of the eggshells of predatory birds which led to less protection for the foetus and a greater mortality rate for young birds. It has been shown that toxic chemicals in humans influence the quality of reproduction cells. Eventually the use of these chemicals will not only affect the insects, its primary target, but the entire food chain, and therefore mankind. Although Rachel Carson's publication attracted a lot of attention, one cannot say that the problems she identified have been solved. On the contrary, the high use of pesticides is still a controversial issue in the implementation process of environmental policies, as agricultural vested interests try to counter efforts to reduce its use.

Concluding remarks

The period described in this part was subject to many changes. Social, cultural, economic and political conditions in society changed enormously. These rapid and sometimes violent changes can be labelled as a revolution. The quality of the natural environment deteriorated dramatically during this period. Effects at a local level became regional problems and soon moved on to a global level. All the classics described stimulated the development of new studies or functioned as eye-openers for groups in society. Themes like the law of diminishing returns, natural selection, externalisation of internal cost, ecology, and the accumulation of toxic substances in food chains are even now frequently used and referred to in books, publications, and speeches.

References
Chiras, D.D., 1994. *Environmental Science; Action for a Sustainable Future*, Redwood City.
Humphrey, C.R. and F.R. Buttel, 1982. *Environment, Energy, and Society*, Belmont.

CHAPTER 1

An Essay on the Principle of Population

T.R. MALTHUS

Statement of the Subject
Ratios of the Increase of Population and Food

In an inquiry concerning the improvement of society, the mode of conducting the subject which naturally presents itself is:

1 To investigate the causes that have hitherto impeded the progress of mankind towards happiness; and,

2 To examine the probability of the total or partial removal of these causes in future.

To enter fully into this question, and to enumerate all the causes that have hitherto influenced human improvement, would be much beyond the power of an individual. The principal object of the present essay is to examine the effects of one great cause intimately united with the very nature of man; which, though it has been constantly and powerfully operating since the commencement of society, has been little noticed by the writers who have treated this subject. The facts which establish the existence of this cause, have, indeed, been repeatedly stated and acknowledged; but its natural and necessary effects have been almost totally overlooked; though probably among these effects may be reckoned a very considerable portion of that vice and misery, and of that unequal distribution of the bounties of nature, which it has been the unceasing object of the enlightened philanthropist in all ages to correct.

The cause to which I allude, is the constant tendency in all animated life to increase beyond the nourishment prepared for it.

It is observed by Dr. Franklin, that there is no bound to the prolific nature of plants or animals but what is made by their crowding and interfering with each other's means of subsistence. Were the face of the earth, he says, vacant of other plants, it might be gradually sowed and overspread with one kind only, as, for instance, with fennel: and were it empty of other inhabitants, it might in a few ages be replenished from one nation only, as, for instance, with Englishmen.[1]

This is incontrovertibly true. Throughout the animal and vegetable kingdoms Nature has scattered the seeds of life abroad with the most profuse and liberal hand; but has been comparatively sparing in the room and the nourishment necessary to rear them. The germs of existence contained in this earth, if they could freely develop themselves, would fill millions of worlds in the course of a few thousand years. Necessity, that imperious, all-pervading law of nature, restrains them within the prescribed bounds. The race of plants and the race of animals shrink under this great restrictive law; and man cannot by any efforts of reason escape from it.

In plants and irrational animals, the view of the subject is simple. They are all impelled by a powerful instinct to the increase of their species, and this instinct is interrupted by no doubts about providing for their offspring. Wherever, therefore, there is liberty, the power of increase is exerted, and the superabundant effects are repressed afterwards by want of room and nourishment.

The effects of this check on man are more complicated. Impelled to the increase of his species by an equally powerful instinct, reason interrupts his career, and asks him whether he may not bring beings into the world for whom he cannot provide the means of support. If he attend to this natural suggestion, the restriction will be constantly endeavouring to increase beyond the means of subsistence. But as, by that law of our nature which makes food necessary to the life of man, population can never actually increase beyond the lowest nourishment capable of supporting it, a strong check on population, from the difficulty of acquiring food, must be constantly in operation. This difficulty must fall somewhere, and must necessarily be severely felt in some or other of the various forms of misery, or the fear of misery, by a large portion of mankind.

That population has this constant tendency to increase beyond the means of subsistence, and that it is kept to its necessary level by these causes, will sufficiently appear from a review of the different states of society in which man has existed. But, before we proceed to this review, the subject will perhaps be seen in a clearer light, if we endeavour to ascertain what would be the natural increase of population, if left to exert itself with perfect freedom; and what might be expected to be the rate of increase in the productions of the earth, under the most favourable circumstances of human industry.

It will be allowed that no country has hitherto been known, where the manners were so pure and simple, and the means of subsistence so abundant, that no check whatever has existed to early marriages from the difficulty of providing for a family, and that no waste of the human species has been occasioned by vicious customs, by towns, by unhealthy occupations, or too severe labour. Consequently in no state that we have yet known, has the power of population been left to exert itself with perfect freedom.

Whether the law of marriage be instituted or not, the dictate of nature and virtue seems to be and early attachment to one woman; and where there were no impediments of any kind in the way of union to which such an attachment would lead, and no causes of depopulation afterwards, the increase of the human species would be evidently much greater than any increase which has been hitherto known.

In the northern states of America, where the means of subsistence have been more ample, the manners of the people more pure, and the checks to early marriages fewer, than in any of the modern states of Europe, the population has been found to double itself, for above a century and a half successively, in less than twenty-five years.[2] Yet, even during these periods, in some of the towns, the deaths exceeded the births,[3] a circumstance which clearly proves that, in those parts of the country which supplied this deficiency, the increase must have been much more rapid than the general advantage.

In the back settlements, where the sole employment is agriculture, and vicious customs and unwholesome occupations are little known, the population has been found to double itself in fifteen years.[4] Even this extraordinary rate of increase is probably short of the utmost power of population. Very severe labour is requisite to clear a fresh country; such situations are not in general considered as particularly healthy; and the inhabitants, probably, are occasionally subject to the incursions of the Indians, which may destroy some lives, or at any rate diminish the fruits of industry.

According to a table of Euler, calculated on a mortality of 1 in 36, if the births be to the deaths in the proportion of 3 to 1, the period of doubling will be only twelve years and four-fifths. And this proportion is not only a possible supposition, but has actually occurred for short periods in more countries than one.

Sir William Petty supposes a doubling possible in so short a time as ten years.[5]

But, to be perfectly sure that we are far within the truth, we will take the slowest of these rates of increase, a rate in which all concurring testimonies agree, and which has been repeatedly ascertained to be from procreation only. It may safely be pronounced, therefore, that population, when unchecked, goes on doubling itself every twenty-five years, or increases in a geometrical ratio.

The rate according to which the productions of the earth may be supposed to increase, will not be so easy to determine. Of this, however, we may be perfectly certain, that the ratio of their increase in a limited territory must be of a totally different nature from the ration of the increase of population. A thousand millions are just as easily doubled every twenty-five years by the power of population as a thousand. But the food to support the increase from the greater number will be no means be obtained with the same facility. Man is necessarily confined in room. When acre has been added to acre till all the fertile land is occupied, the yearly increase of food must depend upon the melioration of the land already in possession. This is a fund, which, from the nature of all soils, instead of increasing, must be gradually diminishing. But population, could it be supplied with food, would go on with unexhausted vigour; and the increase the next, and this without any limit.

From the accounts we have of China and Japan, it may be fairly doubted, whether the best directed efforts of human industry could double the produce of these countries even once in any number of years. There are many parts of the globe, indeed, hitherto uncultivated and almost unoccupied; but the right of exterminating, or driving into a corner where they must starve, even the inhabitants of these thinly-peopled regions, will be questioned in a moral view. The process of improving their minds and directing their industry would necessarily be slow; and during this time, as population would regularly keep pace with the increasing produce, it would rarely happen that a great degree of knowledge and industry would have to operate at once upon rich unappropriated soil. Even when this might take place, as it does sometimes in new colonies, a geometrical ratio increases with such extraordinary rapidity, that the advantage could not last long. If the United States of America continue increasing, which they certainly will do, though not with the same rapidity as formerly, the Indians will be driven farther and farther back into the country, till the whole race is ultimately exterminated, and the territory is incapable of further extension.

These observations are, in a degree, applicable to all the parts of the earth where the soil is imperfectly cultivated. To exterminate the inhabitants of the greatest part of Asia and Africa, is a thought that could not be admitted for a moment. To civilise and direct the industry of the various tribes of Tartars and Negroes, would certainly be a work of considerable time, and of variable and uncertain success.

Europe is by no means so fully peopled as it might be. In Europe there is the fairest chance that human industry may receive its best direction. The science of agriculture has been much studied in England and Scotland; and there is still a great portion of uncultivated land in these countries. Let us consider at what rate the produce of this island (Great Britain) might be supposed to increase under circumstances the most favourable to improvement.

If it be allowed that by the best possible policy, and great encouragements to agriculture, the average produce of the island could be doubled in the first twenty-five years, it will be allowing, probably, a greater increase than could with reason be expected.

In the next twenty-five years, it is impossible to suppose that the produce could be quadrupled. It would be contrary to all our knowledge of the properties of land. The improvement of the barren parts would be a work of time and labour; and it must be evident to those who have the slightest acquaintance with agricultural subjects, that in proportion as cultivation extended, the additions that could yearly be made to the former average produce must be gradually and regularly diminishing. That we may be the better able to compare the increase of population and food, let us make a supposition, which, without pretending to accuracy, is clearly more favourable to the power of production in the earth than any experience we have had of its qualities will warrant.

Let us suppose that the yearly additions which might be made to the former average produce, instead of decreasing, which they certainly would do, were to remain the same; and that the produce of this island might be increased every twenty-five years, by a quantity equal to what it at present produces. The most enthusiastic speculator cannot suppose a greater increase than this. In a few centuries it would make every acre of land in the island like a garden.

If the supposition be applied to the whole earth, and if it be allowed that the subsistence for man which the earth affords might be increased every twenty-five years be a quantity equal to what it at present produces, this will be supposing a rate of increase much greater than we can imagine that any possible exertions of mankind could make it.

It may be fairly pronounced, therefore, that, considering the manner, the dignity of human nature. It cannot be without its effect on men, and nothing can be more obvious than its tendency to degrade the female character, and to destroy all its most amiable and distinguishing characteristics. Add to which, that among those unfortunate females with which all great towns abound, more real distress and aggravated misery are, perhaps, to be found, than in any other department of human life.

When a general corruption of morals, with regard to the sex, pervades all the classes of society, its effects must necessarily to be to poison the springs of domestic happiness, to weaken conjugal and parental affection, and to lessen the united exertions and ardour of parents in the care and education of their children; effects which cannot take place without a decided diminution of the general happiness and virtue of society; particularly as the necessity of art in the accomplishment and conduct of intrigues, and in the concealment of their consequences, necessarily leads to many other vices.

The positive checks to population are extremely various, and include every cause, whether arising from vice or misery, which in any degree contribute to shorten the natural duration of human life. Under this head, therefore, may be enumerated all unwholesome occupations, severe labour and exposure to the seasons, extreme poverty, bad nursing of children, large towns, excesses of all kinds, the whole train of common diseases and epidemics, wars, plague, and famine.

On examining these obstacles to the increase of population which are classed under the heads of preventive and positive checks, it will appear that they are all resolvable into moral restraint, vice, and misery.

Of the preventive checks, the restraint from marriage which is not followed by irregular gratifications may properly be termed moral restraint.[6]
Promiscuous intercourse, unnatural passions, violations of the marriage bed,

and improper arts to conceal the consequences of irregular connections, are preventive checks that clearly come under the head of vice.

Of the positive checks, those which appear to arise unavoidably from the laws of nature, may be called exclusively misery; and those which we obviously bring upon ourselves, such as wars, excesses, and many others which it would be in our power to avoid, are of a mixed nature. They are brought upon us by vice, and their consequences are misery.[7]

The sum of all these preventive and positive checks, taken together, forms the immediate check to population; and it is evident that, in every country where the whole of the procreative power cannot be called into action, the preventive and the positive checks must vary inversely as each other; that is, in countries either naturally unhealthy, or subject to a great mortality, from whatever cause it may arise, the preventive check is found to prevail with considerable force, the positive check will prevail very little, or the mortality be very small.

In every country some of these checks are, with more or less force, in constant operation; yet, notwithstanding their general prevalence, there are few states in which there is not a constant effort in the population to increase beyond the means of subsistence. This constant effort as constantly tends to subject the lower classes of society to distress, and to prevent any great permanent melioration of their condition.

These effects, in the present state of society, seem to be produced in the following manner. We will suppose the means of subsistence in any country just equal to the easy support of its inhabitants. The constant effort towards population, which is found to act even in the most vicious societies, increases the number of people before the means of subsistence are increased. The food, therefore, which before supported eleven millions, must now be divided among eleven millions and a half. The poor consequently must live much worse, and many of them be reduced to severe distress. The number of labourers also being above the proportion of work in the market, the price of labour must tend to fall, while the price of provisions would at the same time tend to rise. The labourer therefore must do more work to earn the same as he did before. During this season of distress, the discouragements to marriage and the difficulty of rearing a family are so great, that the progress of population is retarded. In the meantime, the cheapness of labour, the plenty of labourers, and the necessity of an increased industry among them, encourage cultivators to employ more labour upon their land, to turn up fresh soil, and to manure and improve more completely what is already in tillage, till ultimately the means of subsistence may become in the same proportion to the population as at the period from which we set out. The situation of the labourer being then again tolerable comfortable, the restraints to population are in some degree loosened; and, after a short period, the same retrograde and progressive move-ments with respect to happiness are repeated.

This sort of oscillation will not probably be obvious to common view; and it may be difficult even for the most attentive observer to calculate its periods. Yet that, in the generality of old states, some alternation of this kind does exist, though, in a much less marked, and in a much more irregular manner, than I have described it, no reflecting man, who considers the subject deeply, can well doubt.

One principal reason why this oscillation has been less remarked, and less decidedly confirmed by experience than might naturally be expected, is, that the histories of mankind which we possess are, in general, histories only of the higher classes. We have not many accounts that can be depended upon, of the manners and customs of that part of mankind where these retrograde and pro-gressive movements chiefly take place. A satisfactory history of this kind, of one people and of one period, would require the constant and minute attention of many observing minds in local and general remarks on the state of the lower classes of society, and the causes that influenced it; and, to draw accurate in-ferences upon this subject, a succession of such historians for some centuries would be necessary. This branch of statistical knowledge has, of late years, been attended to in some countries,[8] and we may promise ourselves a clearer insight into the internal structure of human society from the progress of these inquiries. But the science may be said yet to be in its infancy, and many of the objects on which it would be desirable to have information, have been either omitted or not stated with sufficient accuracy. Among these, perhaps, may be reckoned the proportion of the number of adults to the number of marriages; the extent to which vicious customs have prevailed in consequence of the re-straints upon matrimony; the comparative mortality among the children of the most distressed part of the community, and of those who live rather more at their ease; the variations in the real price of labour; the observable differences in the state of the lower classes of society, with respect to ease and happiness, at different times during a certain period; and very accurate registers of births, deaths, and marriages, which are of the utmost importance in this subject.

A faithful history, including such particulars, would tend greatly to elucidate the manner in which the constant check upon population acts; and would probably prove the existence of the retrograde and progressive movements that have been mentioned; though the times of their vibration must necessarily be rendered irregular from the operation of many interrupting causes; such as, the introduction of failure of certain manufactures; a greater or less prevalent spirit of agricultural enterprise; years of plenty, or years of scarcity; wars, sickly sea-sons, poor-laws, emigrations and other causes of a similar nature.

A circumstance which has, perhaps, more than any other, contributed to conceal this oscillation from common view, is the difference between the nomi-

nal and real price of labour. It very rarely happens that the nominal price of labour universally falls; but we well know that it frequently remains the same,
while the nominal price of provisions has been gradually rising. This, indeed,
will generally be the case, if the increase of manufactures and commerce be
sufficient to employ the new labourers that are thrown into the market, and to
prevent the increased supply from lowering the money-price.[9] But an increased
number of labourers receiving the same money-wages will necessarily, by their
competition, increase the money-price of corn. This is, in fact, a real fall in the
price of labour; and during this period, the condition of the lower classes of the
community must be gradually growing worse. But the farmers and capitalists
are growing rich from the real cheapness of labour. Their increasing capitals
enable them to employ a greater number of men; and, as the population had
probably suffered some check from the greater difficulty of supporting a family,
the demand for labour, after a certain period, would be great in proportion to
the supply, and its price would of course rise, if left to find its natural level; and
thus the wages of labour and consequently the condition of the lower classes of
society, might have progressive and retrograde movements, thought the price
of labour might never nominally fall.

In savage life, where there is no regular price of labour, it is little to be
doubted that similar oscillations take place. When population has increased
nearly to the utmost limits of the food, all the preventive and the positive
checks will naturally operate with increased force. Vicious habits with respect
to the sex will be more general, the exposing of children more frequent, and
both the probability and fatality of wars and epidemics will be considerably
greater; and these causes will probably continue their operation till the population is sunk below the level of the food; and then the return to comparative
plenty will again produce an increase, and, after a certain period, its further
progress will again be checked by the same causes.[10]

But without attempting to establish these progressive and retrograde movements in different countries, which would evidently require more minute histories than we possess, and which the progress of civilisation naturally tends to
counteract, the following propositions are intended to be proved:
1 Population is necessarily limited by the means of subsistence.
2 Population invariably increases where the means of subsistence increase, unless prevented by some very powerful and obvious checks.[11]
3 These checks, and the checks which repress the superior power of population, and keep its effects on a level with the means of subsistence, are all resolvable into moral restraints, vice, and misery.

The first of these propositions scarcely needs illustration. The second and third
will be sufficiently established by a review of the immediate checks to population in the past and present state of society.

Malthus, T.R. *An Essay on the Principle of Population*, Murray, London, 1817 (first published, 1798), p. 1-13.

Notes and References

1 Franklin's Miscell. p. 9.

2 It appears, from some recent calculations and estimates, that from the first settlement of America, to the year 1800, the periods of doubling have been but very little above twenty years.

3 Price's Observ. on Revers. Pay. Vol. 1. p. 274, 4th edit.

4 Id. p. 282.

5 Polit. Arith. p. 14.

6 It will be observed, that I here use the term *moral* in its most confined sense. By moral constraint I would be understood to mean a restraint from marriage from prudential motives, with a conduct strictly moral during the period of this restraint; and I have never intentionally deviated from this sense. When I have wished to consider the restraint from marriage unconnected with its consequences, I have either called it prudential restraint, or a part of the preventive check, of which indeed it forms the principal branch.

In my review of the different stages of society, I have been accused of not allowing sufficient weight in the prevention of population to moral restraint; but when the confined sense of the term, which I have here explained, is adverted to, I am fearful that I shall not be found to have erred much in this respect. I should be very glad to believe myself mistaken.

7 As the general consequence of vice is misery, and as this consequence is the precise reason why an action is termed vicious, it may appear that the term *misery* alone would be here sufficient, and that it is superfluous to use both. But the rejection of the term *vice* would introduce a considerable confusion into our language and ideas. We want it particularly to distinguish those actions, the general tendency of which is to produce misery, and which are therefore prohibited by the commands of the creator, and the precepts of the moralist, although, in their immediate or individual effects, they may produce perhaps exactly the contrary. The gratification of all our passions in its immediate effect is happiness, not misery; and, in individual instances, even the remote consequences (at least in this life) may possibly come under the same denomination. There may have been some irregular connections with women, which have added to the happiness of both parties, and have injured no one. These individual actions, therefore, cannot come under the head of misery. But they are still evidently vicious, because an action is so denominated which violated an express precept, founded upon its general tendency to produce misery, whatever may be its individual effect; and no person can doubt the general tendency of an illicit intercourse between the sexes to injure the happiness of society.

8 The judicious questions which Sir John Sinclair circulated in Scotland, and the valuable accounts which he has collected in that part of the island, do him the highest honour; and these accounts will ever remain an extraordinary monument of the learning, good sense, and general information of the clergy of Scotland. It is to be regretted that the adjoining parishes are not put together in the work, which would have assisted the memory both in attaining and recollecting the state of particular districts. The repetitions and contradictory opinions which occur are not, in my opinion, so objectionable; as, to the result of such testimony, more faith may be given than we could possibly give to the testimony of any individual. Even were this result drawn for us by some master hand, though much valuable time would undoubtedly be save, the information would not be so satisfactory. If, with a few subordinate improvements, this work had contained accurate and complete registers for the last 150 years, it would have been estimable, and would have exhibited a better picture of the internal state of a country than has yet been presented to the world. But this last most essential improvement no diligence could have effected.

9   If the new labourers thrown yearly into the market should find no employment but in agriculture, their competition might so lower the money-price of labour as to prevent the increase of population from occasioning an effective demand for more earn; or, in other words, if the landlords and farmers could get nothing but an additional quantity of agricultural labour in exchange for any additional produce which they could raise, they might not be tempted to raise it.

10   Sir James Stuart very justly compares the generative faculty to a spring loaded with a variable weight (Polit. Econ., vol.i., b.i., c.4, p. 20), which would of course produce exactly that kind of oscillation which has been mentioned. In the first part of the subject of population very ably.

11   I have expressed myself in this cautious manner, because I believe there are some instances where population does not keep up to the level of the means of subsistence. But these are extreme cases; and generally speaking, it might be said that:

1   Population is necessarily limited by the means of subsistence.
2   Population always increases where the means of subsistence increase.
3   The checks which repress the superior power of population, and keep its effects on a level with the means of subsistence, are all resolvable into moral restraints, vice and misery.

It should be observed that, by an increase in the means of subsistence, is here meant such an increase as will enable the mass of the society to command more food. An increase might certainly take place which in the actual state of a particular society would not be distributed to the lower classes, and consequently would give no stimulus to population.

CHAPTER 2

On the Origin of Species by Means of Natural Selection

or The Preservation of Favoured Races in the Struggle for Life

CH. DARWIN

Conclusion

I have now recapitulated the facts and considerations which have thoroughly convinced me that species have been modified, during a long course of descent. This has been effected chiefly through the natural selection of numerous successive, slight, favourable variations; aided in an important manner by the inherited effects of the use and disuse of parts; and in an unimportant manner, that is in relation to adaptive structures, whether past or present, by the direct action of external conditions, and by variations which seem to us in our ignorance to arise spontaneously. It appears that I formerly underrated the frequency and value of these latter forms of variation, as leading to permanent modifications of structure independently of natural selection. But as my conclusions have lately been much misrepresented, and it has been stated that I attribute the modification of species exclusively to natural selection, I may be permitted to remark that in the first edition of this work, and subsequently, I placed in a most conspicuous position—namely, at the close of the Introduction—the following words: "I am convinced that natural selection has been the main but not the exclusive means of modification." This has been of no avail. Great is the power of steady misrepresentation; but the history of science shows that fortunately this power does not long endure.

It can hardly be supposed that a false theory would explain, in so satisfactory a manner as does the theory of natural selection, the several large classes of facts above specifies. It has recently been objected that this is an unsafe method of arguing; but it is a method used in judging of the common events of life, and has often been used by the greatest natural philosophers. The undulated theory of light has thus been arrived at; and the belief in the revolution of the earth on its own axis was until lately supported by hardly any evidence. It is no valid objection that science as yet throws no light on the far higher problem of the essence of origin of life. Who can explain what is the essence of the attraction of gravity? No one now objects to following out the results consequent on this unknown element of attraction; notwithstanding that Leibnitz formerly accused Newton of introducing "occult qualities and miracles into philosophy".

I see no good reason why the views given in this volume should shock the religious feelings of any one. It is satisfactory, as showing how transient such impressions are, to remember that the greatest discovery ever made by man, namely the law of the attraction of gravity was also attacked by Leibnitz, "as subversive of natural, and inferentially of revealed, religion". A celebrated author and divine has written to me that he has "gradually learnt to see that it is just as noble a conception of the Deity to believe that He created a few original forms capable of self-development into other and needful forms, as to believe that He required a fresh act of creation to supply the voids caused by the action of His laws".

Why, it may be asked, until recently did nearly all the most eminent living naturalists and geologists disbelieve in the mutability of species. It cannot be asserted that organic beings in a state of nature are subject to no variation; it cannot be proved that the amount of variation in the course of long ages is a limited quantity; no clear distinction has been, or can be, drawn between species and well-marked varieties. It cannot be maintained that species when intercrossed are invariably sterile, and varieties invariably fertile; or that sterility is a special endowment and sign of creation. The belief that species were immutable productions thought to be of short duration; and now that we have acquired some idea of the lapse of time, we are too apt to assume, without proof, that the geological record is so perfect that it would have afforded us plain evidence of the mutation of species, if they had undergone mutations.

But the chief cause of our natural unwillingness to admit that one species has given birth to other and distinct species, is that we are always slow in admitting great changes of which we do no see the steps. The difficulty is the same as that felt by so many geologists, when Lyell first insisted that long lines of inland cliffs had been formed, and great valleys excavated, by the agencies which we see still at work. The mind cannot possibly grasp the full meaning of the term of even a million years; it cannot add up and perceive the full effects of many slight variations, accumulated during an almost infinite number of generations.

Although I am fully convinced of the truth of the views given in this volume under the form of an abstract, I by no means expect to convince experienced naturalists whose minds are stocked with a multitude of facts all viewed, during a long course of years, from a point of view directly opposite to mine. It is so easy to hide our ignorance under such expressions as the "plan of creation", "unity of design", &c., and to think that we give an explanation when we only re-state a fact. Any one whose disposition leads him to attach more weight to unexplained difficulties that to the explanation of a certain number of facts will certainly reject the theory. A few naturalists, endowed with much flexibility of mind, and who have already begun to doubt the immutability of species, may be influenced by this volume; but I look with confidence to the future,—to young and rising naturalists, who will be able to view both sides of the question with impartiality. Whoever is led to believe that species are mutable will do good service by conscientiously expressing his conviction; for thus only can the load of prejudice by which this subject is overwhelmed be removed.

Several eminent naturalists have of late published their belief that a multitude of reputed species in each genus are not real species; but that other species are real, that is, have been independently created. This seems to me a strange conclusion to arrive at. They admit that a multitude of forms, which till lately they themselves thought were special creations, and which are still thus looked at by the majority of naturalists, and which consequently have all the external characteristic features of true species,—they admit that these have been produced by variation, but they refuse to extend the same view to other and slightly different forms. Nevertheless they do not pretend that they can define, or even conjecture, which are the created forms of life, and which are those produced by secondary laws. They admit variation as a *vera causa* in one case, they arbitrarily reject it in another, without assigning any distinction in the two cases. The day will come when this will be given as a curious illustration of the blindness of preconceived opinion. These authors seem no more startled at a miraculous act of creation than at an ordinary birth. But do they really believe that at innumerable periods in the earth's history certain elemental atoms have been commanded suddenly to flash into living tissues? Do they believe that at each supposed act of creation one individual or many were produced? Were all the infinitely numerous kinds of animals and plants created as eggs or seed, or as full grown? And in the case of mammals, were they created bearing the false marks of nourishment from the mother's womb? Undoubtedly some of these same questions cannot be answered by those who believe in the appearance or creation of only a few forms of life, or of some one form alone. It has been maintained by several authors that it is as easy to believe in the creation of a million beings as of one; but Maupertuis' philosophical axiom "of last action" leads the mind more willingly to admit the smaller number; and certainly we ought not to believe that innumerable beings within each great class have been created with plain, but deceptive, marks of descent from a single parent.

As a record of a former state of things, I have retained in the foregoing paragraphs, and elsewhere, several sentences which imply that naturalists believe in the separate creation of each species; and I have been much censured for having thus expressed myself. But undoubtedly this was the general belief when the first edition of the present work appeared. I formerly spoke to very many naturalists on the subject of evolution, and never once met with any sympathetic agreement. It is probable that some did the believe in evolution, but they were either silent, or expressed themselves so ambiguously that it was not easy to understand their meaning. Now things are wholly changed, and almost every naturalist admits the great principle of evolution. There are, however, some who still think that species have suddenly given birth, through quite unexplained means, to new and totally different forms; but as I have attempted to show, weighty evidence can be opposed to the admission of great and abrupt modifications. Under a scientific point of view, and as leading to further investigation, but little advantage is gained by believing that new forms are suddenly developed in an inexplicable manner from old and widely different forms, over the old belief in the creation of species from the dust of the earth.

It may be asked how far I extend the doctrine of the modification of species. The question is difficult to answer, because the arguments in favour of community of descent become fewer in number and less in force. But some arguments of the greatest weight extend very far. All the members of whole classes are connected together by a chain of affinities, and all can be classed on the same principle, in groups subordinate to groups. Fossil remains sometimes tend to fill up very wide intervals between existing orders.

Organs in a rudimentary condition plainly show that an early progenitor had the organ in a fully developed condition; and this in some cases implies an enormous amount of modification in the descendants. Throughout whole classes various structures are formed on the same pattern, and at a very early age the embryos closely resemble each other. Therefore I cannot doubt that the theory of descent with modification embraces all the members of the same great class or kingdom. I believe that animals are descended from at most only four or five progenitors, and plans from an equal or lesser number.

Analogy would lead me one step farther, namely, to the belief that all animals and plants are descended from one prototype. But analogy may be a deceitful guide. Nevertheless all living things have much in common, in their chemical composition, their cellular structure, their laws of growth, and their liability to injurious influences. We see this even in so trifling a fact as that the same poison often similarly affects plants and animals; or that the poison secreted by the gall-fly produces monstrous growths on the wild rose or oak-tree. With all organic beings, excepting perhaps one of the very lowest, sexual reproduction seems to be essentially similar. With all, as far is at present known, the germinal vesicle is the same; so that all organisms start from a common origin. If we look even to the two main divisions—namely, to the animal and vegetable kingdoms—certain low forms are so far intermediate in character that naturalists have disputed to which kingdom they should be referred. As Professor Asa Gray has remarked, "the spores and other reproductive bodies of many of the lower algæ may claim to have first a characteristically animal, and then an unequivocally vegetable existence". Therefore, on the principle of natural selection with divergence of character, it does not seem incredible that, from some such low and intermediate form, both animals and plants may have developed;

and, if we admit this, we must likewise admit that all the organic beings which have ever lived on this earth may be descended from some one primordial form. But this inference is chiefly grounded on analogy, and it is immaterial whether or not it be accepted. No doubt it is possible, as Mr. G.H. Lewes has urged, that at the first commencement of life many different forms were evolved; but if so, we may conclude that only a very few have left modified descendants. For, as I have recently remarked in regard to the members of each great kingdom, such as the Vertebrata, Articulata, &c., we have distinct evidence in their embryological, homologous, and rudimentary structures, that within each kingdom all the members are descended from a single progenitor.

When the views advanced by me in this volume, and by Mr. Wallace, or when analogous views on the origin of species are generally admitted, we can dimly foresee that there will be a considerable revolution in natural history. Systematists will be able to pursue their labours as at present; but they will not be incessantly haunted by the shadowy doubt whether this or that form be a true species. This, I feel sure and I speak after experience, will be no slight relief. The endless disputes whether or not some fifty species of British brambles are good species will cease. Systematists will have only to decide (not that this will be easy) whether any form be sufficiently constant and distinct from other forms, to be capable of definition; and if definable, whether the differences be sufficiently important to deserve a specific name. The latter point will become a far more essential consideration than is at present; for differences, however slight, between any two forms, if not blended by intermediate gradations, are looked at by most naturalists as sufficient to raise both forms to the rank of species.

Hereafter we shall be compelled to acknowledge that the only distinction between species and well-marked varieties is, that the latter are known, or believed, to be connected at the present day by intermediate gradations whereas species were formerly thus connected. Hence, without rejecting the consideration of the present existence of intermediate gradations between any two forms, we shall be led to weigh more carefully and to value higher the actual amount of difference between them. It is quite possible that forms now generally acknowledged to be merely varieties may hereafter be thought worthy of specific names; and in this case scientific and common language will come into accordance. In short, we shall have to treat species in the same manner as those naturalists treat genera, who admit that genera are merely artificial combinations made for convenience. This may not be a cheering prospect; but we shall at least be freed from the vain search for the undiscovered and undiscoverable essence of the term species.

The other and more general departments of natural history will rise greatly in interest. The term used by naturalists, of affinity, rise greatly in interest. The terms used by naturalists, of affinity, relationship, community of type, paternity, morphology, adaptive characters, rudimentary and aborted organs, &c., will cease to be metaphorical, and will have a plain significance. When we no longer look at an organic being as a savage looks at a ship, as something wholly beyond his comprehension; when we regard every production of nature as one which has had a long history; when we contemplate every complex structure and instinct as the summing up of many contrivances, each useful to the possessor, in the same way as any great mechanical invention is the summing up of labour, the experience, the reason, and even the blunders of numerous workmen; when we thus view each organic being, how far more interesting—I speak from experience—does the study of natural history become!

A grand and almost untrodden field of inquiry will be opened, on the causes and laws of variation, on correlation, on the effects of use and disuse, on the direct action of external conditions, and so forth. The study of domestic productions will rise immensely in value. A new variety raised by man will be a more important and interesting subject for study than one more species added to the infinitude of already recorded species. Our classifications will come to be, as far as they can be so made, genealogies; and will then truly give what may be called the plan of creation. The rules for classifying will no doubt become simpler when we have a definite object in view. We posses no pedigrees or armorial bearings; and we have to discover and trace the many diverging lines of descent in our natural genealogies, by characters of any kind which have long been inherited. Rudimentary organs will speak infallibly with respect to the nature of long lost structures. Species and groups of species which are called abberrant, and which may fancifully be called living fossils, will aid us in forming a picture of the ancient forms of life. Embryology will often reveal to us the structure, in some degree obscured, of the prototypes of each great class.

When we can feel assured that all the individuals of the same species, and all the closely allied species of most genera, have within a not very remote period descended from one parent, and have migrated from some one birth-place; and when we better know the many means of migration, then, by the light which geology now throws, and will continue to throw, on former changes of climate and of the level of the land, we shall surely be enabled to trace in an admirable manner the former migration of the inhabitants of the whole world. Even at present, by comparing the differences between the inhabitants of the sea on the opposite sides of a continent, and the relation to their apparent means of immigration, some light can be thrown on ancient geography.

The noble science of Geology loses glory from the extreme imperfection of the record. The crust of the earth with its imbedded remains must not be looked at as a well-filled museum, but as a poor collection made at hazard and at rare intervals. The accumulation of each great fossiliferous formation will be recognised as having depended on an unusual concurrence of favourable circumstances, and the blank intervals between the successive stages as having been of vast duration. But we shall be able to gauge with some security the duration of these intervals by a comparison of the preceding and succeeding organic forms. We must be cautious in attempting to correlate as strictly contemporaneous two formations, which do not include many identical species, by the general succession of the forms of life. As species are produced and exterminated by slowly acting and still existing causes, and not by miraculous acts of creation; and as the most important of all causes of organic change is one which is almost independent of altered and perhaps suddenly altered physical conditions, namely, the mutual relation of organism to organism,—the improvement of one organism entailing the improvement or the extermination of others; it follows, that the amount of organic change in the fossils of consecutive formations probably serves as a fair measure of the relative, though not actual lapse of time. A number of species, however, keeping in a body might remain for a long period unchanged, whilst within the same period, several of these species by migrating into new countries and coming into competition with foreign associates, might become modified; so that we must not overrate the accuracy of organic change as a measure of time.

In the future I see open fields for far more important researches. Psychology will be securely based on the foundation already well laid by Mr. Herbert Spencer, that of the necessary acquirement of each mental power and capacity by gradation. Much light will be thrown on the origin of man and his history.

Authors of the highest eminence seem to be fully satisfied with the view that each species has been independently created. To my mind it accords better with what we know of the laws impressed on matter by the Creator, that the production and extinction of the past and present inhabitants of the world should have been due to secondary causes, like those determining the birth and death of the individual. When I view all beings not as special creations, but as the lineal descendants of some few beings which lived long before the first bed of the Cambrian system was deposited, they seem to me to become ennobled. Judging from the past, we may safely infer that not one living species will transmit its unaltered likeness to a distant futurity. And of the species now living very few will transmit progeny of any kind to a far distant futurity; for the manner in which all organic beings are grouped, shows that the greater number of species in each genus, and all the species in many genera, have left no descendants, but have become utterly extinct. We can so far take a prophetic glance into futurity as to foretell that it will be the common and widely-spread species, belonging to the larger and dominant groups within each class, which will ultimately prevail and procreate new and dominant species. As all the living forms of life are the lineal descendants of those which lived long before the Cambrian epoch, we may feel certain that the ordinary succession by generation has never once been broken, and that no cataclysm has desolated the whole world. Hence we may look with some confidence to a secure future of great length.

And as natural selection works solely by and for the good of each being, all corporeal and mental endowments will tend to progress towards perfection.

It is interesting to contemplate a tangled bank, clothed with many plants of many kinds, with birds singing on the bushes with various insects flitting about, and with worms crawling through the damp earth, and to reflect that these elaborately constructed forms, so different from each other, and dependent upon each other in so complex a manner, have all been produced by laws acting around us. These laws, taken in the largest sense, being Growth with Reproduction; Inheritance from the indirect and direct action of the conditions of life, and from use and disuse: a Ratio of Increase so high as to lead to a Struggle for Life, and as a consequence to Natural Selection, entailing Divergence of Character and the Extinction of less-improved forms. Thus, from the war of nature, from famine and death, the most exalted object which we are capable of conceiving, namely, the production of the higher animals, directly follows. There is grandeur in this view of life, with its several powers, having been originally breathed by the Creator into a few forms or into one; and that, whilst this planet has gone cycling on according to the fixed law of gravity, from so simple a beginning endless forms most beautiful and most wonderful have been, and are being evolved.

Darwin, Ch., *On the Origin of Species by Means of Natural Selection, or the Preservation of Favoured Races in the Struggle for Life*, Murray, London, 1859, p. 395-403.

CHAPTER 3

The Economics of Welfare

A.C. PIGOU

Social and Private Net Product

I now turn to the second class of divergence between social and private net product. Here the essence of the matter is that one person A, in the course of rendering some service, for which payment is made, to a second person B, incidentally also renders services or disservices to other persons (not producers of like services) of such a sort that payment cannot be exacted from the benefited parties or compensation enforced on behalf of the injured parties. If we were to be pedantically loyal to the definition of the national dividend it would be necessary to distinguish further between given in Chapter III. of Part I., industries in which the uncompensated benefit or burden respectively is and is not one that can be readily brought into relation with the measuring rod of money. This distinction, however, would be of formal rather than of real importance, and would obscure rather than illuminate the main issues. I shall, therefore, in the examples I am about to give, deliberately pass it over.

Among these examples we may set out first a number of instances in which marginal private net product falls short of marginal social net product, because incidental services are performed to third parties from whom it is technically difficult to exact payment. Thus, as Sidgwick observes, "it may easily happen that the benefits of a well-placed lighthouse must be largely enjoyed by ships on which no toll could be conveniently levied".[1] Again, uncompensated services are rendered when resources are invested in private parks in cities; for these, even though the public is not admitted to them, improve the air of the neighbourhood. The same thing is true though here allowance should be made for a detriment elsewhere of resources invested in roads and tramways that increase the value of the adjoining land—except, indeed, where a special betterment rate, corresponding to the improvements they enjoy, is levied on the owners of this land. It is true, in like manner, of resources devoted to afforestation, since the beneficial effect on climate often extends beyond the borders of the estates owned by the person responsible for the forest. It is true also of resources invested in lamps erected at the doors of private houses, for these necessarily throw light also on the streets.[2] It is true of resources devoted to the prevention of smoke from factory chimneys:[3] for this smoke in large towns inflicts a heavy uncharged loss on the community, in injury to buildings and vegetables, expenses for washing, clothes and cleaning rooms, expenses for the provision of extra artificial light, and in many other ways.[4] Lastly and most important of all, it is true of resources devoted alike to the fundamental problems of scientific research, out of which, in unexpected ways, discoveries of high practical utility often grow, and also to the perfecting of inventions and improvements in industrial processes. These latter are often of such a nature that they can neither be patented nor kept secret, and, therefore, the whole of the extra reward, which they at first bring to their inventor, is very quickly transferred from him to the general public in the form of reduced prices. The patent laws aim, in effect, at bringing marginal private net product and marginal social net product more closely together. By offering the prospect of reward for certain types of invention they do not, indeed, appreciably stimulate inventive activity, which is, for the most part, spontaneous, but they do direct it into channels of general usefulness.[5]

Corresponding to the above investments in which marginal private net product falls short of marginal social net product, there are a number of others, in which, owing to the technical difficulty of enforcing compensation for incidental disservices, marginal private net product is greater than marginal social net product. Thus, incidental uncharged disservices are rendered to third parties when the game-preserving activities of one occupier involve the overrunning of a neighbouring occupier's land by rabbits—unless, indeed, the two occupiers stand in the relation of landlord and tenant, so that compensation is given in an adjustment of the rent. They are rendered, again, when the owner of a site in a residential quarter of a site builds a factory there and so destroys a great part of the amenities of the neighbouring sites; or, in a less degree, when he uses his site in such a way as to spoil the lighting of the houses opposite:[6] or when he invests resources in erecting buildings in a crowded centre, which, by contracting the air space and the playing-room of the neighbourhood, tend to injure the health and efficiency of the families living there. Yet again, third parties—this time the public in general—suffer incidental uncharged disservices from resources invested in the running of motor cars that wear out the surface of the roads. The case is similar—the conditions of public taste being assumed—with resources devoted to the production and sale of intoxicants. To enable the social net product to be inferred from the private net product of the marginal pound invested in this form of production, the investment should, as Mr. Bernard Shaw observes, be debited with the extra costs in policemen and prisons which it indirectly makes necessary.[7] Exactly similar considerations hold good in some measure of foreign investment in general. For, if foreigners can obtain some of the exports they need from us by selling promises, they will not have to send so many goods; which implies that the ratio of interchange between our exports and our imports will become slightly less favourable to us. For certain sorts of foreign investments more serious reactions come into account. Thus, when the indirect effect of an increment of investment made abroad, or of the diplomatic manoeuvres employed in securing the concession for it, is an actual war or preparations to guard against war, the cost of these things ought to be deducted from any interest that the increment yields before its net contribution to the national dividend is calculated. When this is done, the marginal social net product even of investments, which, as may often happen in countries where highly profitable openings are still unworked and hard bargains can be driven with corrupt officials, yield a very high return to the investors, may easily turn out to be negative. Yet again, when the investment consists in a loan to a foreign government and makes it possible for that government to engage in a war which otherwise would not have taken place, the indirect loss which Englishmen in general suffer, in consequence of the world impoverishment caused by the war, should be debited against the interest which English financiers receive. Here, too, the marginal social net product may well be negative. Perhaps, however, the crowning illustration of this order of excess of private over social net product is afforded by the work done by women in factories, particularly during the periods immediately preceding and succeeding confinement; for there can be no doubt that this work often carries with it, besides the earnings of the women themselves, grave injury to the health of their children.[8] The reality of this evil is not disproved by the low, even negative, correlation which sometimes is found to exist between the factory work of mothers and the rate of infantile mortality. For in districts where women's work of this kind prevails there is presumably—and this is the cause of the women's work—great poverty. This poverty, which is obviously injurious to children's health, is likely, other things being equal, to be greater than elsewhere in families where the mother declines factory work, and it may be that the evil of the extra poverty is greater than that of the factory work.[9] This consideration explains the statistical facts that are known. They, therefore, militate in no way against the view that, *other things equal*, the factory work of mothers is injurious. All that they tend to show is that prohibition of such work should be accompanied by relief to those families whom the prohibition renders necessitous.[10]

At this point it is desirable to call attention to a somewhat specious fallacy. Some writers unaccustomed to mathematical analysis have imagined that, when improved methods of producing some commodities are introduced, the value of the marginal social net product of the resources invested in developing these methods is less than the value of the marginal private net product, because there is not included in the latter any allowance for the depreciation which the improvement causes in the value of existing plant; and as they hold, in order to arrive at the value of the marginal social net product, such allowance ought to be included.[11] If this view were correct, reason would be shown for attempts to make the authorisation of railways dependent on the railway

companies compensating existing canals, for refusals to license motor om-
nibuses in the interests of municipal tramways, and for the placing of hind-
rances in the way of electric lighting enterprises in order to conserve the contribution made to the rates by municipal gas companies. But in fact the view is
not correct. The marginal social net product of resources devoted to *improved
methods of producing a given commodity* is not, in general, different from the marginal private net product; for whatever loss the old producers suffer during a
reduction in the price of their products is balanced by the gain which the reduction confers upon the purchasers of their products. This is obvious if, after
the new investment have been made, the old machines continue to produce the
same output as before at reduced prices. If the production of the old machines
is diminished on account of the change, it seems at first sight doubtful. Reflection, however, makes it plain that no unit formerly produced by the old machinery will be supplanted by one produced by the new machinery, except
when the new machinery can produce it at a *total cost* smaller than the *prime cost*
that would have been involved in its production with the old machinery: except, that is to say, when it can produce it at a price so low that the old machinery would have earned nothing by producing it at that price. This implies that
every unit taken over by the new machinery from the old is sold to the public at
a price *reduced* by as much as the whole of the net receipts, after discharging
prime costs, which the old machinery would have obtained from it if it had produced that unit. It is thus proved that there is no loss to the owners of the old
machineries, in respect of any unit of their former output, that is not offset by
an equivalent gain to consumers. It follows that to count the loss to these
owners, in respect of any unit taken over from them by the new machinery, as a
part of the social cost of producing that unit would be incorrect.

An attempt to avoid this conclusion may, indeed, still be made. It may be
granted that, so far as direct effects are concerned, ordinary commercial policy,
under which investment in improved processes is not restrained by consideration for the earnings of other people's established plant, stands vindicated.
There remain, however, indirect effects. If expensive plant is liable to have its
earnings reduced at short notice by new inventions, will not the building of
such plant be hindered ? Would not the introduction of improved processes on
the whole be stimulated, if they were in some way guaranteed against too rapid
obsolescence through the competition of processes yet further improved ? The
direct answer to this question is, undoubtedly, yes. On the other side, however,
has to be set the fact that the policy proposed would retain inferior methods in
use when superior methods were available. Whether gain or loss on the whole
would result from these two influences in combination, is a question to which
it seems difficult to give any confident answer. But this impotent conclusion is
not the last word. The argument so far has assumed that the rapidity with
which improvements are invented is independent of the rapidity of their practi-

cal adoption; and it is on the basis of that assumption that our comparison of rival policies fails to attain a definite result. As a matter of fact, however improvements are much more likely to be made at any time, if the best methods previously discovered are being employed and, therefore, watched in actual operation, than if they are being held up in the interest of established plant. Hence the holding-up policy indirectly delays, not merely the adoption of improvements that have been invented, but also the invention of new improvements. This circumstance almost certainly turns the balance. The policy proper to ordinary competitive industry is, therefore, in general and on the whole, of greater social advantage than the rival policy. It is not to the interest of the community that business men, contemplating the introduction of improved methods, should take account of the loss which forward action on their part threatens to other business men. The example of some municipalities in postponing the erection of electric-lighting plant till their gas plant is worn out is not one that should be imitated, nor one that can be successfully defended by reference to the distinction between social and private net products. The danger that beneficial advances may be checked by unwise resistance on the part of interested municipal councils is recognised in this country in the rules empowering the central authority to override attempts at local vetoes against private electrical enterprise. The policy followed by the Board of Trade is illustrated by the following extract from their report on the Ardrossan, Saltcoats and District Electric Lighting Order of 1910: "As the policy of the Board has been to hold that objection on the grounds of competition with a gas undertaking, even when belonging to a local authority, is not sufficient reason to justify them in refusing to grant an Electric Lighting Order, the Board decided to dispense with the consent of the Corporation of Ardrossan."[12]

So far we have considered only those divergences between private and social net products that come about through the existence of uncompensated services and uncharged disservices, the general conditions of popular taste being tacitly assumed to remain unchanged. This is in accordance with the definition of social net product given in Chapter II. As was there indicated, however, it is, for some purposes, desirable to adopt a wider definition. When this is done, we observe that a further element of divergence between social and private net products, important to economic welfare though not to the actual substance of the national dividend, may emerge in the form of uncompensated or uncharged effects upon the *satisfaction that consumers derive from the consumption of things other than the one directly affected.* For the fact that some people are now able to consume the new commodity may set up a psychological reaction in other people, directly changing the amount of satisfaction that they get from their consumption of the old commodity. It is conceivable that the reaction may lead to an increase in the satisfaction they obtain from this commodity, since it may please them to make use of a thing just because it is superseded and more or less archaic. But, in general, the reaction will be in the other direction. For, in some measure, people's affection for the best quality of anything is due simply to the fact that it is the best quality; and, when a new best, superior to the old best, is created, that element of value in the old best is destroyed. Thus, if an improved form of motor car is invented, an enthusiast who desires above all "the very latest thing" will, for the future, derive scarcely any satisfaction from a car, the possession of which, before this new invention, afforded him intense pleasure. In these circumstances the marginal social net product of resources invested in producing the improved type is somewhat smaller than the marginal private net product.[13] It is *possible* that the introduction of electric lighting into a town may, in some very slight degree, bring about this sort of psychological reaction in regard to gas: and this possibility may provide a real defence, supplementary to the fallacious defence described in the preceding section, for the policy of municipalities in delaying the introduction of electricity. This valid defence, however, is almost certainly inadequate. The arguments actually employed in support of the view that municipalities should not permit competition with their gas plant are those described in the preceding section. They are, in general, independent of any reference to psychological reactions, and are, therefore, like the arguments which persons interested in canals brought against the authorisation of the early railways, wholly fallacious.

It is plain that divergences between private and social net product of the kinds we have so far been considering cannot, like divergences due to tenancy laws, be mitigated by a modification of the contractual relation between any two contracting parties, because the divergence arises out of a service or disservice rendered to persons other than the contracting parties. It is, however, possible for the State, if it so chooses, to remove the divergence in any field by "extraordinary encouragements" or "extraordinary restraints" upon investments in that field. The most obvious forms which those encouragements and restraints may assume are, of course those of bounties and taxes. Broad illustrations of the policy of intervention in both its negative and positive aspects are easily provided.

The private net product of any unit of investment is unduly large relatively to the social net product in the businesses of producing, and distributing alcoholic drinks. Consequently, in nearly all countries, special taxes are placed upon these businesses. Marshall was in favour of treating in the same way resources devoted to the erection of buildings in crowded areas. He suggested to a witness before the Royal Commission on Labour, "that every person putting up a house in a district that has got as closely populated as is good should be compelled to contribute towards providing free playgrounds".[14] The principle is susceptible of general application. It is employed, though in a very incomplete and partial manner, in the British levy of a petrol duty and a motor-car licence tax upon the users of motor cars, the proceeds of which are devoted to the service of the roads.[15] It is employed again in an ingenious way in the National Insurance Act. When the sickness rate in any district is exceptionally high, provision is made for throwing the consequent abnormal expenses upon employers, local authorities or water companies, if the high rate can be shown to be due to neglect or carelessness on the part of any of these bodies. Some writers have thought that it might be employed in the form of a discriminating tax upon income derived from foreign investments. But, since the element of disadvantage described in § 10 only belongs to some of these investments and not to others, this arrangement would not be a satisfactory one. Moreover, foreign investment is already penalised to a considerable extent both by general ignorance of foreign conditions and by the fact that income earned abroad is frequently subjected to foreign income tax as well as to British income tax.

The private net product of any unit of investment is unduly small in industries, such as agriculture, which are supposed to yield the indirect service of developing citizens suitable for military training. Partly for this reason agriculture in Germany was accorded the indirect bounty of protection. A more extreme form of bounty, in which a governmental authority provides *all* the funds required, is given upon such services as the planning of towns, police administration, and, sometimes, the clearing of slum areas. This type of bounty is also not infrequently given upon the work of spreading information about improved processes of production in occupations where, owing to lack of appreciation on the part of potential beneficiaries it would be difficult to collect a fee for undertaking that task. Thus the Canadian Government has established a system, "by means of which any farmer can make inquiry, without even the cost of postage, about any matter relating to his business";[16] and the Department of the Interior also sometimes provides, for a time, actual instruction in farming.[17] Many Governments adopt the same principle in respect of information about Labour, by providing the services of Exchanges free of charge. In the United Kingdom the various Agricultural Organisation Societies are voluntary organisations, providing a kindred type of bounty at their subscribers' expense. An important part of their purpose is, in Sir Horace Plunkett's words, to bring freely "to the help of those whose life is passed in the quiet of the field the experience, which belongs to wider opportunities of observation and a larger acquaintance with commercial and industrial affairs".[18] The Development act of 1909, with its provision for grants towards scientific research, instruction, and experiment in agricultural science, follows the same lines.

It should be added that sometimes, when the interrelations of the various private persons affected are highly complex, the Government may find it necessary to exercise some means of authoritative control in addition to providing a bounty. Thus it is coming to be recognised as an axiom of government that, in every town, power must be held by some authority to limit the quantity of building permitted to a given area, to restrict the height to which houses may be carried,—for the erection of barrack dwellings may cause great overcrowding of area even though there is no overcrowding of rooms,[19]—and generally to control the building activities of individuals. It is as idle to expect a well planned town to result from the independent activities of isolated speculators as it would be to expect a satisfactory picture to result if each separate square inch were painted by an independent artist. No "invisible hand" can be relied on to produce a good arrangement of the whole from a combination of separate treatments of the parts. It is, therefore, necessary that an authority of wider reach should intervene and should tackle the collective problems of beauty, of air and of light, as those other collective problems of gas and water have been tackled. Hence, shortly before the war, there came into being, on the pattern of long previous German practice, Mr. Burns's extremely important townplanning Act. In this Act, for the first time, control over individual buildings, from the standpoint, not of individual structure, but of the structure of the town as a whole, was definitely conferred upon those town councils that are willing to accept the powers offered to them. Part II. of the Act begins: "A town-planning scheme may be made in accordance with the provisions of this Part of the Act as respects any land which is in course of development, or appears likely to be used for building purposes, with the general object of securing proper sanitary conditions, amenity, and convenience in connection with the laying out and use of the land, and of any neighbouring lands." The scheme may be worked out, as is the custom in Germany, many years in advance of actual building, thus laying down beforehand the lines of future development. Furthermore, it may, if desired, be extended to include land on which buildings have already been put up, and may provide "for the demolition or alteration of any buildings thereon, so far as may be necessary for carrying the scheme into effect". Finally where local authorities are remiss in preparing a plan on their own initiative, power is given to the appropriate department of the central Government to order them to take action. There is ground for hope, however, that, so soon as people become thoroughly familiarised with town-planning, local patriotism and inter-local emulation will make resort to pressure from above less and less necessary.

Pigou, A.C., *The Economics of Welfare*, Macmillan, London, 1920, p. 183-196.

Notes and References

1   *Principles of Political Economy*, p. 406.

2   Cf. Smart, *Studies in Economics*, p. 314.

3   It has been said that in London, owing to the smoke, there is only 12 per cent as much sunlight as is astronomically possible, and that one fog in five is directly caused by smoke alone, while all the fogs are befouled and prolonged by it (J.W. Graham, *The Destruction of Daylight*, pp. 6 and 24). It would seem that mere ignorance and inertia prevent the adoption of smoke-preventing appliances in many instances where, through the addition they would make to the efficiency of fuel, they would be directly profitable to the users. The general interest, however, requires that these devices should be employed beyond the point at which they "pay". There seems no doubt that by means of mechanical stokers, hot-air blasts and other arrangements, factory chimneys can be made practically smokeless. Noxious fumes from alkali works are suppressed by the law more vigorously than smoke (*ibid.* p. 12136).

4   Thus the Interim Report of the Departmental Committee on smoke and Noxious Vapours Abatement 1920 contains the following passages:

"17. *Actual economic loss Due to Coal Smoke.*—It is impossible to arrive at any complete and exact statistical statement of the amount of damage occasioned to the whole community by smoke. We may, however, quote the following investigations.

"A report on an exhaustive investigation conducted by an expert Committee of engineers, architects, and scientists in 1912 in Pittsburgh, USA, estimated the cost of the smoke nuisance to Pittsburgh at approximately £4 per head of the population per annum.

"18. A valuable investigation was made in 1918 by the Manchester Air Pollution Advisory Boars into the comparative cost of household washing in Manchester—a smoky town—as compared with Harrogate—a clean town. The investigator obtained 100 properly comparable statements for Manchester and Harrogate respectively as to the cost of the weekly washing in working-class houses. These showed an extra cost in Manchester of 71/2d. a week per household for fuel and washing material alone, disregarding the extra labour involved, and assuming no greater loss for middle-class than for working-class households (a considerable under-statement), works out at over £290,000 a year for over a populations of three quarters of a million."

5   Cf. Taussig, *Inventors and Money Makers*, p. 51.

6   In Germany the town-planning schemes of most cities render anti-social action of this kind impossible; but in America individual site-owners appear to be entirely free, and in England to be largely free, to do what they will with their land. (cf. Howe, *European Cities at Work*, pp. 46, 95 and 346.)

7   *The Common Sense of Municipal Trading*, pp. 19-20.

8   Cf. Hutchins, *Economic Journal*, 1908, p. 227.

9   Cf. Newsholme, *Second Report on Infant and Child Mortality* [d. 6909] p. 56. Similar considerations to the above hold good of night work by boys. *The Departmental Committee on Night Employment* did not, indeed, obtain any strong evidence that this work injures the boys' health. But they found that it reacts injuriously on their efficiency in another way, *i.e.* by practically pre-eluding them from going on with their education in continuation classes and so forth. The *theory* of our factory laws appears to be that boys between 14 and 18 should only be permitted at night on unnecessary non-continuous processes which are carried out in the same factory as continuous processes of such a kind that great loss would result if they did not so. The *practice* of these laws, however, permits them to be employed at night on unnecessary non-continuous processes which are carried out in the same factory as continuous processes. Consequently, the Committee recommended that in future "such permits should be granted in terms of processes, and not of premises, factories, or parts of factories without reference to processes" ([Cd. 6503], p. 17).

10 Cf. *Annual Report of the Local Government Board*, 1909-10, p. 57. The suggestion that the injurious consequences of the factory work of mothers can be done away with, if the factory worker gets some unmarried woman to look after her home in factory hours, is mistaken, because it ignores the fact that a woman's work has a special personal value in respect of her own children. In Birmingham this fact seems to be recognised, for, after a little experience of the bad results of putting their children out to "mind", married women are apt, it was said before the war, to leave the factory and take to home work. (Cf. Cadbury, *Women's Work*, p. 175.)

11 For example, J.A. Hobson, *Sociological Review*, July 1911, p. 197, and *Gold, Prices and Wages*, pp. 107-8. Even Sidgwick might be suspected of countenancing the argument set out in the text (cf. *Principles of Political Economy*, p. 408). It does not seem to have been noticed that this argument, if valid, would justify the State in prohibiting the use of new machinery that dispenses with the services of skilled mechanics until the generation of mechanics possessing that skill has been depleted by death.

12 Cf. Knoop, *Principles and Methods of Municipal Trading*, p. 35.

13 It should be noted that the argument of the text may be applicable even where the product formerly consumed is wholly superseded by the new rival, and where, therefore, nobody is actually deriving diminished satisfaction from the old product: for it may be that complete supersession would not have come about unless people's desire for the old product had been reduced by the psychological reaction we have been contemplating. Furthermore, the preceding argument shows that inventions may actually diminish aggregate economic welfare; for they *may* cause labour to be withdrawn from other forms of productive service to make a new variety of some article to supersede an old one, whereas, if there had been no invention, the old one would continued in use would have yielded as much economic satisfaction as the new one yields now. This is true, broadly speaking, of inventions of new weapons of war, so far as these are known to all nations, because it is of no advantage to one country to have improved armaments if its rivals have them also.

14 *Royal Commission on Labour*, Q. 8665.

15 The application of the principle is incomplete, because the revenue from these taxes, administered through the Road Board, must be devoted, "Not to the ordinary road maintenance are all, however onerous it might be, but exclusively to the execution of new and specific road improvements" (Webb, *The King's Highway*, p. 250). Thus, in the main, the motorist does not pay for the damage he does to the ordinary roads, but obtains in return for his payment an additional services useful to him rather than to the general public.

16 Mavor, *Report on the Canadian North-West*, p. 36.

17 *Ibid.* p. 78.

18 C. Webb, *Industrial Co-operation*, p. 149.

19 Mr. Dawson believes that this type of overcrowding prevails to a considerable extent in German Towns. He writes: " The excessive width of the streets, insisted on by cast-iron regulations, adds greatly to the cost of housing-building, and in order to recoup himself, and make the most of his profits, the builder begins to extend his house vertically instead of horizontally" (*Municipal Life and Government in Germany*, pp. 163-4). Hence German municipalities now often control the height of buildings, providing a scale of permitted heights which decreases on passing from the centre to the outlying parts of a town.

CHAPTER 4

The Sociology of Nature

L. REID

The Elements

Perhaps the most striking, the most moving, thing about the wild creatures in all their widespread diversity, still teeming over the surface of the earth, is the simple fact that they do so, that in the worlds of Mallory explaining his desire to climb Mount Everest, they are there. We, or some of us in our egocentric way, are sometimes heard to ask: 'What good are they?' But this is a question unlikely even to occur to anyone who has taken the trouble to examine them and their ways. The student of natural history knows that they exist in their own right, and that this is an important part of their fascination. They live their own lives sublimely independent of humanity, except when humanity intervenes to extirpate them because they get in the way of its feverish activities. They were tenants of this planet long before man ceased to be an ape-like creature, many of them by hundreds of millions of years. Their frail yet marvellously enduring generation have succeeded one another over a lapse of time for which the hackneyed adjective immemorial is an absurd understatement.

Think of *Volvox*, that minute, emerald-green plant-animal, rolling now in serene detachment through a drop of water on the slide of a microscope, then of all the hosts of *Volvox* as having rolled in identical serenity for almost as long as life has existed. Think of all the creatures inhabiting a coral reef in kaleidoscopic variety, of beetles in the grass, ants in their thronged cities, butterflies lapping up the moisture of a sandbank flanking some tropical river, ripplemarked fishes in rippling movement, gannets nesting in their thousands on some unvisited island. All these and uncountably more rely wholly on themselves and on one another. They exist for us to wonder at.

But after all there is much more to move us than this fact of self-sufficient existence. Their diversity is equally a subject for wonder, as is their irrepressible urge to propagate their kind, their grip upon life, and perhaps above all the way their structure and habits are adapted to the environment in which they live. To wish to understand something of their lives and ways, and to pursue this quest in a disinterested spirit, believing that such knowledge is worth having for its own sake, is an admirable thing. It has resulted in the accumulation of an imposing corpus of theory and conclusion which is among the greatest achievements of human mind, but and achievement almost solely of recent times. At the beginning there was bewilderment, an attempt with many a false start to grasp the significance of the wood without first examining the trees. Before long it became clear that no single intelligence could hope to cover the whole field. For this reason there arose the specialist, whose business it became to isolate circumscribed departments of knowledge and give them his whole attention. His method was analytical, a breaking down, a dividing. This went on in spite of a growing realization that it was the wood as a whole that mattered, that specialization can never be an end in itself. Analysis must give way to synthesis, since the further specialization continues the clearer does it become that a close dependence exists, not only between the different branches of a science, but between one science and another. Science therefore, having given detailed attention to the trees, is now turning its attention on the wood, is coming back in fact to the viewpoint of the earliest inquirers, with the all-important difference that it is now provided with the raw material for synthesis, the varied contributions of the specialists.

There is little doubt that this is true of science as a whole. It is certainly true of the science as a whole. It is certainly true of the science of biology, in which perhaps there has been more specialization than in any other. A synthesis is now well under way, and we call it ecology, which can be defined as the study of living things in relation to their surroundings and to one another. The more deeply ecology is studied, and it is still in its early stages, the clearer does it become that mutual dependence is a governing principle, that animals are bound to one another by unbreakable ties of dependence. But that is no more than a fraction of the truth, for the ties of dependence linking animals with one another are no closer than those linking animals with plants, while plants in their turn cannot be dissociated from climate, nor from the earth itself in the form of rocks and soil. Without plants there could be no animals, without soil and without rainfall there could be no plants. Nor are the several ties traceable in one direction only. If animals depend on plants, so in a number of vitally important ways do plants depend on animals. Similarly, while plants cannot exist without soil, neither can soil exist as such without contributions made to it by plants. Yet again, rainfall is a vital necessity for plants, and plants in the form of trees are not without their influence in the making of rain, while playing a part of the greatest importance in helping to conserve the rain for their own use after it has soaked into the soil. If trees can be said to increase rainfall, so to a very much greater extent do mountains. So the earth, which cannot support life without the rain that the plants require, plays its part in adding to the supply.

What has just been written is no more than a bare summary of a fundamental principle governing the existence of life. I shall return to it again and again. There is another fundamental principle that the modern study of ecology brings out, and that is change. It was the Greek philosopher Heraclitus who, as long ago as the sixth century B.C., summed up his teaching in the sentence 'all things flow'. Denying the reality of static existence, he declared that eternal flux was the condition of the universe. With all things in nature, according to him, it is not a matter of being but of becoming, a dynamic process of transition. He declared further that the primordial substance of the universe was fire. Today we know better than to agree with the second of these conclusions, but the ecologist of modern times will certainly agree whole-heartedly with the first. Both biologists and geologists of today are well aware of the soundness of that doctrine and can offer proof entirely unknown to Heraclitus, who after all was no scientist as we understand the term.

Much of the evidence in favour of Heraclitus' dictum is simple and apparent. We can see it for ourselves, can hardly help seeing it, for indeed it governs our lives. The sun rises and sets as the earth rotates. The seasons succeed one another as the earth speeds on its orbit round the sun, its axis making a never-changing angle with the plane of that orbit. The seed germinates, the seedling proliferates into leaves, flowers, and fruit. Every animal, from the most primitive up to man himself, undergoes a life-cycle which is one continuous change. These and a hundred others like them are appreciable to us because the phase of their fulfilment falls well within the span of human life. They are the short-term, obvious changes, and the very first men of all were certainly aware of them. They have the utmost relevance for ecology. But there are others, equally relevant, but far less easily perceptible because slower, more gradual, some requiring a span equivalent to many human lives, others of such gradualness that we are quite incapable of seeing them directly and can infer their existence only from the traces they have left. Such are the slow, secular changes of climate, the onset and recession of glaciers, the transgression and withdrawal of seas, the deposition of sediments to form rocks, the erosion of mountains. It is only within the past 150 years that we have become aware of fluxations of that order. Their realization was one of the major revolutions in human thought; and intimately linked with them was the realization of changing, developing life, the doctrine of evolution, which indeed was one of them, perhaps the most important.

We see further that it is not merely the notion of change that matters, but the inevitable gradualness of change. In nature there are few abrupt transitions, for on the whole she abhors them. This applies even to the transitions in space that we can see for ourselves. Abrupt enough to a casual glance, they are seen to be less so when looked at more closely. That between sea and land for instance is traceable, not as a line but as a zone, that between tidemarks, which in truth is a zone involving both space and time. Apart from this, the things of the land birds, plants, insects—invade to some extent the kingdom of the sea; while the influence of the sea penetrates landwards to a notable and varying extent. In a similar way the snow-line on mountains ebbs and flows with the seasons, forests grade into savanna or steppe, savanna and steppe into desert. Broadly speaking we can say that it is only when the normal gradualness of a change has in some way and for some time been prevented that it takes place suddenly and with violence.

This conception of the gradualness of change is bound up with certain primary, and to us sharply defined distinctions. We make one of this sort for instance between the two kinds of life, plant and animal; and while there is no denying that the distinction is a valid one, it has long been recognized that there are many unicellular organisms having the attributes of both. We cannot say that they are exclusively either the one or the other. The change from plant to animal, in origin at least, was a gradual one. Another and even more sharply defined distinction, to us the most important of all, is that between the living and the non-living. We are convinced that every particle of matter in the universe is either the one or the other. But even that distinction has lost a large measure of its rigidity in recent times. Crystals are known to possess almost all the attributes of living things; while those still mysterious and infinitesimally small organisms, the filter-passing viruses, behave at one time like inanimate crystals, at another like living things, according to whether they are detached from or are part of the hosts on which they prey. This is a comparatively recent discovery, amounting to, or looking as though it might in time amount to a revolution in thought in itself. It may be that life evolved from non-life as gradually and by as continuous a process as multicellular from unicellular organisms, or birds from reptiles. It is highly probable that the change from the non-living to the living was as gradual as that from plant to animal. It may even be that death itself, to us the most violent change of all, is less violent than we suppose.

These two distinctions, between the plant-cell and the animal-cell, between the organic and inorganic; and these two primary principles, mutual dependence and gradual change, cannot fail to be implicit in any survey of the whole complex web of life. The distinctions can be taken for granted, with the reservations already made. As for the two principles they will have to be explicit and that continually, since they are far less easily realized, and since both of them determine the very nature, weave the essential fabric, of the web of life.

The four spheres

The main concern of this book is with animal life, but since plants sustain animals, they must play an important part in it as well. Further, since plants depend utterly on soil on the one hand and on climate on the other, these two also must be given a place in this introductory chapter at least. With this in mind and taking the broadest of views, we can distinguish four great entities or spheres: first the lithosphere which is the outer crust of the earth; secondly the hydrosphere, comprising all the salt waters; thirdly the atmosphere; and lastly the biosphere, a word signifying the network of all animated things, both plant and animal. Of these, where all land-surfaces are concerned, the first and the third provide the indispensable substratum for the maintenance of life, both plant and animal, in all its wonderful diversity.

The lithosphere

The deeper parts of the earth's crust are made up of igneous rocks, those that is to say that have cooled and solidified from the molten condition. Over much of the surface as well, rocks of this kind are to be found, granite, basalt and the like; but overlying them elsewhere and covering an enormous area of continent and island, are the rocks of sedimentary origin, laid down for the most part in shallow seas and lakes particle by particle. The connexion between the two is an intimate one, since the sedimentary rocks—sandstones, limestones, clays—were built up originally from the piecemeal destruction of the igneous. They are in a special sense secondary as well as sedimentary; and the slow, subsequent fragmentation of both kinds is brought about by atmospheric weathering. Then comes erosion, the transportation and the final deposition of the particles by the four great agents of erosion, rivers, glaciers, wind, and sea. Of these four rivers perhaps are the most important, at least in those parts of the earth where rainfall is sufficient for rivers to flow perennially.

Contemplate imaginatively a rock section where beds say of sandstone are exposed in some quarry. Grasp the fact that their presence is accounted for by geographical conditions obtaining at some remote period, a river flowing into a sea, winding perhaps over a swampy delta, that as it meandered and bifurcated over the delta, and still more when it came to the sea and the current slackened to immobility, a slow rain of particles fell and very gradually accumulated. Realize that those beds of sandstone are the direct result of those conditions, and finally that they represent a period of time during which the conditions held good and that work was done.

In time the conditions are bound to change. The sea at that particular point may deepen, alternatively it may ebb away. If it deepens, perhaps in response to some warping of the crust the river continues to deposit its particles, but the shallows at the mouth of the river are now at a distance, and the particles continuing to be deposited in the deepening sea are finer because conveyed further offshore. It is only the finer particles that can be so conveyed. For this reason, if the particles are very fine and the sea comparatively deep, the rock being formed will be of a different kind, a day rather than a sandstone. Here is change and gradual change, as perhaps may be shown in the rock-section exposed in the quarry from the way in which the coarse-textured sandstone shades off into a fine-textured day. Broadly speaking the finer the particles the deeper the sea and the greater the distance from the landmass more an assemblage of trees, discharges moisture previously sucked from the soil. This is known as transpiration, and it adds inevitably to the humidity of the air, increasing its burden of water-vapour, increasing the likelihood of condensation. So trees, themselves vitally dependent on rainfall, play their part in bringing it about. This is a good example of those intimate, mutual associations of which there are so many in the natural world. Break the association, as man has done in many parts of the world, and the result can be catastrophic where life is concerned, including human life. Cut down trees and you help to make a desert.

The hydrosphere
So two of the great spheres, the lithosphere and the atmosphere, the one pertaining to the earth, the other ultimately to the sun, in their mutual association provide the conditions needed for the maintenance of life on land, whether plant or animal. But there is a third, the hydrosphere, the great oceans, covering more than seventy per cent of the earth's surface, with the continents after all no more than islands in their midst. Quite apart from this claim to our consideration 'the great salt deep' teems with living things to an extent surpassing dry land, since the world of the waters is a three-dimensional world in a way that the land, even including the air above it, can never be. Since there are animals uncountable in the great waters, there must also be plants; but the distribution of plants on dry land and in the sea is restricted, though for different reasons; on dry land because parts of it are too dry, in the sea because parts of it, indeed the major part, are too dark. Plants must have sunlight to synthesize their food and that of animals. For this reason the larger fixed plants are found in the sea only to the depth penetrable by sunlight, which means the continental shelves, and the plants are seaweeds exclusively. The deeps of the sea, not the ultimate abysses only, but the great oceanic basins below a depth of a hundred fathoms, are devoid of seaweeds. On the other hand the surface waters everywhere, again to the depth penetrable by sunlight, swarm with floating plants of microscopic size the diatoms in their uncountable millions. Seaweeds contribute comparatively little to the food of creatures inhabiting the great waters. Diatoms on the other hand are the basis of all the food that there is.

   The sea has another claim on our attention, for there is little doubt that it became both the womb and the nursery of life itself. It was in the primeval oceans of the world that life must have originated, perhaps somewhere in the shallows accessible to what little sunlight could penetrate the dense cloudbanks, from which torrential rain fell ceaselessly for months, even years at a time. Somewhere in that dim natural laboratory protoplasm, the stuff of life, was prepared, kneaded, given coordinated shape. We can only guess at the nature of the first living organisms, but it is reasonable to suppose that they stood at that border-line between plant and animal, where even today some organisms are to be found. But perhaps it is more likely that they were strictly neither the one nor the other, organisms of a simpler structure than anything we know today, re-lated remotely to the bacteria and capable, like some of them, of ingesting inor-ganic substances directly.

At a much later date, in response to augmented sunshine, there took place that great parting of the roads, one by way of simple unicellular organisms, green with the first chlorophyll, to all the plants that have since appeared; the other by way of equally simple organisms, lacking chlorophyll and capable of devouring those that possessed it, to all the animals. It was then, along those two associated but diverging paths, that there began that evolution of living things which has resulted in so wonderful a diversity. The oceans began to swarm, first with invertebrate then with vertebrated creatures, but for a long time the oceans, and no doubt the fresh waters, alone. For millions of years life was confined to the waters, while the continents showed nothing but naked rock, were the playground of wind and rain, of glaciers, volcanoes, and blown desert sand, devoid of life, devoid of true soil.

The biosphere

Then at a period comparatively advanced in across which the river flowed. This conveys a relationship in space between the two. A relationship in time is simultaneously shown by the fact that the fine-textured clay overlies the coarse-textured sandstone. The first was laid down at a later date than the sec-ond.

If the sea on the other hand is slowly ebbing and becoming less deep, the shoreline advances, a widening belt is won from the sea and a fresh land sur-face will be laid bare, with all that that means in the way of an increasing mantle of living things. In that event no sedimentation can occur. Picture a change of yet another kind. The river may be diverted or dry up altogether be-cause of some alteration of climate. The sea is then clear of sediment, and there may occur a slow precipitation of lime steadily accumulating over the floor. This, together with the mingling deposition of the shells of dead creatures, or of countless millions of the tests of minute organisms, will give rise to beds of limestone, or of chalk which is a form of limestone.

Changes of this sort, slow, rhythmic pulsations, responding to warpings and heavings of the underlying substratum, or others of a more violent sort, such as oozings of lava from fissures in the crust, explosive eruptions also from time to time, scattering rock fragments and clouds of comminuted dust, have all gone on at intervals succeeding one another, repeating themselves over and over again since the earth cooled from a molten condition some thousands of millions of years ago. In this way there has evolved the crust of the earth as we know it, the lithosphere, layer upon layer, each sooner or later to be exposed, and when exposed acted upon instantly by the agents of erosion and destroyed, so that fresh layers can be laid down elsewhere. The major sort of rock are comparatively few: granites, basalts, conglomerates, sandstones, limestones, clays, and shales. These, with local variations differing in some sort chemically and in another sort organically, as shown by the fossils they contain, have come into existence over and over again in endless repetition throughout the long history of the earth. They constitute the crust and are of primary importance for living things, since from them has been formed the soil, the platform for plants and so for animals, that soil which varies from place to place largely according to the size of the particles originally laid down, that size of the particles determining the capacity of the soil to retain the moisture on which the plants depend for their sustenance.

The atmosphere
Moisture calls to mind climate, and so to the second great sphere, the atmosphere. Temperature, pressure, rainfall are the things that matter most where climate is concerned. Temperature, varying in the widest sense of all with latitude, is affected also by the angle that the sun makes with the horizon, that is to say according to the seasons, by the distance of any one part of the earth from the sea, by altitude above sea-level. Warm air thins away, cold air is compressed. Thus it is variations in temperature that cause variations in atmospheric pressure measured by the barometer. Variations in pressure cause winds to blow in an attempt to equalize the pressure. Winds bring rain. All the airs that blow hold water-vapour to an extent that varies with the temperature of the air. But there is a limit to the holding capacity, and the point of saturation depends also upon temperature. Reduce the temperature of the saturated air and the water-vapour condenses as rain. The rain-making agency, one of the fundamentally important power-producing engines of the world, is that which cools air. Cooling occurs on a stupendous scale over the great oceans when cold, dry air from polar regions mingles in great swirls with warm, moist air from equatorial latitudes. The moist air is driven upwards by the drier air, cools as it rises and lets fall torrents of rain. These ascending spirals are the depressions we hear about so frequently, and when they reach our continental coasts they are the more likely to bring rain because of the friction of land against air. On a local scale mountains cause further ascent and so further precipitation. Similarly, on a yet more local scale, will forests with the friction they can provide. Apart from this, a tree, and still the history of the earth, there took place that invasion of the land from the sea, that milestone in history if ever there was one. What creatures they were that, in Silurian[1] times, undertook this

laudable enterprise is far from certain, but it is supposed that they were arthropods, perhaps scorpion-like things. Considering what the consequences were to be we are justified in calling this an invasion, but as such there is little doubt that it was a compromising, tentative business, for a time at least far from a wholehearted abandonment of the one element for the other, rather as always a gradual transition, resulting in an amphibious existence. There could be no complete exodus and settlement of animal life on the land unless plants took part at the same time. But this must have happened as an equally gradual transition, perhaps of some of the seaweeds taking possession of a zone above high-water mark, so that in time they and their descendants could get to work on the inhospitable rocks of the continental margins, hasten their decay, and contribute humus in such a way that soil was at last produced. Only then could there be true land plants and true land animals.

Even at that remote time, measured in terms of the history of life rather than that of the earth itself, evolution had progressed to a surprising extent. The earliest fishes appeared in late Silurian times, and during the succeeding period, the Devonian, both salt waters and fresh fairly swarmed with them. But in the sea, apart from land animals that returned to it much later to become the whales and seals we know today, evolution went not much further, no further than the fishes of the present time. The sea, though an admirable medium both then and now for life, has its limits where further progress is concerned; and the tremendous significance of the migration of living things from the sea to the land is to be found in the fact that not until it had occurred could opportunities be found for the gradual development of the great succeeding classes—amphibians, reptiles, birds, mammals, and finally man himself. Much the same is true of plants. In the sea there are unicellular plants innumerable, but apart from them seaweeds only. It was not until the land had become widely and securely colonized that the higher forms could arise—lichens, fungi, mosses, ferns, and finally the great host of flowering plants.

The ecologist can divide the peopled earth into the two great habitats of land and water, and to a certain extent he must. But in spite of the obvious differences between them, there are also fundamental similarities and these on the whole are of eater importance. In both elements it is plants that make animal life possible. In both it is oxygen that animals must breathe in order to live, whether the oxygen is dissolved in the water and taken in by means of gills, as with truly aquatic creatures, or as a free constituent of the air, as with land creatures making use of lungs. This does not mean that all animals to be found in or on the earth's waters breathe dissolved oxygen by means of gills. In the oceans there are the whales, the dolphins and the seals depending on oxygen taken in at the surface our rivers and lakes there are scores of creatures breathing free oxygen either for a part or for the whole of their life-cycle here is a third parallel between life in the two elements. On land soil is a vital necessity because of the dependence of animals on plants, but soil does not enter into the lives of aquatic animals and this seems at first sight like a paramount distinction. But once again a close similarity underlies the difference. The creatures of both habitats require certain chemical substances, mainly phosphates and nitrates, in order to live, and these are conveyed to them through the agency of plants. The sea contains these substances in solution, available a varying extent everywhere. But the soil also contains them, again in solution in the water that the soil is enabled to hold, and this soil water is the primary justification for the existence of soil so far as plants, and consequently animals, are concerned. A fertile soil holds water containing these substances, and the plants draw them into their tissues by means of their roots, converting them by an elaborate process into more complex stances available as food for animals. An infertile soil is deficient in or devoid of them. So in both elements we find water containing mineral substances playing a part of primary importance.

It is clear, therefore, that ecology must begin with the great spheres of this planet, the primary envelopes of crust, water, air, and finally its ultimate concern, the envelope of life itself. These are the fundamental four, and though distinct, though of contrasted constitution in many important respects, they are nevertheless dependent on one another, renewing one another, constantly giving and taking. The atmosphere gives oxygen to the great waters both salt and fresh, gives rain, gives variations in temperature. Without the atmosphere there could be no hydrosphere as we know it. But the air also takes from the water, since the sea is by far the most important reservoir of that water-vapour which the air takes into its embrace, only to return sooner or later as rain. But if the rain has its importance where the sea is concerned, it has more for the land, for the crust of weathered and eroded rock where all the higher forms of life have their origin and their existence. The atmosphere makes possible the formation of soil, transforms it into the basis of life, but cannot do so on the one hand without the rain which it contributes, on the other without the humus provided by the plants. Nor is that all, since plants, themselves unable to thrive without both soil and rain, are vitally necessary to animals; while animals, as though mindful of their debt, provide not only the waste products of their bodies to make more nutriment for plants, but in dramatic and wholehearted fashion yield up those bodies themselves to replenish yet more the universal repository of mineral salts locked up in the soil.

So it goes on, this dynamism of give and take, this omni-present nexus of mutual dependence. It goes on not merely between crust, air, water and life, not merely between plants and animals, but also between one plant and other plants, between one animal and other animals, until the strands of the web cross, bifurcate, and return upon themselves with a complexity to excite wonder and baffle analysis. This mutualism is the very stuff of ecology, a principle;

one of two principles, of which the other is change. The web of things and the flow of things.

Reid, L., *The Sociology of Nature*, Penguin Books, Harmondsworth, 1962, p. 13-26.

Note

1   See table of geological ages, page 279.

CHAPTER 5

Silent Spring

R. CARSON

The Obligation to Endure

The history of life on earth has been a history of interaction between living things and their surroundings. To a large extent, the physical form and the habits of the earth's vegetation and its animal life have been moulded by the environment. Considering the whole span of earthly time, the opposite effect, in which life actually modifies its surroundings, has been relatively slight. Only within the moment of time represented by the present century has one species—man—acquired significant power to alter the nature of his world.

During the past quarter-century this power has not only increased to one of disturbing magnitude but it has changed in character. The most alarming of all man's assaults upon the environment is the contamination of air, earth, rivers, and sea with dangerous and even lethal materials. This pollution is for the most part irrecoverable; the chain of evil it initiates not only in the world that must support life but in living tissues is for the most part irreversible. In this now universal contamination of the environment, chemicals are the sinister and little-recognized partners of radiation in changing the very nature of the world the very nature of its life. Strontium 90, released through nuclear explosions into the air, comes to earth in rain or drifts down as fallout, lodges in soil, enters into the grass or corn or wheat grown there, and in time takes up its abode in the bones of a human being, there to remain until his death. Similarly, chemicals sprayed on croplands or forests or gardens lie long in soil, entering into living organisms, passing from one to another in a chain of poisoning and death. Or they pass mysteriously by underground streams until they emerge and, through the alchemy of air and sunlight, combine into new forms that kill vegetation, sicken cattle, and work unknown harm on those who drink from once pure wells. As Albert Schweitzer has said, 'Man can hardly even recognize the devils of his own creation.'

It took hundreds of millions of years to produce the life that inhabits the earth—aeons of time in which that developing and evolving and diversifying life reached a state of adjustment and balance with its surroundings. The environment, rigorously shaping and directing the life it supported, contained elements that were hostile as well as supporting. Certain rocks gave out dangerous radiation even within the light of the sun, from which all life draws its energy, there were short-wave radiations with power to injure. Given time—time not in years but in millennia—life adjusts, and a balance has been reached. For time is the essential ingredient; but in the modern world there is no time.

The rapidity of change and the speed with which new situations are created follow the impetuous and heedless pace of man rather than the deliberate pace of nature. Radiation is no longer merely the background radiation of rocks, the bombardment of cosmic rays, the ultra-violet of the sun that have existed before there was any life on earth; radiation is now the unnatural creation of man's tampering with the atom. The chemicals to which life is asked to make its adjustments are no longer merely the calcium and silica and copper and all the rest of the minerals washed out of the rocks and carried in rivers to the sea; they are the synthetic creations of man's inventive mind, brewed in his laboratories and having no counterparts in nature.

To adjust to these chemicals would require time on the scale that is nature's; it would require not merely the years of a man's life but the life of generations. And even this, were it by some miracle possible, would be futile, for the new chemicals come from our laboratories in an endless stream; almost five hundred annually find their way into actual use in the United States alone. The figure is staggering and its implications are not easily grasped—five hundred new chemicals to which the bodies of men and animals are required somehow to adapt each year, chemicals totally outside the limits of biologic experience.

Among them are many that are used in man's war against nature. Since the mid 1940s over two hundred basic chemicals have been created for use in killing insects, weeds, rodents, and other organisms described in the modern vernacular as 'pests'; and they are sold under several thousand different brand names.

These sprays, dusts and aerosols are now applied almost universally to farms, gardens forests, and homes—non-selective chemicals that have the power to kill every insect, the 'good' and the 'bad', to still the song of birds and the leaping of fish in the streams, to coat the leaves with a deadly film, and to linger on in soil—all this though the intended target may be only a few weeds or insects. Can anyone believe it is possible to lay down such a barrage of poisons on the surface of the earth without making it unfit for all life? They should not be called 'insecticides', but biocides.

The whole process of spraying seems caught up in an endless spiral. Since DDT was released for civilian use, a process of escalation has been going on in which ever more toxic materials must be found. This has happened because insects, in a triumphant vindication of Darwin's principle of the survival of the fittest, have evolved super races immune to the particular insecticide used, hence a deadlier one has always to be developed—and then a deadlier one than that. It has happened also because, for reasons to be described later, destructive insects often undergo a 'flareback', or resurgence, after spraying, in numbers greater than before. Thus the chemical war is never won, and all life is caught in its violent crossfire.

Along with the possibility of the extinction of mankind by nuclear war, the central problem of our age has therefore become the contamination of man's total environment with such substances of incredible potential for harm—substances that accumulate in the tissues of plants and animals and even penetrate the germ cells to shatter or alter the very material of heredity upon which the shape of the future depends.

Some would-be architects of our future look towards a time when it will be possible to alter the human germ plasm by design. But we may easily be doing so now by inadvertence, for many chemicals, like radiation, bring about gene mutations. It is ironic to think that man might determine his own future by something so seemingly trivial as the choice of an insect spray.

All this has been risked—for what? Future historians may well be amazed by our distorted sense of proportion. How could intelligent beings seek to control a few unwanted species by a methods that contaminated the entire environment and brought the threat of disease and death even to their own kind? Yet this is precisely what we have done. We have done it, moreover, for reasons that collapse the moment we examine them. We are told that the enormous and expanding use of pesticides is necessary to maintain farm production. Yet is our real problem not one of *over-production*? Our farms, despite measures to remove acreages from production and to pay farmers *not* to produce, have yielded such a staggering excess of crops that the American taxpayer in 1962 is paying out more than one billion dollars a year as the total carrying cost of the surplus-food storage programme. And the situation is not helped when one branch of the Agriculture Department tries to reduce production while another states, as it did in 1958,

> It is believed generally that reduction of crop acreages under provisions of the Soil Bank will stimulate interest in use of chemicals to obtain maximum production on the land retained in crops.

All this is not to say there is no insect problem and no need of control. I am saying, rather, that control must be geared to realities, not to mythical situations, and that the methods employed must be such that they do not destroy us along with the insects.

The problem whose attempted solution has brought such a train of disaster in its wake is an accompaniment of our modern way of life. Long before the age of man, insects inhabited the earth—a group of extraordinarily varied and adaptable beings. Over the course of time since man's advent, a small percentage of the more than half a million species of insects have come into conflict with human welfare in two principal ways: as competitors for the food supply and as carriers of human disease.

Disease-carrying insects become important where human beings are crowded together, especially under conditions where sanitation is poor, as in time of natural disaster or war or in situations of extreme poverty and deprivation. Then control of some sort becomes necessary. It is a sobering fact, however, as we shall presently see, that the method of massive chemical control has had only limited success, and also threatens to worsen the very conditions it is intended to curb.

Under primitive agricultural conditions the farmer had few insect problems. These arose with the intensification of agriculture—the devotion of immense acreages to a single crop. Such a system set the stage for explosive increases in specific insect populations. Single-crop farming does not take advantage of the principles by which nature works; it is agriculture as an engineer might conceive it to be. Nature has introduced great variety into the landscape, but man has displayed a passion for simplifying it. Thus he undoes the built-in checks and balances by which nature holds the species within bounds. One important natural check is a limit on the amount of suitable habitat for each species. Obviously then, an insect that lives on wheat can build up its population to much higher levels on a farm devoted to wheat than on one in which wheat is intermingled with other crops to which the insect is not adapted.

The same thing happens in other situations. A generation or more ago, the towns of large areas of the United States lined their streets with the noble elm tree. Now the beauty they hopefully created is threatened with complete destruction as disease sweeps through the elms, carried by a beetle that would have only a limited chance to build up large populations and to spread from tree to tree if the elms were only occasional trees in a richly diversified planting.

Another factor in the modern insect problem is one that must be viewed against a background of geologic and human history: the spreading of thousands of different kinds of organisms from their native homes to invade new territories. This world-wide migration has been studied and graphically described by the British ecologist Charles Elton in his recent book *The Ecology of Invasions*. During the Cretaceous Period, some hundred million years ago, flooding seas cut many land bridges between continents and living things found themselves confined in what Elton calls 'colossal separate nature reserves'. There, isolated from others of their kind, they developed many new species. When some of the land masses were joined again, about fifteen million years ago, these species began to move out into new territories—a movement that is not only still in progress but is now receiving considerable assistance from man.

The importation of plants is the primary agent in the modern spread of species, for animals have almost invariably gone along with the plants, quarantine being a comparatively recent and not completely effective innovation. The United States Office of Plant Introduction alone has introduced almost 200,000 species and varieties of plants from all over the world. Nearly half of the 180 or so major insect enemies of plants in the United States are accidental imports from abroad, and most of them have come as hitch-hikers on plants.

In new territory, out of reach of the restraining hand of the natural enemies that kept down its numbers in its native land, an invading plant or animal is able to become enormously abundant. Thus it is no accident that our most troublesome insects are introduced species.

These invasions, both the naturally occurring and those dependent on human assistance, are likely to continue indefinitely. Quarantine and massive chemical campaigns are only extremely expensive ways of buying time. We are faced, according to Dr Elton, 'with a life-and-death need not just to find new technological means of suppressing this plant or that animal'; instead we need the basic knowledge of animal populations and their relations to their surroundings that will promote an even balance and damp down the explosive power of outbreaks and new invasions'.

Much of the necessary knowledge is now available but we do not use it. We train ecologists in our universities and even employ them in our governmental agencies but we seldom take their advice. We allow the chemical death rain to fall as though there were no alternative, whereas in fact there are many, and our ingenuity could soon discover many more if given opportunity.

Have we fallen into a mesmerized state that makes us accept as inevitable that which is inferior or detrimental, as though having lost the will or the vision to demand that which is good? Such thinking, in the words of the ecologist Paul Shepard,

> idealizes life with only its head out of water, inches above the limits of toleration of
> the corruption of its own environment . . . Why should we tolerate a diet of weak
> poisons, a home in insipid surroundings, a circle of acquaintances who are not
> quite our enemies, the noise of motors with just enough relief to prevent insanity?
> Who would want to live in a world which is just not quite fatal?

Yet such a world is pressed upon us. The crusade to create a chemically sterile, insect-free world seems to have engendered a fanatic zeal on the part of many specialists and most of the so-called control agencies. On every hand there is evidence that those engaged in spraying operations exercise a ruthless power. 'The regulatory entomologists . . . function as prosecutor, judge and jury, tax assessor and collector and sheriff to enforce their own orders', said Connecticut entomologist Neely Turner. The most flagrant abuses go unchecked in both state and federal agencies.

It is not my contention that chemical insecticides must never be used. I do contend that we have put poisonous and biologically potent chemicals indiscriminately into the hands of persons largely or wholly ignorant of their potentials for harm. We have subjected enormous numbers of people to contact with these poisons, without their consent and often without their knowledge. If the Bill of Rights contains no guarantee that a citizen shall be secure against lethal poisons distributed other by private individuals or by public officials, it is surely only because our forefathers, despite their considerable wisdom and foresight, could conceive of no such problem.

I contend, furthermore, that we have allowed these chemicals to be used with little or no advance investigation of their effect on soil, water, wildlife, and man himself. Future generations are unlikely to condone our lack of prudent concern for the integrity of the natural world that supports all life.

There is still very limited awareness of the nature of the threat. This is an era of specialists, each of whom sees his own problem and is unaware of or intolerant of the larger frame into which it fits. It is also an era dominated by industry, in which the right to make a dollar at whatever cost is seldom challenged. When the public protests, confronted with some obvious evidence of damaging results of pesticide applications, it is fed little tranquillizing pills of half truth. We urgently need an end to these false assurances, to the sugar coating of unpalatable facts. It is the public that is being asked to assume the risks that the insect controllers calculate. The public must decide whether it wishes to continue on the present road, and it can do so only when in full possession of the facts. In the words of Jean Rostand, 'The obligation to endure gives us the right to know.'

Carson, R., *Silent Spring*, Houghton Mifflin, Boston, 1962, p. 23-30.

Part II

THE REVIVAL

Introduction to Part II: The Revival

After the Second World War, most European countries were completely absorbed in rebuilding their shattered society. Owing to a lack of industrial capacity, many basic goods were in short supply. The growth of production, or economic growth as it was often called, could therefore be identified as the main goal of Western societies at that time. This idea of economic growth was strongly supported by the memory of mass unemployment in the 1930s. It was often argued that this situation of mass unemployment had contributed to the rise of nazi and fascist powers in many European countries. 'Never again' was the slogan.

The Cold War

In the second half of the 1940s, the relationship between Western countries and the communist regime in the Soviet Union worsened. Communist parties assumed power in various Central European countries, which led to the Cold War. The Second World War had made it clear that industrial power was an essential part of military strength. Germany and Japan were defeated by the industrial power of the Allies. Nobody could guess to what extent the Cold War might develop into a military conflict in which military power would again be an essential factor. This situation created a political climate in which the development of industrial capacity was propagated as an instrument of military defence against the presumed bad intentions of the Soviet Union. Comparisons of the growth of the GNP were often used in the political debate to demonstrate the ideological superiority of the West.

The first environmental problems

These factors made the growth of industrial capacity, or the growth of GNP, an indisputably important phenomenon. In the late 1940s and 1950s, the growth of the national product was never criticised, even in more progressive quarters. Environmental problems were barely given consideration and, compared with the current situation, one could argue that environmental problems did not exist. This changed rapidly in the course of the 1960s when environmental problems suddenly became a part of everyday life as a result of the ongoing ex-

pansion of industrial capacity without environmental protection. The first
large-scale environmental problems were evaluated as problems of public
health. Water pollution and air pollution, which caused respiratory problems
for people living next to polluting industries, attracted the attention of many.
People living and working in the Rijnmond area in the Netherlands, the Ruhrgebiet in Germany, and the Midlands in England were suffering greatly from
air pollution, especially during certain climatic conditions. Rather suddenly,
environmental problems became a part of everyday life.

The welfare state

Society, however, did not react directly, nor was the concept of industrial expansion criticised. Rather, all the arguments in favour of economic growth
were considered as valid as before. Furthermore, an additional argument was
formulated by those who defended the building of the welfare state. In the latter half of the 1960s, the notion of the welfare state was propagated by parties
across the political spectrum. However, building a welfare state is highly dependent on the continuous growth of GNP. The welfare state can be paid for
from the surplus created by the growth of GNP, which implies that distributional problems can be neutralised as long as the growth of GNP is a reality.

Traditional science

As environmental scientists in the actual sense of the word did not exist at that
time, there could be no reaction from them. There were, of course, some scientists who saw that environmental pollution could be detrimental to society.
Consequently, they looked for paradigms, approaches, concepts, and instruments which could provide a scientific argument against environmental pollution. They were inclined to use the concepts of their own disciplines as arguments to defend a sound environment. Some economists, who would later be
called environmental economists, and some biologists and ecologists published
their work on environmental problems.

The environmental movement

Most of these scientists were more or less involved with the environmental
movement, which became a significant political factor in the course of the
1960s. Their arguments were often used to demonstrate that 'things were
going in the wrong direction'. The increasing influence of the environmental
movement cannot be separated from general developments in society in the
latter half of the 1960s. Without going into detail, it can be argued that the environmental movement, the student movement, the second wave of women's

liberation, the antinuclear weapons movement and the anti-Vietnam movement were in one way or another parts of the same social movement. This created a climate in which the idea that 'things could be changed' became a common belief, in any case in these groups.

Revival

The social and political climate influenced scientific attempts to cope with environmental problems. Therefore, this period is termed the Revival. It is the revival of public awareness of environmental problems accompanied by the arguments of many scientists looking for new concepts which could be found in their disciplines. This revival can also be seen as the societal answer to a sharp increase in environmental problems resulting from unlimited industrial expansion, supported by the plea for economic growth by the elite in society.

Monodisciplines

As was stated above, there were no environmental scientists in the actual sense of the word in this period. University educations were monodisciplinary. Thus, when environmental problems were first discussed, there were only scientists who had been educated in a certain discipline. Environmental problems, however, did not belong to any discipline in particular, and to further complicate the situation, politics and public opinion did not recognize environmental problems in society. What then was the reaction of scientists when they realised that environmental problems could no longer be ignored as these problems had become a part of 'normal' life?

A new approach

In some disciplines, like economics, the first general reaction was that environmental problems did not belong to science. The general view was that economics should deal with scarce goods and services, produced by using production factors, to be sold on a market for the benefit of consumers. Environmental issues were not a factor. Some economists were aware of the fundamental shortcomings of this argument. They argued that scarcity was a more central issue in economics than price. And, in this view, the production process created an increasing scarcity of environmental goods. Hence, one may conclude that economists, being aware of environmental threats, tried to demonstrate that environmental problems did indeed belong to economic science.

Generally speaking, two concepts were used to demonstrate this point. Firstly, they used the generally accepted argument given by Robins (1935), who said that economics deals with alternative uses of scarce goods. As free environmental goods were becoming scarce, this new scarcity belonged to economic science. Secondly, Pigou's concept of externalities (1920/1954) was fairly well known among economists. However, it did not play a significant role in the economic literature. Blaug (1978) pointed out that the concept of externalities could only be found in the footnotes of textbooks that had been published over the previous fifty years. Nevertheless, when environmental problems became relevant, the economists dealing with them, having been trained in the externality argument, recognised the possibilities of using this in the environmental debate.

Sociologists traditionally were involved in the process of changes in society. Authors such as Marx, Durkheim, and Weber were not aware of the changing relationship between man, society, and the environment. They studied society from the point of view of changing relationships between individuals and social groups. They pointed to the disappearance of traditional human interactions and the rise of functional, formal interactions. They studied society completely isolated from the natural context. Only a few sociologists began to see that one of the most important social questions of the future would be the conflicts between man, society, and the environment.

Biologists played a significant role in the Revival period. Their position was completely different from that of economists and sociologists, as they had never been trained in the study of societal questions. However, their subject of study being nature, they had been, from an early stage, confronted with the influence of human behaviour on nature and the environment. They were able to demonstrate how negative this development was. Some biologists restricted their arguments to this negative development while others included societal elements as well. As they had not been trained in economics and sociology, some of their arguments could be easily criticised by social scientists. Human societies are not the same as animal societies, nor can the functioning of ecosystems be applied to human societies without any 'translation'. We may conclude that the most significant contribution from biologists in the Revival period can be found in their valuable arguments demonstrating the negative influence of human economic behaviour on nature. These arguments in particular were used by economists and sociologists who warned that these signals could no longer be ignored.

Mishan: The Costs of Economic Growth

Mishan's most significant contribution to the discussion of environmental problems can perhaps be found in the fact that he, being a fairly well-known economist, attacked traditional economics, in which environmental issues were ignored. Owing to his position, other economists could not ignore his arguments. However, most of them eventually rejected his line of reasoning which attacked traditional economics and, consequently, paid little attention to his arguments. However, Mishan's influence reached beyond the field of economics. His book, *The Costs of Economic Growth* (1967), full of scientific economic arguments, could be read by every academic, even those not trained in economics. The book was read by 'everybody' who wanted to contribute to the societal debate about environmental issues.

The concept of economic growth is strongly criticised in the book as it is based on easily measured economic parameters such as the level of employment, the level of income, public debt, the exchange rate, etc. By concentrating on these easily measurable economic parameters, other relevant economic issues are overlooked and neglected. Mishan brought the measurement of economic value, in particular when environmental issues are at stake, back to the core of the economic debate. On this point, his approach can be evaluated as fairly dualistic.

On the one hand, Mishan argued that economic value is often difficult to measure, and he propagated Pigou's concept of negative external effects: the cost price of products does not really reflect the total economic scarcity which has been sacrificed by producing the good, as environmental costs are not included in the cost price. In line with Pigou's argument, these environmental costs have to be included in the price of the product. On the other hand Mishan argued that it is often very difficult or even impossible to quantify the economic costs of environmental disruption. Furthermore, he argued that marginal cost pricing is a normal practice based on well-established neoclassical paradigms. However, this statement can only be relevant in the absence of substantial environmental costs. Hence, we may conclude that, in the case of an increasing relevance of negative external effects, marginal cost pricing is based on substantial shortcomings, as we do not know what the 'real' costs of environmental use are. These examples demonstrate how difficult the search for scientific arguments was in this period.

Mishan's criticism of 'economic growth' is consistent with these arguments. He argued that it is not certain that an increase in the level of production results in an increase in welfare, as is assumed by traditional economists. If an increase in production creates a more than proportional increase in negative external effects, the increase in production will result in a decrease of welfare. Welfare could be increased by lowering the level of production, as this would lower the environmental costs so sharply that an increase in welfare would be the result. It goes without saying that such ideas were completely at odds with the plea for expansion of the GNP, which was, as we saw in the previous sections, one of the main aims of society in that period. Mishan was fully aware how controversial his arguments were, responding that economic growth was society's goal mainly because of its effects on distributional problems: 'With economic growth we will remedy all social evils', was society's answer to these problems.

In traditional neoclassical approaches, generally speaking, private economic goods are given more attention than public goods because of the focus on market forces. Nature and the environment—in particular the loss of nature and environment—have more to do with public goods than with private goods. Mishan claimed that this focus on private goods would lead to an overestimation of environmental costs in the case of private goods and an underestimation of these costs in the case of public goods, as the costs to the latter are difficult to determine.

He demonstrated this point (1970) when discussing the location of the third London Airport, which was then a hot issue. A report on this problem was published by the Roskill Commission. After an in-depth cost-benefit analysis, the Commission had concluded that the best location would be in the neighbourhood of London itself. Mishan argued that this choice had been strongly influenced by the fact that, on the one hand, there was no market for the environmental costs of the disturbance in residential areas due to air traffic, whereas on the other hand, there was a market for travel time which is relevant when the airport is located far away from the London conurbation. So, Mishan's conclusion was that the outcome of the cost-benefit analysis was based on the fact that there is no market for environmental disruption due to air traffic. The definition of the property rights of travel time is not the same as that of the absence of noise in residential areas. Hence, the property rights of environmental goods are not well described, and this will be reflected in the outcome of every cost-benefit analysis in which environmental goods play a role, as economists have to conform to the definition of property rights in society. It goes without saying that such ideas could hardly be accepted by a societal elite responsible for the definition of property rights.

These ideas about property rights are strongly reflected in Mishan's ideas about 'solutions' to the environmental problem. One instrument could, in Mishan's view be, the implementation of amenity rights. These amenity rights can be considered to be a kind of property rights with regard to environmental goods. As soon as these property rights or amenity rights are established, people who experience a loss of amenity may go to court to force the economic actor to stop frustrating the amenity right, or the victim may start a negotiation process with the polluter aimed at reaching a certain acceptable level of pollution with or without compensation. This can only function if amenity rights are well established through governmental action. Compensation may work as an instrument to calculate environmental costs, as the level of compensation has to be in line with the welfare loss of the victim.

This approach demonstrates the lack of knowledge of environmental problems in that period. The amenity rights argument can only work if certain assumptions are met. It is assumed, for instance, that dose-effect relationships of environmental problems can be described and that the victim can negotiate his problem with a limited number of polluters. Of course, this assumption is seldom met in modern societies as polluters are numerous; transboundary pollution is more common than pollution by local industry; synergetic effects and thresholds are common in environmental problems; and many effects are unknown. All these factors complicate Mishan's solution considerably.

A second way of 'solving' the environmental problems mentioned by Mishan is the idea of separate facilities: certain parts of a country can be used for polluting activities while others are excluded. In such a case, people can choose the location of their houses according to their evaluation of environmental pollution. People who do not want to live in the neighbourhood of a polluting industry can look for other localities where pollution is absent. Thus, the market process will provide an optimal pollution situation, as pollution is only present in locations where environmental costs are low. Here again, we have the assumption that dose-effect relationships are known and that it is possible to restrict pollution to a certain geographical area. In this model, national and global pollution, causing problems in all locations, is a non-issue. More generally speaking, the implementation of separate facilities can be seen as a form of spatial planning, which has been common practice throughout Europe for many decades.

There is another dualistic point in Mishan's book. In the Revival period, his arguments about economic growth and environmental problems were considered very progressive. They were very often used by alternative environmental groups in their fight against the well-established economic forces in society which benefited, in this view, from environmental pollution and were not willing to tackle these problems. On the other hand, Mishan's description of all the terrible things in modern society and the call for 'the good old days' contained a strong reactionary and conservative element.

Finally, it has to be stressed that the value of Mishan's work can be found in the polemical character of his arguments in a period in which environmental problems were increasingly becoming part of normal life. In this climate, such a polemical publication by a well-known economics professor about the dangers and the high costs of environmental problems was an eye-opener for many people confronted with environmental issues in everyday life. His arguments were used by many alternative environmental groups that were convinced of the need to create a better world in which it was worth living. In this respect too, Mishan's work can be regarded as a milestone of the Revival period.

Hardin: The Tragedy of the Commons

Hardin's article, *The Tragedy of the Commons* (1968), attracted the attention of many scientists because it made it clear that 'something has to be done' to save the world from overusing its natural resources. Hardin is an American biologist; no close reading is needed to recognise this fact. The article is difficult to evaluate as it contains several levels of argument. Perhaps the best starting point for understanding the article is the population problem. The implicit assumption is often made that, regarding future generations, human populations behave just as animal populations do. From this point of view, it makes sense to argue about an optimal and a maximum population. Hardin argued that there are too many people, since people can have as many children as they are willing to generate. There are no limitations in human society whith regard to the acceptable upper level of population. Hardin is of the opinion that since natural resources are finite, human populations cannot increase forever. Of course, the question is how the human population can be limited.

Hardin's article does not provide an explicit answer to this question. The implicit answer is found in rather long deliberations on the arguments of Malthus and Darwin. This leaves the impression that the limitations should be realised in the manner suggested in Darwin's approach: the survival of the fittest. Although this is not explicitly stated in the article, the implication is clearly there.

This point led to many controversies. According to Hardin, all human beings have the same characteristics with regard to the use of natural resources. However, the use of food, energy and natural resources by an American or a West European citizen is completely different from the use of these resources by people from India, Ghana, or Honduras. This implies that the optimal and maximum population of the world can only be defined by taking these differences into account. Hence, the optimal population of the world could be 30 billion Indians or 2 billion West Europeans and North Americans. Hardin's approach has been attacked by Third World countries and by the American Left because it excluded the unequal use of natural resources from his analysis.

Another point of controversy is the position of technology. Hardin states that the problem he analyses can be defined as a 'no technical solution problem'. This implies that technical questions can be ignored. This paradigm is in fact similar to that of Malthus, who argued that technological development could not change the outcome of the economic process. In Malthus' model, however, the idea was that technology would not change substantially in the future as the relationships between the level of production and the level of resources are fixed. In Hardin's model, technology does not matter, which overlooks the point that there are very substantial differences in the use of natural resources with significant differences in output. Commoner (1971) was particularly critical of these points.

The problem of overpopulation is discussed in the context of *The Tragedy of the Commons*. Hardin argued that the commons (viz., the common property resources) are always frustrated by the same types of problems. He used the example of shepherds using a common pasture. Every shepherd wants to increase the number of sheep he takes to the common pasture land. By doing so, he has the benefit of a substantial increase in personal income from the higher output of the animals, whereas the negative effect of potential overgrazing is shifted to all the other shepherds using the common pasture land. In Hardin's analysis, this process will continue until the commons are completely used up, resulting in collapse.

This metaphor is used to extend the argument to the population problem. The only way to stop overgrazing is to control the commons. The same has to be done in the case of the number of children people have. Hardin indicated only implicitly how this could be done: 'The only way we can preserve and nurture other and more precious freedoms is by relinquishing the freedom to breed, and that very soon. Freedom is the recognition of necessity—and it is the role of education to reveal to all the necessity of abandoning the freedom to breed. Only so, can we put an end to this aspect of the tragedy of the commons.' Education can solve the problem. It is often concluded that people in Third World countries need to be educated in order to reduce the number of children they have. They do not know their responsibility in this respect, and Americans can explain to them how to behave by educating them. This point has triggered many reactions from fundamental critics as it completely overlooks why people in Third World countries have more children than Americans; additionally, this kind of reasoning serves to secure the high consumption levels of Americans.

When we take the environmental point into account again, a final essential problem has to be discussed. In Hardin's approach, common property resources are always overexploited and overused until the resource is exhausted. However, particularly in the case of the common pasture land, historical evidence countradicts his view. In many countries of the world, there are common property pasture lands. Let us take the Alps as an example. For many centuries, common pasture lands (the Alps) have been the normal practice. However, common property can only exist if the common property is defined. Without a definition of common property, it does not exist, as was clearly demonstrated by Bromly (1991). This means that access to common property is not open to everybody, which is the case in Hardin's example. The most typical characteristic of a common property right is that some persons (in most cases, people from one village) are allowed to use the land to the exclusion of others. If everybody is allowed to use the resource, there is no clear definition of property rights and no ownership, which means open access for everybody. In all cases of traditional common property, there are regulations stating what the users of the resource may and may not do. We can, therefore, conclude that Hardin did not realise the difference between common property and open access.

Hardin's article is often misused to argue that common property rights lead to overuse of the resource; thus, the solution to environmental problems is the privatisation of common property. This is not the point in Hardin's article. Furthermore, this frustrates the real possibilities, namely, public policies to limit the use of common resources and to protect the environment.

Finally, there is the question of the relevance of Hardin's article in the Revival period. Various authors come to the conclusion that there is no publication in environmental science which has attracted so much unnecessary attention without reason as Hardin's article. This opinion is, of course, based on the criticism we discussed above. Nonetheless, it can be concluded that Hardin put the population problem on the agenda as a problem related to the limited resources of our planet. On the other hand, Hardin's approach clearly demonstrates how many pitfalls and barriers a biologist has to confront when trying to explain human societies and human behaviour using the behaviour of animals and ecosystems as a guideline. Of course, the central question is how the relationship between the ecosystem and human society can be defined and analysed.

Ehrlich and Ehrlich: The Population Bomb

The book *The Population Bomb* (1969) is a typical example of biologists using their discipline to explain society. In this view, everything that is true for animals is also true for human beings. Man is seen as a typical mammal. The level of human populations will be influenced by the same factors that influence those of other animals. Food is seen as a dominant factor. Without the availability of a sufficient quantity of food, people will die. The number of children people will have is related to this food supply. An increase in population is seen as a dangerous development which will result in starvation on a massive scale.

The book starts with an empirical example of the multiplication of human population in the course of time. The figures are impressive. The doubling time of the world population has been 1,000,000 years, 1,000 years, 200 years, 80 years, and 37 years. The conclusion is clear: population growth can no longer be sustained. Additional information is given about the doubling time of human populations in underdeveloped countries, where the doubling time is much shorter than in developed countries.

In the book we find some ideas about the influence of technological development on the relevance of these figures. It is said, for example, that medical care lowered the death rate in some underdeveloped countries such as Costa Rica and Sri Lanka. Additionally, it is said, without any proof that rich people have more children than poor people. This phenomenon has to be recognised in Western countries.

Ehrlich's approach is strongly based on Malthus and Darwin: the Law of Diminishing Returns and the survival of the fittest. In Malthus' approach, the only way out of the dilemma of overpopulation is birth control; a higher yield of agricultural products is not seen as a possibility. Production functions are fixed in the Malthusian way of thinking, which implies that technology cannot significantly change the outcome of agricultural processes. Indeed, in such a situation, birth control is the only option. In contrast to Malthus, Ehrlich argues that birth control *and* a sharp increase in the level of food production is the only solution to this complicated problem. When these projects are carried out, optimum population-environment goals for the world should be defined. Additionally, methods for reaching these goals must be elaborated. Ehrlich recognised that this is a difficult job, but argued that it could be realised.

In an animal population, information about the level of food leads to the development and implementation of mechanisms appropriate to the adaptation of this new information. This implies that any type of public policy is a non-issue in these types of ecosystems. Information automatically leads to the adaptation of the system to the new situation. Clearly, Ehrlich had these ideas in mind when discussing the problems of human populations as, in the book, hardly any idea or model is elaborated that deals with the question of needing public policy to implement birth control measures and the expansion of food production.

This Darwinian approach can also be recognised when Ehrlich discusses the differences between developed and underdeveloped countries. The most significant difference in his view is the difference in population growth, which is much higher in underdeveloped countries. The Ehrlichs concluded that the topic of birth control is much more relevant in underdeveloped countries where people are dying because of a lack of food. The distribution of power and access to natural resources is not given attention. Owing to the concentration on food supply, pollution and the exhaustion of non-renewable resources are not evaluated as a threat to mankind. This implies that it is not the high quantities of natural resources used by Western developed countries that have to be brought down; in Ehrlich's model, these issues are not relevant. The food supply, as in Malthus' publications, is the main issue. This leads, of course, to the conclusion that population growth, particularly in underdeveloped countries, needs to be given priority.

Ehrlich's book was especially popular in the USA and in some other Western developed countries. The book gave strong arguments for safeguarding the American way of life and blamed the underdeveloped countries for causing problems by creating too much human life. By using a simple Malthusian/Darwinian approach, the policy of blaming the victims of exploitation in the under-developed countries received pseudo-scientific support.

It goes without saying that the Ehrlichs took a peculiar position in the Revival period. On the one hand, they were able to put the problem of overpopulation on the political agenda. On the other hand, this approach was often recognised as a neo-imperialistic attempt to guarantee control of the underdeveloped countries. Of course, the elites in the underdeveloped countries and the Critical Left in Western countries did not support these ideas. Nowadays, the Ehrlich publication is generally seen as a very dated attempt to discuss population problems without paying attention to the real problems of underdeveloped countries. The Brundtland Commission (1987) took a different starting point by including many other arguments in the debate, which led to the conclusion that developed countries should decrease their level of economic growth and create new opportunities for economic growth in underdeveloped countries. This is a completely different conclusion than that presented by Ehrlich.

Odum: The Strategy of Ecosystem Development

The article *The Strategy of Ecosystem Development* (1969) is another example of an attempt to define the relationship between human society and the ecosystem more precisely. Odum is a biologist who studied the functioning of ecosystems and particularly the energy flows in ecosystems. His aim was to use this experience to deal with environmental problems in society.

Odum argues that there are, generally speaking, two types of ecosystems. The first is a young one in which production is high; in this stage, the ecosystem is specialised in growth, the number of species being more relevant than the quantity. These types of ecosystems are interesting for mankind as they can provide us with the food, timber, and resources we need. On the other hand, we have the highly stable mature ecosystems in which the quality of species is more relevant than the quantity. These ecosystems do not play a significant role in production as this level is low; they provide us, however, with other features relevant to mankind. Watershed protection and mature forests give us a type of protection which can only be provided by these types of ecosystems.

We can learn two lessons from the strategies used by ecosystems. First, we need both types of ecosystems; young ones for the production of food, timber, and other resources, and mature ecosystems for protection and ecological stability. This means that we have to arrange our policies to use ecosystems in such a fashion that we take these starting points into account. This idea is similar to the concept of Sustainable Development developed by the World Commission on Environment and Development two decades later (1987).

The second lesson is that we have to become a mature society. The changes which are necessary in the use of ecosystems can only be realised when we organise societal institutions such as education, planning, and environmental policies to meet the demands of ecosystem development.

The relevance of Odum's contribution to the Revival period can be found in the clear and understandable lessons of ecosystem development applied to human society. Traditionally, economists were able to give a value to nature and the environment only in as far as these resources had a price on the market. Odum shows that this is a short-sighted view in which the unpriced values of nature and the environment are overlooked. In other, more contemporary, words: Odum makes it clear that nature and the environment have an intrinsic value which should be defined outside the traditional realm of economics where values can only be defined as far as prices can be defined. In this way Odum has strongly influenced recent thinking on the value of nature and the environment.

White: The Historical Roots of our Ecological Crisis

White's aim in his article *The Historical Roots of our Ecological Crisis* (1967) was to define the historical antecedents of our environmental problems, i.e., to explain the high level of disruption of nature and the environment which is a common practice in industrialised Western countries. From this perspective as a historian, this can only be explained by paying attention to the way thinking and science developed in the Western world in the last two millennia.

White's starting point is the definitions given by both medieval and more recent scholars. A common paradigm found in their publications is that science should be affected by the demands of God. Science can provide a better understanding of God as well as giving God his proper place in our way of doing things. Such ideas are based on natural theology, which is the religious study of nature to achieve a better understanding of God. Western Christianity, rooted in Latin cultures, is the basis for these ideas. In Christianity, man is defined as a creature with divine origins, who has been given sovereignty over animal and plant life. By being sovereign, mankind is carrying out the will of God.

These ideas are found only in Western countries where science and production were developed in the context of this natural theology. The relationship of mankind with nature was based on the notion that man is master over animals and plants. This philosophical and religious climate gave full reign to all activities which strengthened this master relationship. The development of science and technology therefore became particularly possible in Western countries. When the general starting point is that nature and the environment should be used and exploited by mankind, it is difficult to define misuse and over-exploitation. In this religious and philosophical climate, no connotations are found to define these situations.

Though White's arguments are firmly based on empirical evidence, these arguments cannot explain why in other parts of the globe the same behaviour vis-a-vis nature and the environment can be recognised. In Eastern Europe, where Eastern orthodox religion is found, which is in White's view completely different from Western Latin-dominated Christianity, environmental disruption is not at a lower level than in Western Europe. On the contrary, in countries such as Greece, Bulgaria, and Russia, pollution is at least as severe as in Western countries. Furthermore, environmental disruption in Third World Countries with a low level of Western influences is often very high. One can conclude that factors other than Western Christianity influence the production and consumption behaviour of people with regard to nature and the environment.

White discusses the approach of Francis of Assisi, who believed that animals and plants are the brothers and sisters of human beings. In other words, nature does have an intrinsic value. Nature is, in this view, no longer evaluated as something which can be used for all purposes beneficial to mankind. Therefore, White suggests at the end of his article that Francis should be the patron saint of ecologists. He is of the opinion that Western societies need to be aware of the shortcomings of approaches based on traditional Christian views, which needs reconstruction when they are confronted with the massive negative outcomes of these starting points.

The relevance of White's contribution can be defined on two levels. First, White made it clear that environmental disruption is not accidental, but is based on well-established ideas and religious traditions which have always influenced Western societies. This view was particularly important as many traditional scientists in the Revival period saw environmental issues as minor problems which could be solved easily by taking certain measures. Even if we do not agree with White's conclusions, the deeply rooted origins of ecological disruption cannot be overlooked. White's contribution made discussion of the roots and significance of environmental disruption possible. This was of great importance in a period when many vested interests and traditional scientists tried to dismiss every effort ecologists made to define ecological problems as an ecological crisis.

Additionally, White opened the discussion on the value of nature itself. Should we see the value of nature only insofar as it could be exploited and therefore used in the production process, or do nature and the environment have a value of their own, as articulated by Francis of Assisi? These questions make the problem central in the societal debate.

Stone: Should Trees Have Standing?

Stone's article, *Should Trees Have Standing?* (1972), is a major contribution to the debate on the place nature and the environment should have in society. His arguments are mainly based on legal arguments. In his detailed discussion, he examines the pros and cons of the idea of legal rights for natural objects. This discussion is particularly relevant and interesting for legal scientists trying to determine the extent to which the structure of American legislation can support such a plea for legal rights. On the other hand, one should not overlook the point that legislation is different in many countries. In several European countries, for example, current legislation makes it possible for an environmental group to be a stakeholder if the quality and the quantity of nature and the environment are affected by others. In other countries, such as Belgium, environmental groups do not have access to information on the conditions of licenses and permits given to polluting industries, which means that these groups can never go to court as they are not privy to the necessary information.

In our view, advocating legal rights for natural objects is not just a legal question. More relevant are the social views which are inherent in these ideas. Generally speaking, people can start a case against others when they are of the opinion that their interests are being hurt or diminished by these persons. The most common case is when someone's property, let us say a car or a house, is damaged as the result of the actions of others. In that case, it is the owner who has the right to go to court. It is private property rights which determine the outcome of this process. When there is no private property right, but a common or societal one, it is still possible to go to court as soon as the common property rights can be recognised in society.

However, in the case of natural objects, the definition of property rights is not as clear as in the cases mentioned above. Are the property rights of birds, trees, grasshoppers, and butterflies recognised in society and therefore described and legislated? One could argue that there are, in some ways, common property rights, though that is not so easy to prove in a legal sense. Stone's 1972 contribution made it clear that, regarding natural objects, this was a weakness of the traditional legal system.

Private property rights of natural objects deal, by definition only, with the values relevant in a private situation. They are related to use values. Traditional common property rights can deal with use values such as the quality of the water to be used as drinking water. In the case of legal rights for natural objects, we are more in the realm of the intrinsic value of nature, which means that nature and the environment have a value of their own, regardless of other stakeholders who might believe that their traditional use values will be frustrated as soon as nature is protected for its own sake.

We can conclude that at a very early stage, Stone initiated a discussion about the relevance of the intrinsic value of nature, which has, in this view, priority over all other use values related to the position of economic agents on a market which by definition excludes negative effects outside the market. In this market situation, the intrinsic values of nature and the environment which are not measured by traditional market processes are traditionally not relevant.

Komarov: The Destruction of Nature in the Soviet Union

Before the fall of the Communist regimes in Central and Eastern Europe, information about the state of nature and the environment in these countries was limited. In Western Europe, the environmental issues in these countries were seldom considered; ideological and economic issues were given much more attention. Various authors have periodically criticised aspects of the environmental situation in the Soviet Union. Komarov was the first Soviet scientist to make an overall analysis of the state of the environment in his country in his book *The Destruction of Nature in the Soviet Union* (1980).

On the other hand, one could argue that even Komarov's publication did not direct much attention to the situation in the East. In Western Europe, it was the Critical Left who put the disruption of nature and the environment on the public agenda. Generally speaking, these groups did not intend to criticise the Soviet Union; this was done by the right wing of the political spectrum which was not willing to pay much attention to environmental issues. This brought Western Europe to a point where the disruption of nature and the environment was not given any societal priority.

In this climate, more publications appeared about the benefits of Marxist approaches to environmental issues than about the shortcomings of these approaches. Komarov underlines the power element when discussing the possibility of a critical review in the Soviet Union itself. However, he and many other scientists in the West overlooked the influence of Marxist theories on environmental policies. An in-depth perception of the relation between nature and society was hardly an issue in the period in which Marx lived and worked. The rise of the Industrial Revolution and the accompanying technical development generated a widely held belief in the potential of technical progress. This idea can also be found in the work of Marx. The development of the forces of production were, in his view, a necessary condition for realizing a socialist society. Marx never developed systematic ideas regarding the relation between nature and society.

We have to be aware that both traditional Western neoclassical and Marxist theories originated and were elaborated in the second half of the nineteenth century, describing and analysing the same market process. Neoclassical economists emphasized the issue of the efficient use of the production factors of labour and capital, while Marxist economists stressed what they considered to be an unfair distribution of power and income between labour and capital, resulting from the same market processes by which, according to neoclassical economists, an optimal allocation of production factors is realised. This brings us to the conclusion that in both theoretical systems, nature and the environment are irrelevant.

Additionally, the two systems in Europe focused on production for political reasons. The experience of the Second World War made it clear that in the event of war, industrial power would be a very significant factor in modern societies. This gave a strong impetus to industrial vested interests to argue that the only aim in Soviet society should be a high level of industrial production, disregarding nature and the environment.

After the fall of the Communist regimes, the West was surprised by the high levels of environmental disruption in these countries. Even the Chernobyl incident was seen as an accident and not as a structural situation in Communist countries. The relevance of Komarov's work can be found in the fact that he wrote about the deplorable situation in the Soviet Union in 1980, a period in which hardly anybody in the West had the slightest idea about the level of environmental disruption in that area.

Concluding Remarks

The Revival period presented new ways of looking at the relationship between man, society and the environment. Economists, sociologists, biologists, and legal scholars introduced concepts and paradigms to reinterpret the traditional view of this relationship. The starting point of their reasoning was the discovery of the problematic character of this relationship and its fatal consequences for the world's future. The causes of the problem were to be found in an uneconomic economy, in dated value systems, in legal rights systems, and in autonomous processes of population growth. These approaches can be considered one-factor theories in that they (over)emphasized the importance of one causal factor and overlooked other (important) factors. Nevertheless, the classics of the Revival period have been of great importance in promoting an awareness of environmental problems and in necessitating the development of environmental policy programmes.

References

Blaug, M., 1978. *Economic Theory in Retrospect*. Cambridge University Press, Cambridge.

Bromley, Daniel W., *1991*. *Environment and Economy; Property Rights and Public Policy*. Basil Blackwell, Oxford.

Dietz, F.J. and J. Van der Straaten, 1992. Rethinking Environmental Economics: Missing Links between Economic Theory and Environmental Policy. *Journal of Economic Issues*, Vol. 26, Nr. 1, pp. 27-51.

Dietz, F.J. and J. Van der Straaten, 1993. Economic Theories and the Necessary Integration of Ecological Insights. In: A. Dobson and P. Lucardie (Eds.), *The Politics of Nature*, Routledge, London/New York, pp. 118-144.

Hueting, R., 1970. *Nieuwe Schaarste en Economische Groei*. Agon Elsevier, Amsterdam/Brussel.

Hueting, R., 1980. *New Scarcity and Economic Growth*. North-Holland, Amsterdam.

Mishan, E.J., 1970. What is Wrong with Roskill? *Journal of Transport Economics and Policy*, September 1970, pp. 221-234.

Pigou, A.C., 1920/1954. *The Economics of Welfare*. MacMillan, London.

Robins, L., 1935. *An Essay on the Nature and Significance of Economic Science*. Second Edition, London.

CHAPTER 6

The Costs of Economic Growth

E.J. MISHAN

External Diseconomies and Property Rights: 2

The three considerations discussed in the preceding chapter, the existing distribution of wealth, the incentives to improved allocation within the existing institutional framework, and the justice or equity of that framework, combine to reveal the deficiencies of the *status quo*, in particular its inability to respond properly to significant external diseconomies. Suppose a private airport is to be built, or expanded, close enough to a large residential area as to disturb the peace of the inhabitants. In order to simplify the issues we may suppose that a fixed number of flights are involved all of which, if the airport is to be profitable, must be undertaken. The optimal outcome is, therefore, either that of siting the airport near this residential area or that of not siting it there. We may dismiss the simplest *laissez faire* argument that all is well, that an optimal outcome must be realized since either (1) the inhabitants are able to compensate the airport authorities to move elsewhere, and if so it will be in their own interests to bring this about, or else (2) they are unable to compensate the airport authorities, in which case the establishment of the airport is the optimal outcome. This argument, as we have seen, neglects the problem of initiative in organizing the protest and the costs incurred (in terms of time, effort and money) of large numbers of people arriving at a decision not only of the largest sum they can collectively offer but also of the contribution to be made by each family. The larger the population affected the smaller is the likelihood of effective initiative and the higher the sum of the costs in reaching a decision. As suggested, however, these are not inevitable obstacles. If the existing law were such that the airport authorities were compelled to compensate the inhabitants, the costs of reaching a decision would be likely to be very much lower.[1] Even if these costs were nil, however, it is altogether possible that the airport authorities could not afford to compensate the inhabitants. The optimal outcome would then require their plans to be changed: the airport would have to be sited elsewhere.

Again, there is the consideration that the individual resident families are poorer than the airport-owners, from which two things follow: First, the simple point that if we are interested in a more equal distribution of income and/or social justice, then in the event that the airport company *could* compensate all the victims of aircraft noise—implying, by definition that the siting of the airport there is the optimal outcome—we should require that the company actually compensate the victims. Secondly in consequence of *subjective* (non-market) estimates of damages sustained by one or both of the opposing parties the optimal outcome may very easily be ambiguous even if we ignore all costs of reaching voluntary agreements and of implementing them. The *maximum* that any of the inhabitants is prepared to pay to the airport company to avoid aircraft noise is limited by his wealth, by his prospective income and assets. No matter how excruciating his suffering will be, his contribution is perforce limited. The *minimum* sum that he is willing to accept to bear with the aircraft noise is not, however, subject to such constraint. In fact such a sum will exceed the maximum he is prepared to pay by a margin that is wider the larger is the sum of money required to present him with feasible alternatives.[2] And the relevant alternative is not the sum that would suffice to soundproof his house—or suffice to soundproof a room of his house, as the government is apt to think—unless he is indifferent to being shut in his house all the year round. Indeed such a sum should not be less than the amount necessary to compensate him fully for the inconvenience and expense of moving to a quiet area similar in other respects. And if alternative quiet areas have disappeared so that it is not worth his while to move house, the minimal compensation that would enable him to feel no worse off than he was before being exposed to aircraft noise would be larger than this.[3] Thus, if we reckon the maximum sum that the inhabitants could pay to be rid of aircraft disturbance as £10 million, and the minimum they would be prepared to accept to bear with it as £20 million, and compare these figures with the sum, say of, £15 million which is the most the company is willing to pay to operate on this site, and, also, the least it will accept to move elsewhere—this £15 million being the capital value of the estimate of their excess future profits from operating on the present site over the next best alternative site the company could not be bribed by the inhabitants to move their airport elsewhere. Even if costs (i), (ii) and (iii) are all zero, then the decision already taken to establish the airport is the optimal one. If, on the other hand, the law found in favour of the inhabitants (and all costs again were zero) the company would be unable to compensate the inhabitants to put up with the disturbance. Once again, then, whichever situation the law brings about is the optimal one. Considerations both of the distribution of wealth and of plain justice, however, suggest that the victims of aircraft disturbance be given legal rights to full compensation.

In so far as the activities of private or public industry are in question, the alteration required of the existing law is clear. For private industry, when it bothers at all to justify its existence to society, is prone to do so just on the grounds that the value of what it produces exceeds the cost it incurs—gains exceed losses, in other words. But what are costs under the existing law and what ought to count as costs is just what is in issue. A great impetus would doubtless take place in the expansion of certain industries if they were allowed freely to appropriate or trespass on the land or properties of others. Even where they were effectively bought off by the victims, the owners of such favoured industries would thereby become the richer. And one could be sure that if, after the elapse of some years, the Government sought to revoke this licence there would be an outcry that such arbitrary infringement of liberties would inevitably 'stifle progress', 'jeopardize employment' and, of course, 'lose us valuable export markets'. Such an example though admittedly farfetched is distinctly relevant. For private property in this country has been regarded as inviolate for centuries. Even if the Government during a national emergency or in pursuit of national policy takes over the ownership or management of private property it is obliged to compensate owners. It may well be alleged that in any instance the Government paid too little or too much, but it would not occur to a British Government merely to confiscate private property.

In extending this principle of compensation, largely on the grounds of equity, the law should explicitly recognize also the facts of allocation. Privacy and quiet and clean air are scarce goods—far scarcer than they were before the war—and sure to become scarcer still in the foreseeable future. They are becoming more highly valued by millions of people, most of them anxious to find a quiet place to live not too far from their work. There is no warrant, therefore, for allowing them to be treated as though they were free goods, as though they were so abundant that a bit more or less made not the slightest difference to anyone. Clearly if the world were so fashioned that clean air and quiet took on a physically identifiable form, and one that allowed it to be transferred as between people, we should be able to observe wether a man's quantum of the stuff had been appropriated, or damaged, and institute legal proceedings accordingly. The fact that the universe has not been so accommodating in this respect does not in the least detract from the principle of justice involved, or from the principle of economy regarding the allocation of scarce resources. One has but to imagine a country in which men were invested by law with property rights in privacy, quiet, and in clean air simple things, but for many indispensable to the enjoyment of life—to recognize that the extent of the compensatory payments that would perforce accompany the operation of industries, motorized traffic, and airlines, would constrain many of them to close down or to operate at levels far below those which would prevail in the absence of such a law, at least until industry and transport discovered economical ways of controlling their own noxious by-products.

The consequence of recognizing such rights in one form or another, let us call them *amenity rights*, would be far-reaching. Such innovations as the invisible electronic bugging devices currently popular in the US among people eager to 'peep in' on other people's conversations could be legally prohibited in recognition of such rights.[4] The case against their use would rest simply on the fact that the users of such devices would be unable to compensate the victims, including all the potential victims, to continue living in a state of unease or anxiety. So humble an invention as the petrol-powered lawn-mower, and other petrol-driven garden implements would come also into conflict with such rights. The din produced by any one man is invariably heard by dozens of families who, of course, may be enthusiastic gardeners also. If they are all satisfied with the current situation or could come to agreement with one another, well and good. But once amenity rights were enacted, at least no man could be forced against his will to absorb these noxious byproducts of the activity of others. Of course, compensation that would satisfy the victim (always assuming he tells the truth) may exceed what the offender could pay. In the circumstances, the enthusiast would have to make do with a hand lawn-mower until the manufacturer discovered means of effectively silencing the din. The manufacturer would, of course, have every incentive to do so, for under such legislation the degree of noise-elimination would be regarded as a factor in the measurement of technical efficiency. The commercial prospects of the product would then vary with the degree of noise-elimination achieved.

Admittedly there are difficulties whenever actual compensation payments have to be made, say, to thousands of families disturbed by aircraft noise. Yet once the principle of amenity rights is recognized in law, a rough estimate of the magnitude of compensation payments necessary to maintain the welfare of the number of families affected would be entered as a matter of course into the social cost calculus. And unless these compensatory payments could also be somehow covered by the proceeds of the air service there would be no *prima facie* case for maintaining the air service.[5] If, on the other hand, compensatory payments could be paid (and their payment costs the company less than any technical device that would effectively eliminate the noise) some method of compensation must be devised. It is true that the courts, from time to time, have enunciated the doctrine that in the ordinary pursuit of industry a reasonable amount of inconvenience must be borne with. The recognition of amenity rights, however, does no more than impose an economic interpretation on the word 'reasonable', and therefore also on the word 'unreasonable', by transferring the cost of the inconvenience on to the shoulders of those who cause it. If by actually compensating the victims—or by paying to eliminate the disamenity by the cheapest technical method available—an existing service cannot be continued (the market being unwilling to pay the increased cost) the inconvenience that is currently being borne with is to be deemed unreasonal. And since those who cause the inconvenience are now compelled to shoulder the increased costs there should be no trouble in convincing them that the inconvenience is unreasonable and, therefore, in withdrawing the service in question.

A law recognizing this principle would have drastic effects on private enterprise which, for too long, has neglected the damage inflicted on society at large in producing its wares. For many decades now private firms have, without giving it a thought, polluted the air we breathe, poisoned lakes and rivers with their effluence, and produced gadgets that have destroyed the quiet of millions of families, gadgets that range from motorized lawn-mowers and motor-cycles to transistors and private planes. What is being proposed therefore may be regarded as an alteration of the legal framework within which private firms operate in order to direct their enterprise towards ends that accord more closely with the interests of society. More specifically, it would provide industry with the incentive necessary to undertake prolonged research into methods of removing the potential amenity-destroying features of so many of today's existing products and services.

The social advantage of enacting legislation embodying amenity rights is further reinforced by a consideration of the regressive nature of many existing external diseconomies. The rich have legal protection of their property and have less need, at present, of protection from the disamenity created by others. The richer a man is the wider is his choice of neighbourhood. If the area he happened to choose appears to be sinking in the scale of amenity he can move, if at some inconvenience, to a quieter area. He can select a suitable town house, secluded perhaps, or made soundproof throughout, and spend his leisure in the country or abroad at times of his own choosing. *Per contra*, the poorer the family the less opportunity there is for moving from its present locality. To all intents it is stuck in the area and must put up with whatever disamenity is inflicted upon it. And, generalizing from the experience of the last ten years or so, one may depend upon it that it will be the neighbourhoods of the working and lower middle classes that will suffer most from the increased construction of fly-overs and fly-unders and roadwidening schemes intended to speed up the accumulating road traffic that all but poison the air. Thus the recognition of amenity rights has favourable distributive effects also. It would promote not only a rise in the standards of environment generally, it would raise them most for the lower income groups that have suffered more than any other group from unchecked 'development' and the growth of motorized traffic since the war.

Mishan, E.J., *The Costs of Economic Growth*, Staples Press, London, 1967, p. 67-73.

Notes

1   Since we are concerned here to illustrate principles, not to offer practical proposals, we shall assume that all parties affected by the siting of the airport have the relevant knowledge on which to estimate their potential compensation payments, and, moreover, that they consistently tell the truth. The cost of collecting information about compensation payments, and any costs of enforcing the optimal arrangements—costs (i) and (ii)—we continue to assume the same whichever party is to be compensated.

2   An extreme case of a man in the desert dying of thirst brings out this point convincingly. The maximum he would pay for a bucket of drinking water which would ensure his survival is limited by his prospective wealth. He could sign the lot away, but no more. The minimum sum he would be willing to take instead of the bucket of water—assuming he wished to live and was not stupid—would approach infinity; or rather, there could be no sum large enough to induce him to part with the life-saving bucket of water. For in the circumstance, a sum of money, no matter how large, would be worth nothing to him since there are no alternative means at any price of keeping himself alive. Only as such means become available at a price does this minimum sum become finite.

3   The reader may now appreciate how recent calculations of the differences in the market value between houses, alike in all other relevant respects, at different distances from an airport, understate the loss suffered from aircraft noise for two reasons, (1) they represent an estimate of the maximum loss that house-owners in the noisier area are able and willing to bear to move out of the area; not the larger estimate of the minimum sum they would accept to put up with the noise. And, as alternative quiet zones become harder to find, this minimum sum they would accept grows relative to the maximum loss they are able and willing to bear. Indeed, even if there were several currently quiet areas into which a family might move, the lack of any announced government plan of maintaining noise-free zones leaves open a risk that effectively reduces the attraction of these areas. (2) If the Government's existing policy continues and noise-free inhabitable areas gradually disappear, an increased level of noise throughout the country as a whole is accompanied by a narrowing of differentials between areas. To regard such a calculation as an index of disamenity is absurd, since it will ultimately reveal zero disamenity for any area whenever all areas are subject to the same amount of aerial disturbance, no matter how great.

4   According to *Life International* (June 13, 1966): 'As manufacturers leap-frog each other turning out ingenious new refinements, the components they sell have been getting smaller and more efficient . . . So rapidly is the field developing that today's devices may be soon outmoded by systems using microcircuits so tiny that a transmitter made of them would be thinner and smaller than a postage stamp, and could be slipped undetected virtually anywhere . . . How to safeguard individual rights in a world suddenly turned into a peep-hole and listening-post has become the toughest legal problem facing the US today.'
Whether the law could be made effective is, of course, a problem. To the extent it could not, one would have to recognize a loss of welfare arising directly from technological progress.

5   It is always open to the Government to claim that a certain air service should be maintained even though it cannot cover its social costs for reasons connected with the defence of the realm. However, it would have to vindicate its claims about the high value to the nation of this particular air service by a willingness to pay a direct subsidy to the company, from the taxpayers' money, in order to cover the costs of compensating the victims.

CHAPTER 7

The Tragedy of the Commons

G. HARDIN

> The population problem has no technical solution;
> it requires a fundamental extension in morality.

At the end of a thoughtful article on the future of nuclear war, Wiesner and York[1] concluded that: "Both sides in the arms race are . . . confronted by the dilemma of steadily increasing military power and steadily decreasing national security. *It is our considered professional judgment that this dilemma has no technical solution.* If the great powers continue to look for solutions in the area of science and technology only, the result will be to worsen the situation."

I would like to focus your attention not on the subject of the article (national security in a nuclear world) but on the kind of conclusion they reached, namely that there is no technical solution to the problem. An implicit and almost universal assumption of discussions published in professional and semipopular scientific journals is that the problem under discussion has a technical solution. A technical solution may be defined as one that requires a change only in the techniques of the natural sciences, demanding little or nothing in the way of change in human values or ideas of morality.

In our day (though not in earlier times) technical solutions are always welcome. Because of precious failures in prophecy, it takes courage to assert that a desired technical solution is not possible. Wiesner and York exhibited this courage; publishing in a science journal, they insisted that the solution to the problem was not to be found in the natural sciences. They cautiously qualified their statement with the phrase, "It is our considered professional judgment . . ." Whether they were right or not is not the concern of the present article. Rather, the concern here is with the important concept of a class of human problems which can be called "no technical solution problems", and, more specifically, with the identification and discussion of one of these.

It is easy to show that the class is not a null class. Recall the game of tick-tack-toe. Consider the problem, "How can I win the game of tick-tack-toe?" It is well known that I cannot, if I assume (in keeping with the conventions of game theory) that my opponent understands the game perfectly. Put another way, there is no "technical solution" to the problem. I can win only by giving a radical meaning to the word "win". I can hit my opponent over the head; or I can drug him; or I can falsify the records. Every way in which I "win" involves, in some sense, an abandonment of the game, as we intuitively understand it (I can also, of course, openly abandon the game—refuse to play it. This is what most adults do).

The class of "No technical solution problems" has members. My thesis is that the "population problem", as conventionally conceived, is a member of this class. How it is conventionally conceived needs some comment. It is fair to say that most people who anguish over the population problem are trying to find a way to avoid the evils of overpopulation without relinquishing any of the privileges they now enjoy. They think that farming the seas or developing new strains of wheat will solve the problem—technologically. I try to show here that the solution they seek cannot be found. The population problem cannot be solved in a technical way, any more than can the problem of winning the game of tick-tack-toe.

What Shall We Maximize?

Population, as Malthus said, naturally tends to grow "geometrically", or, as we would now say, exponentially. In a finite world this means that the per capita share of the world's goods must steadily decrease. Is ours a finite world?

A fair defense can be put forward for the view that the world is infinite; or that we do not know that it is not. But, in terms of the practical problems that we must face in the next few generations with the foreseeable technology, it is clear that we will greatly increase human misery if we do not, during the immediate future, assume that the world available to the terrestrial human population is finite. "Space" is no escape.[2]

A finite world can support only a finite population; therefore, population growth must eventually equal zero. (The case of perpetual wide fluctuations above and below zero is a trivial variant that need not be discussed.) When this condition is met, what will be the situation of mankind? Specifically, can Bentham's goal of "the greatest good for the greatest number" be realized?

No—for two reasons, each sufficient by itself. The first is a theoretical one. It is not mathematically possible to maximize for two (or more) variables at the same time. This was clearly stated by von Neumann and Morgenstern,[3] by giving a radical meaning to the word but the principle is implicit in the theory of partial differential equations, dating back at least to D'Alembert (1717-1783).

The second reason springs directly from biological facts. To live, any organism must have a source of energy (for example, food). This energy is utilized for two purposes: mere maintenance and work. For man, maintenance of life requires about 1600 kilocalories a day ("maintenance calories"). Anything that he does over and above merely staying alive will be defined as work, and is supported by "work calories" which he takes in. Work calories are used not only for what we call work in common speech; they are also required for all forms of enjoyment, from swimming and automobile racing to playing music and writing poetry. If our goal is to maximize population it is obvious what we must do: We must make the work calories per person approach as close to zero as possible. No gourmet meals, no vacations, no sports, no music, no literature, no art . . . I think that everyone will grant, without argument or proof, that maximizing population does not maximize goods. Bentham's goal is impossible.

In reaching this conclusion I have made the usual assumption that it is the acquisition of energy that is the problem. The appearance of atomic energy has led some to question this assumption. However, given an infinite source of energy, population growth still produces an inescapable problem. The problem of the acquisition of energy is replaced by the problem of its dissipation, as J. H. Fremlin has so wittily shown.[4] The arithmetic signs in the analysis are, as it were, reversed; but Bentham's goal is still unobtainable.

The optimum population is, then, less than the maximum. The difficulty of defining the optimum is enormous; so far as I know, no one has seriously tackled this problem. Reaching an acceptable and stable solution will surely require more than one generation of hard analytical work—and much persuasion.

We want the maximum good per person; but what is good? To one person it is wilderness, to another it is ski lodges for thousands. To one it is estuaries to nourish ducks for hunters to shoot; to another it is factory land. Comparing one good with another is, we usually say, impossible because goods are incommensurable. Incommensurables cannot be compared.

Theoretically this may be true; but in real life incommensurables are commensurable. Only a criterion of judgment and a system of weighting are needed. In nature the criterion is survival. Is it better for a species to be small and hideable, or large and powerful? Natural selection commensurates the incommensurables. The compromise achieved depends on a natural weighting of the values of the variables.

Man must imitate this process. There is no doubt that in fact he already does, but unconsciously. It is when the hidden decisions are made explicit that the arguments begin. The problem for the years ahead is to work out an acceptable theory of weighting. Synergistic effects, nonlinear variation, and difficulties in discounting the future make the intellectual problem difficult, but not (in principle) insoluble.

Has any cultural group solved this practical problem at the present time, even on an intuitive level? One simple fact proves that none has: there is no prosperous population in the world today that has, and has had for some time, a growth rate of zero. Any people that has intuitively identified its optimum point will soon reach it, after which its growth rate becomes and remains zero.

Of course, a positive growth rate might be taken as evidence that a population is below its optimum. However, by any reasonable standards, the most rapidly growing populations on earth today are (in general) the most miserable. This association (which need not be invariable) casts doubt on the optimistic assumption that the positive growth rate of a population is evidence that it has yet to reach its optimum.

We can make little progress in working toward optimum population size until we explicitly exorcize the spirit of Adam Smith in the field of practical demography. In economic affairs, *The Wealth of Nations* (1776) popularized the "invisible hand", the idea that an individual who "intends only his own gain", is, as it were, "led by an invisible hand to promote . . . the public interest".[5] Adam Smith did not assert that this was invariably true, and perhaps neither did any of his followers. But he contributed to a dominant tendency of thought that has ever since interfered with positive action based on rational analysis, namely, the tendency to assume that decisions reached individually will, in fact, be the best decisions for an entire society. If this assumption is correct it justifies the continuance of our present policy of laissez-faire in reproduction. If it is correct we can assume that men will control their individual fecundity so as to product the optimum population. If the assumption is not correct, we need to reexamine our individual freedoms to see which ones are defensible.

Tragedy of Freedom in a Commons

The rebuttal to the invisible hand in population control is to be found in a scenario first sketched in a little-known pamphlet[6] in 1833 by a mathematical amateur named William Forster Lloyd (1794-1852). We may well call it "the tragedy of the commons", using the word "tragedy" as the philosopher Whitehead used it:[7] "The essence of dramatic tragedy is not unhappiness. It resides in the solemnity of the remorseless working of things." He then goes on to say, "This inevitableness of destiny can only be illustrated in terms of human life by incidents which in fact involve unhappiness. For it is only by them that the futility of escape can be made evident in the drama."

The tragedy of the commons develops in this way. Picture a pasture open to all. It is to be expected that each herdsman will try to keep as many cattle possible on the commons. Such an arrangement may work reasonably satisfactorily for centuries because tribal wars, poaching, and disease keep numbers of both man and beast well below the carrying capacity of the land. Finally, however, comes the day reckoning, that is, the day when the long desired goal of social stability becomes a reality. At this point, the inherent logic of the commons remorselessly generates tragedy.

As a rational being, each herdsman seeks to maximize his gain. Explicitly, or implicitly, more or less consciously he asks, "What is the utility *to me* of adding one more animal to my herd?" This utility has one negative and a positive component.

1 The positive component is a function of the increment of one animal. Since the herdsman receives all the proceeds from the sale of the additional animal, the positive utility is nearly +1.

2 The negative component is a function of the additional overgrazing created by one more animal. Since, however, the effects of overgrazing are shared by all the herdsmen, the negative utility for any particular decisionmaking herdsman is only a fraction of -1.

Adding together the component partial utilities, the rational herdsman concludes that the only sensible course for him to pursue is to add another animal to his herd. And another; and another. . . . But this is the conclusion reached by each and every ration herdsman sharing a commons. There is the tragedy. Each man is locked in a system that compels him to increase his herd without limit—in a world that is limited. Ruin is the destination toward which all men rush, each pursuing his own best interest in a society that believes in the freedom of the commons. Freedom in a commons brings ruin to all.

Some would say that this is a platitude. Would that it were! In a sense, it was learned thousands of years ago, but natural selection favours the forces psychological denial.[8] The individual benefits as an individual from his ability to deny the truth even though society a whole, of which he is a part, suffers. Education can counteract the natural tendency to do the wrong thing, but the exorable succession of generations requires that the basis for this knowledge be constantly refreshed.

A simple incident that occurred a few years ago in Leominster, Massachusetts, shows how perishable the knowledge is. During the Christmas shopping season the parking meters downtown were covered with plastic bags that bore tags reading: "Do not open until after Christmas. Free parking courtesy of the mayor and city council." In other words, facing the prospect of an increased demand for already scarce space, the city fathers reinstituted the system of the commons. (Cynically, we suspect that they gained more votes than they lost by this retrogressive act.)

In an approximate way, the logic of the commons has been understood for a long time, perhaps since the discovery of agriculture or the invention of private property in real estate. But it is understood mostly only in special cases which are not sufficiently generalized. Even at this late date, cattlemen leasing national land on the western ranges demonstrate no more than an ambivalent understanding, in constantly pressuring the authorities to increase the head count to the point where overgrazing produces erosion and weed dominance. Likewise, the oceans of the world continue to suffer from the survival of the philosophy of the commons. Maritime nations still respond automatically to the shibboleth of the "freedom of the seas". Professing to believe in the "inexhaustible resources of the oceans", they bring species after species of fish and whales closer to extinction.[9]

The National Parks present another instance of the working out of the tragedy of the commons. At present, they are all open to all, without limit. The parks themselves are limited in extent—there is only one Yosemite Valley—whereas population seems to grow without limit. The value that visitors seek in the parks are steadily eroded. Plainly, we must soon cease to treat the parks as commons or they will be of no value to anyone.

What shall we do? We have several options. We might sell them off as private property. We might keep them as public property, but allocate the right to enter them. The allocation might be on the basis of wealth, by the use of an auction system. It might be on the basis of merit, as defined by some agreed upon standards. It might be by lottery. Or it might be on a first-come, first-served basis, administered to long queues. These, I think, are all the reasonable possibilities. They are all objectionable. But we must choose—or acquiesce in the destruction of the commons that we call our National Parks.

Pollution

In a reverse way, the tragedy of the commons reappears in problems of pollution. Here it is not a question of taking something out of the commons, but of putting something in—sewage, or chemical, radioactive, and heat wastes into water; noxious and dangerous fumes into the air; and distracting and unpleasant advertising signs into the line of sight. The calculations of utility are much the same as before. The rational man finds that his share of the cost of the wastes he discharges into the commons is less than the cost of purifying his wastes before releasing them. Since this is true for everyone, we are locked into a system of "fouling our own nest", so long as we behave only as independent, rational, free-enterprisers.

The tragedy of the commons as a food basket is averted by private property, or something formally like it. But the air and water surrounding us cannot readily be fenced, and so the tragedy of the commons as a cesspool must be prevented by different means, by coercive laws or taxing devices that make it cheaper for the polluter to treat his pollutants than to discharge them untreated. We have not progressed as far with the solution of this problem as we have with the first. Indeed, our particular concept of private property, which deters us from exhausting the positive resources of the earth, favours pollution. The owner of a factory on the bank of a stream—whose property extends to the middle of the stream—often has difficulty seeing why it is not his natural right to muddy the waters flowing past his door. The law, always behind the times, requires elaborate stitching and fitting to adapt it to this newly perceived aspect of the commons.

The pollution problem is a consequence of population. It did not much matter how a lonely American frontiers-man disposed of his waste. "Flowing water purifies itself every 10 miles", my grandfather used to say, and the myth was near enough to the truth when he was a boy, for there were not too many people. But as population became denser, the natural chemical and biological recycling processes became overloaded, calling for a redefinition of property rights.

How To Legislate Temperance?

Analysis of the pollution problem as a function of population density uncovers a not generally recognized principle of morality, namely: *the morality of an act is a function of the state of the system at the time it is performed.*[10] Using the commons as a cesspool does not harm the general public under frontier conditions, because there is no public; the same behaviour in a metropolis is unbearable. A hundred and fifty years ago a plainsman could kill an American bison, cut out only the tongue for his dinner, and discard the rest of the animal. He was not in any important sense being wasteful. Today, with only a few thousand bison left, we would be appalled at such behaviour.

In passing, it is worth noting that the morality of an act cannot be determined from a photograph. One does not know wether a man killing an elephant or setting a fire to the grassland is harming others until one knows the total system in which his act appears. "One picture is worth a thousand words", said an ancient Chinese; but it may take 10,000 words to validate it. It is as tempting to ecologists as it is to reformers in general to try to persuade others by way of the photographic shortcut. But the essence of an argument cannot be photographed: it must be presented rationally—in words.

That morality is system-sensitive escaped the attention of most codifiers of ethics in the past. "Thou shalt not . . ." is the form of traditional ethics, and therefore are poorly suited to governing a complex, crowded, changeable world. Our epicyclic solution is to augment statutory law with administrative law. Since it is practically impossible to spell out all the conditions under which it is safe to run an automobile without smog-control, by law we delegate the details to bureaus. The result is administrative law, which is rightly feared for an ancient reason—*Quis custodiet ipsos custodes?*—"Who shall watch the watchers themselves?" John Adams said that we must have "a government of laws and not men". Bureau administrators, trying to evaluate the morality of acts in the total system, are singularly liable to corruption, producing a government by men, not laws.

Prohibition is easy to legislate (though not necessarily to enforce); but how do we legislate temperance? Experience indicates that it can be accomplished best through the mediation of administrative law. We limit possibilities unnecessarily if we suppose that the sentiment of *Quis custodiet* denies us the use of administrative law. We should rather retain the phrase as a perpetual reminder of fearful dangers we cannot avoid. The great challenge facing us now is to invent the corrective feedbacks that are needed to keep custodians honest. We must find ways to legitimate the needed authority of both the custodians and the corrective feedbacks.

Freedom To Breed Is Intolerable

The tragedy of the commons is involved in population problems in another way. In a world governed solely by the principle of "dog eat dog" —if indeed there ever was such a world—how many children a family had would not be a matter of public concern. Parents who bred too exuberantly would leave fewer descendants, not more, because they would be unable to care adequately for their children. David Lack and others have found that such a negative feedback demonstrably controls the fecundity of birds.[11] But men are not birds, and have not acted like them for millenniums, at least.

If each human family were dependent only on its own resources; *if* the children of improvident parents starved to death; *if*, thus, overbreeding brought its own "punishment" to the germ line—*then* there would be no public interest in controlling the breeding of families. But our society is deeply committed to the welfare state,[12] and hence is confronted with another aspect of the tragedy of the commons.

In a welfare state, how shall we deal with the family, the religion, the race, or the class (or indeed any distinguishable and cohesive group) that adopts overbreeding as a policy to secure its own aggrandizement?[13] To couple the concept of freedom to breed with the belief that everyone born has an equal right to the commons is to lock the world into a tragic course of action.

Unfortunately this is just the course of action that is being pursued by the United Nations. In late 1967, some 30 nations agreed to the following:[14]

> The Universal Declaration of Human Rights describes the family as the natural and fundamental unit of society. It follows that any choice and decision with regard to the size of the family must irrevocably rest with the family itself, and cannot be made by anyone else.

It is painful to have to deny categorically the validity of this right; denying it, one feels as uncomfortable as a resident of Salem, Massachusetts, who denied the reality of witches in the 17th century. At the present time, in liberal quarters, something like a taboo acts to inhibit criticism of the United Nations. There is a feeling that the United Nations is "our last and best hope", that we shouldn't find fault with it; we shouldn't play into the hands of the archconservatives. However, let us not forget what Robert Louis Stevenson said: "The truth that is suppressed by friends is the readiest weapon of the enemy." If we love the truth we must openly deny the validity of the Universal Declaration of Human Rights, even though it is promoted by the United Nations. We should also join with Kingsley Davis[15] in attempting to get Planned Parenthood-World Population to see the error of its ways in embracing the same tragic ideal.

Conscience Is Self-Eliminating

It is a mistake to think that we can control the breeding of mankind in the long run by an appeal to conscience. Charles Galton Darwin made this point when he spoke on the centennial of the publication of his grandfather's great book. The argument is straightforward and Darwinian.

People vary. Confronted with appeals to limit breeding, some people will undoubtedly respond to the plea more than others. Those who have more children will produce a larger fraction of the next generation than those with more susceptible consciences. The difference will be accentuated, generation by generation.

In C. G. Darwin's words: "It may well be that it would take hundreds of generations for the progenitive instinct to develop in this way, but if it should do so, nature would have taken her revenge, and the variety *Homo contracipiens* would become extinct would be replaced by the variety *Homo progenitivus*".[16]

The argument assumes that science or the desire for children (no matter which) is hereditary—but hereditary only in the most general for sense. The result will be the same whether the attitude is transmitted through germ cells, or exosomatically to use A. J. Lotka's term. (If one denies the latter possibility as well as former, then what's the point of education?) The argument has here been stated in the context of the population problem, but it applies equally well any instance in which society appeals to an individual exploiting a commons to restrain himself for the general good—by means of his conscience. To make such an appeal is to set up a selective system that works toward the elimination of conscience from the race.

Pathogenic Effects of Conscience

The long-term disadvantage of an appeal to conscience should be enough to condemn it; but has serious short term disadvantages as well. If we ask a man who is exploiting a commons desist "in the name of conscience", what are we saying to him? What does he hear? —not only at the moment but also in the wee small hours of the night when, half asleep, he remembers not merely the words we used but also the nonverbal communication cues gave him unawares? Sooner or later, consciously or subconsciously, he senses that he has received two communications, and that they are contradictory: (i) (intended communication) "If you don't do as we ask, we will openly condemn you for not acting like a responsible citizen"; (ii) (the unintended communication) "If you *do* behave we ask, we will secretly condemn you for a simpleton who can be shamed into standing aside while the rest of exploit the commons."

Everyman then is caught in what Bateson has called a "double bind". Bateson and his co-workers have made a plausible case for viewing the double bind as an important causative factor in the genesis of schizophrenia.[17] The double bind may not always be so damaging, but it always endangers the mental health of anyone to whom it is applied. "A bad conscience", said Nietzsche, "is a kind of illness".

To conjure up a conscience in other is tempting to anyone who wishes to extend his control beyond the legal limits. Leaders at the highest level succumb to this temptation. Has any President during the past generation failed to call on labour unions to moderate voluntarily their demands for higher wages, or to steel companies to honor voluntary guidelines on prices? I can recall none. The rhetoric used on such occasions is designed to produce feelings of guilt in non-cooperators.

For centuries it was assumed without proof that guilt was a valuable, perhaps even an indispensable, ingredient of the civilized life. Now, in this post-Freudian world, we doubt it.

Paul Goodman speaks from the modern point of view when he says:

"No good has ever come from feeling guilty, neither intelligence, policy, nor compassion. The guilty do not pay attention to the object but only to themselves, and not even to their own interests, which might make sense, but to their anxieties".[18]

One does not have to be a professional psychiatrist to see the consequences of anxiety. We in the Western world are just emerging from a dreadful two-centuries-long Dark Ages of Eros that was sustained partly by prohibition laws, but perhaps more effectively by the anxiety-generating mechanisms of education. Alex Comfort has told the story well in *The Anxiety Makers;*[19] it is not a pretty one.

Since proof is difficult, we may even concede that the results of anxiety may sometimes, from certain points of view, be desirable. The larger question we should ask is whether, as a matter of policy, we should ever encourage the use of a technique the tendency (if not the intention) of which is psychologically pathogenic. We hear much talk these days of responsible parenthood; the coupled words are incorporated into the titles of some organizations devoted to birth control. Some people have proposed massive propaganda campaigns to instill responsibility into the nation's (or the world's) breeders. But what is the meaning of the word responsibility in this context? Is it not merely a synonym for the word conscience? When we use the word responsibility in the absence of substantial sanctions are we not trying to browbeat a free man in a commons into acting against his own interest? Responsibility is a verbal counterfeit for a substantial quid pro quo. It is an attempt to get something for nothing.

If the word responsibility is to be used at all, I suggest that it be in the sense Charles Frankel uses it.[20] "Responsibility", says this philosopher, "is the product of definite social arrangements". Notice that Frankel calls for social arrangements—not propaganda.

Mutual Coercion
Mutually Agreed upon

The social arrangements that produce responsibility are arrangements that create coercion, of some sort. Consider bank-robbing. The man who takes money from a bank acts as if the bank were a commons. How do we prevent such action? Certainly not by trying to control his behaviour solely by a verbal appeal to his sense of responsibility. Rather than rely on propaganda we follow Frankel's lead and insist that a bank is not a commons; we seek the definite social arrangements that will keep it from becoming a commons. That we thereby infringe on the freedom of would-be robbers we neither deny nor regret.

The morality of bank-robbing is particularly easy to understand because we accept complete prohibition of this activity. We are willing to say "Thou shalt not rob banks", without providing for exceptions. But temperance also can be created by coercion. Taxing is a good coercive device. To keep downtown shoppers temperate in their use of parking space we introduce parking meters for short periods, and traffic fines for longer ones. We need not actually forbid a citizen to park as long as he wants to; we need merely make it increasingly expensive for him to do so. Not prohibition, but carefully biased options are what we offer him. A Madison Avenue man might call this persuasion; I prefer the greater candor of the word coercion.

Coercion is a dirty word to most liberals now, but it need not forever be so. As with the four-letter words, its dirtiness can be cleansed away by exposure to the light, by saying it over and over without apology or embarrassment. To many, the word coercion implies arbitrary decisions of distant and irresponsible bureaucrats; but this is not a necessary part of its meaning. The only kind of coercion I recommend is mutual coercion, mutually agreed upon by the majority of the people affected.

To say that we mutually agree to coercion is not to say that we are required to enjoy it, or even to pretend we enjoy it. Who enjoys taxes? We all grumble about them. But we accept compulsory taxes because we recognize that voluntary taxes would favor the conscienceless. We institute and (grumblingly) support taxes and other coercive devices to escape the horror of the commons.

An alternative to the commons need not be perfectly just to be preferable. With real estate and other material goods, the alternative we have chosen is the institution of private property coupled with legal inheritance. Is this system perfectly just? As a genetically trained biologist I deny that it is. It seems to me that, if there are to be differences in individual inheritance, legal possession should be perfectly correlated with biological inheritance—that those who are biologically more fit to be the custodians of property and power should legally inherit more. But genetic recombination continually makes a mockery of the doctrine of "like father, like son" implicit in our laws of legal inheritance. An idiot can inherit millions, and a trust fund can keep his estate intact. We must admit that our legal system of private property plus inheritance is unjust—but we put up with it because we are not convinced, at the moment, that anyone has invented a better system. The alternative of the commons is too horrifying to contemplate. Injustice is preferable to total ruin.

It is one of the peculiarities of the warfare between reform and the status quo that it is thoughtlessly governed by a double standard. Whenever a reform measure is proposed it is often defeated when its opponents triumphantly discover a flaw in it. As Kingsley Davis has pointed out,[21] worshippers of the status quo sometimes imply that no reform is possible without unanimous agreement, an implication contrary to historical fact. As nearly as I can make out, automatic rejection of proposed reforms is based on one of two unconscious assumptions: (i) that the status quo is perfect; or (ii) that the choice we face is between reform and no action; if the proposed reform is imperfect, we presumably should take no action at all, while we wait for a perfect proposal.

But we can never do nothing. That which we have done for thousands of years is also action. It also produces evils. Once we are aware that the status quo is action, we can then compare its discoverable advantages and disadvantages with the predicted advantages and disadvantages of the proposed reform, discounting as best we can for our lack of experience. On the basis of such a comparison, we can make a rational decision which will not involve the unworkable assumption that only perfect systems are tolerable.

Recognition of Necessity

Perhaps the simplest summary of this analysis of man's population problems is this: the commons, if justifiable at all, is justifiable only under conditions of low-population density. As the human population has increased, the commons has had to be abandoned in one aspect after another.

First we abandoned the commons in food gathering, enclosing farm land and restricting pastures and hunting and fishing areas. These restrictions are still not complete throughout the world.

Somewhat later we saw that the commons as a place for waste disposal would also have to be abandoned. Restrictions on the disposal of domestic sewage are widely accepted in the Western world; we are still struggling to close the commons to pollution by automobiles, factories, insecticide sprayers, fertilizing operations, and atomic energy installations.

In a still more embryonic state is our recognition of the evils of the commons in matters of pleasure. There is almost no restriction on the propagation of sound waves in the public medium. The shopping public is assaulted with mindless music, without its consent. Our government is paying out billions of dollars to create supersonic transport which will disturb 50,000 people for every one person who is whisked from coast to coast 3 hours faster. Advertisers muddy the airwaves of radio and television and pollute the view of travellers. We are a long way from outlawing the commons in matters of pleasure. Is this because our Puritan inheritance makes us view pleasure as something of a sin, and pain (that is, the pollution of advertising) as the sign of virtue?

Every new enclosure of the commons involves the infringement of somebody's personal liberty. Infringements made in the distant past are accepted because no contemporary complains of a loss. It is the newly proposed infringements that we vigorously oppose; cries of "rights" and "freedom" fill the air. But what does "freedom" mean? When men mutually agreed to pass laws against robbing, mankind became more free, not less so. Individuals locked into the logic of the commons are free only to bring on universal ruin; once they see the necessity of mutual coercion, they become free to pursue other goals. I believe it was Hegel who said, "Freedom is the recognition of necessity."

The most important aspect of necessity that we must now recognize, is the necessity of abandoning the commons in breeding. No technical solution can rescue us from the misery of overpopulation. Freedom to breed will bring ruin to all. At the moment, to avoid hard decisions many of us are tempted to propagandize for conscience and responsible parenthood. The temptation must be resisted, because an appeal to independently acting consciences selects for the disappearance of all conscience in the long run, and an increase in anxiety in the short.

The only way we can preserve and nurture other and more precious freedoms is by relinquishing the freedom to breed, and that very soon. "Freedom is the recognition of necessity"—and is the role of education to reveal to a the necessity of abandoning the freedom to breed. Only so, can we put a end to this aspect of the tragedy of the commons.

Hardin, G., The Tragedy of the Commons, in: *Science* 162, 1968, p. 1243-1248.

Notes and References

1 J. B. Wiesner and H. F. York, *Sci. Amer.* 211 (No. 4), 27 (1964).

2 G. Hardin, *J. Hered.* 50, 68 (1959); S. von Hoernor, Science 137, 18 (1962).

3 J. von Neumann and O. Morgenstern, *Theory of Games and Economic Behaviour* (Princeton Univ. Press, Princeton, N.J., 1947), p. 11.

4 J. H. Fremlin, *New Sci.*, No. 415 (1964), p. 285.

5 A. Smith, *The Wealth of Nations* (Modern Library, New York, 1937), p. 423.

6 W. F. Lloyd, *Two Lectures on the Checks to Population* (Oxford Univ. Press, Oxford, England, 1833), reprinted (in part) in *Population, Evolution, and Birth Control*, G. Hardin Ed. (Freeman, San Francisco, 1964), p. 37.

7 A. N. Whitehead, *Science and the Modern World* (Mentor, New York, 1948), p. 17.

8 G. Hardin, Ed. *Population, Evolution and Birth Control* (Freeman, San Francisco, 1964), p. 56.

9 S. McVay, *Sci. Amer.* 216 (no. 8), 13 (1966).

10 J. Fletcher, *Situation Ethics* (Westminster, Philadelphia, 1966).

11 D. Lack, *The Natural Regulation of Animal Numbers* (Clarendon Press, Oxford, 1954).

12 H. Girvetz, *From Wealth to Welfare* (Stanford Univ. Press, Stanford, Calif., 1950).

13 G. Hardin, *Perspec. Biol. Med.* 6, 366 (1963).

14 U. Thant, *Int. Planned Parenthood News*, No. 168 (February 1968), p. 3.

15 K. Davis, *Science 158*, 730 (1967).

16 S. Tax, Ed., *Evolution after Darwin* (Univ. of Chicago Press, Chicago, 1960), vol. 2, p. 469.

17 G. Bateson, D. D. Jackson, J. Haley, J. Weakland, *Behav. Sci.* 1, 251 (1956).

18 P. Goodman, *New York Rev. Books* 10(8), 22 (23 May 1968).

19 A. Comfort, *The Anxiety Makers* (Nelson, London, 1967).

20 C. Frankel, *The Case for Modern Man* (Harper, New York, 1955), p. 203.

21 J. D. Roslansky, *Genetics and the Future of Man* (Appleton-Century-Crofts, New York, 1966), p. 177.

CHAPTER 8

The Population Bomb

P. AND A. EHRLICH

Too Many People

Americans are beginning to realize that the undeveloped countries of the world face an inevitable population-food crisis: Each year food production in underdeveloped countries falls a bit further behind burgeoning population growth, and people go to bed a little bit hungrier. While there are temporary or local reversals of this trend, it now seems inevitable that it will continue to its logical conclusion: mass starvation. The rich are going to get richer, but the more numerous poor are going to get poorer. Of these poor, a minimum of three and one-half million will starve to death this year, mostly children. But this is a mere handful compared to the numbers that will be starving in a decade or so. And it is now too late to take action to save many of those people.

In a book about population there is a temptation to stun the reader with an avalanche of statistics. I'll spare you most, but not all, of that. After all, no matter how you slice it, population is a number game. Perhaps the best way to impress you with numbers is to tell you about the "doubling time"—the time necessary for the population to double in size.

It has been estimated that the human population of 6000 BC was about five million people, taking perhaps one million years to get there from two and a half million. The population did not reach 500 million until almost 8,000 years later—about 1650 AD. This means it doubled roughly once every thousand years or so. It reached a billion people around 1850, doubling in some 200 years. It took only 80 years or so for the next doubling, as the population reached two billion around 1930. We have not completed the next doubling to four billion yet, but we now have well over three billion people. The doubling time at present seems to be about 37 years.[1] Quite a reduction in doubling times, 1,000,000 years, 1,000 years, 200 years, 80 years, 37 years. Perhaps the meaning of a doubling time of around 37 years is best brought home by a theoretical exercise. Let's examine what might happen on the absurd assumption that the population continued to double every 37 years into the indefinite future.

If growth continued at that rate for about 900 years, there would be some 60,000,000,000,000,000 people on the face of the earth. Sixty million billion people. This is about 100 persons for each square yard of the Earth's surface, land and sea. A British physicist, J.H. Fremlin[2], guessed that such a multitude might be housed in a continuous 2,000-story building covering our entire planet. The upper 1,000 stories would contain only the apparatus for running this gigantic warren. Ducts, pipes, wires, elevator shafts, etc., would occupy about half of the space in the bottom 1,000 stories. This would leave three of four yards of floor space for each person. I will leave to your imagination the physical details of existence in this ant heap, except to point out that all would not be black. Probably each person would be limited in his travel. Perhaps he could take elevators through all 1,000 residential stories but could travel only within a circle of a few hundred yards' radius on any floor. This would permit, however, each person to choose his friends among some ten million people! And, as Fremlin points out, entertainment on the worldwide TV should be excellent, for at any time "one could expect some ten million Shakespeares and rather more Beatles to be alive".

Could growth of the human population of the Earth continue beyond that point? Not according to Fremlin. We would have reached a "heat limit". People themselves, as well as their activities convert other forms of energy into heat which must be dissipated. In order to permit this excess heat to radiate directly from the top of the "world building" directly into space, the atmosphere would have been pumped into flasks under the sea well before the limiting population size was reached. The precise limit would depend on the technology of the day. At a population size of one billion people, the temperature of the "world roof" would be kept around the melting point of iron to radiate away the human heat generated.

But, you say, surely Science (with a capital "S") will find a way for us to occupy the other planets of our solar system and eventually of other stars before we get all that crowded. Skip for a moment the virtual certainty that those planets are inhabitable. Forget also the insurmountable logistic problems of moving billions of people off the Earth. Fremlin has made some interesting calculations on how much time we could buy by occupying the planets of the solar system. For instance, at any given time it would take only about 50 years to populate Venus, Mercury, Mars, the moon, and the moons of Jupiter and Saturn to the same population density as Earth.[3]

What if the fantastic problems of reaching and colonizing the other planets of the solar system, such as Jupiter and Uranus, can be solved? It would take only about 200 years to fill them "Earth-full". So we could perhaps gain 250 years of time for population growth in the solar system after we had reached an absolute limit on Earth. What then? We can't ship our surplus to the stars. Professor Garrett Hardin[4] of the University of California at Santa Barbara has dealt effectively with fantasy. Using extremely optimistic assumptions, he has calculated that Americans, by cutting their standard of living down to 18 per cent of its present level, could in *one year* set aside enough capital to finance the exportation to the start of *one day's* increase in the population of the world.

Interstellar transport for surplus people presents an amusing prospect. Since the ships would take generation to reach most stars, the only people who could be transported would be those willing to exercise strict birth control. Population explosions on space ships would be disastrous. Thus we would have to export our responsible people, leaving the irresponsible at home on Earth to breed.

Enough of fantasy. Hopefully, you are convinced that the population will have to stop growing sooner or later and that the extremely remote possibility of expanding into outer space offers no escape from the laws of population growth. If you still want to hope for the stars, just remember that, at the current growth rate, in a few thousand years everything in the visible universe would be converted into people, and the ball of people would be expanding with the speed of light![5] Unfortunately, even 900 years is much too far in the future for those of us concerned with the population explosion. As you shall see, the next *nine* years will probably tell the story.

Of course, population growth is not occurring uniformly over the face of the Earth. Indeed, countries are divided rather neatly into two groups: those with rapid growth rates, and those with relatively slow growth rates. The first group, making up about two-thirds of the world population, coincides closely with what are known as the "undeveloped countries" (UDCs). The UDCs are not industrialized, tend to have inefficient agriculture, very small gross national products, high illiteracy rates and related problems. That's what UDCs are technically, but a short definition of undeveloped is "starving". Most Latin American, African, and Asian countries fall into this category. The second group consists, in essence, of the "developed countries" (DCs). The DCs are modern, industrial nations, such as the United States, Canada, most European countries, Israel, Russia, Japan, and Australia. Most people in these countries are adequately nourished.

Doubling times in the UDCs range around 20 to 35 years. Examples of these times (from the 1968 figures just released by the Population Reference Bureau) are Kenya, 24 years; Nigeria, 28; Turkey, 24; Indonesia, 31; Philippines, 20; Brazil, 22; Costa Rica, 20; and El Salvador, 19. Think of what it means for the population of a country to double in 25 years. In order to just keep living standards at the present inadequate level, the food available for the people must be doubled. Every structure and road must be duplicated. The amount of power must be doubled. The capacity of the transport system must be doubled. The number of trained doctors, nurses, teachers, and administrators must be doubled. This would be a fantastically difficult job in the United States—a rich country with a fine agricultural system, immense industries, and rich natural resources. Think of what it means to a country with none of these.

Remember also that in virtually all UDCs, people have gotten the word about the better life it is possible to have. They have seen colored pictures in magazines of the miracles of Western technology. They have seen automobiles and airplanes. They have seen American and European movies. Many have seen refrigerators, tractors, and even TV sets. Almost all have heard transistor radios. They *know* that a better life is possible. They have what we like to call "rising expectations". If twice as many people are to be happy, the miracle of doubling what they now have will not be enough. It will only maintain today's standard of living. There will have to be a tripling or better. Needless to say, they are not going to be happy.

Doubling times for the populations of the DCs tend to be in the 50 to 200-year range. Examples of 1968 doubling times are the United States, 63 years; Austria, 175; Denmark, 88; Norway, 88; United Kingdom, 140; Poland, 88; Russia, 63; Italy, 117; Spain, 88; and Japan, 63. These are industrialized countries that have undergone the so-called demographic transition—a transition from high to low growth rate. As industrialization progressed, children became less important to parents as extra hands to work on the farm and as support in old age. At the same time they became a financial drag—expensive to raise and educate. Presumably these are the reasons for a slowing of population growth after industrialization. They boil down to a simple fact—people just want to have fewer children.

This is not to say, however, that population is not a problem for the DCs. First of all, most of them are overpopulated. They are overpopulated by the simple criterion that they are not able to produce enough food to feed their populations. It is true that they have the money to buy food, but when food is no longer available for sale they will find the money rather indigestible. Then, too, they share with the UDCs a serious problem of population distribution. Their urban centers are getting more and more crowded relative to the countryside. This problem is not as severe as it is in the UDCs (if current trends should continue, which they cannot, Calcutta could have 66 million inhabitants in the year 2000). As you are well aware, however, urban concentrations are creating serious problems even in America. In the United States, one of the more rapidly growing DCs, we hear constantly of the headaches caused by growing population: not just garbage in our environment, but overcrowded highways, burgeoning slums, deteriorating schools systems, rising crime rates, riots, and other related problems.

From the point of view of a demographer, the whole problem is quite simple. A population will continue to grow as long as the birth rate exceeds the death rate—if immigration and emigration are not occurring. It is, of course, the balance between birth rate and death rate that is critical. The birth rate is the number of births per thousand people per year in the population. The death rate is the number of deaths per thousand people per year.[6] Subtracting the death rate from the birth rate, and ignoring migration, gives the rate of increase. If the birth rate is 30 per thousand per year, and the death rate is 10 per thousand per year, then the rate of increase is 20 per thousand per year (30-10=20). Expressed as a per cent (rate per hundred people), the rate of 20 per thousand becomes 2 per cent. If the rate of increase is 2 per cent, then the doubling time will be 35 years. Note that if you simply added 20 people per thousand per year to the population, it would take 50 years to add a second thousand people (20x50=1,000). But the doubling time is actually much less because populations grow at compound interest rates. Just as interest dollars themselves earn interest, so people added to populations produce more people. It's growing at compound interest that makes populations double so much more rapidly than seems possible. Look at the relationship between the annual per cent increase (interest rate) and the doubling time of the population (time for your money to double):

Annual per cent increase	Doubling time
1.0	70
2.0	35
3.0	24
4.0	17

Those are all the calculations—I promise. If you are interested in more details on how demographic figuring is done, you may enjoy reading Thompson and Lewis's excellent book, *Population Problems*.[7]

There are some professional optimists around who like to greet every sign of dropping birth rates with wild pronouncements about the end of the population explosion. They are a little like a person who, after a low temperature of five below zero on December 21, interprets a low of only three below zero on December 22 as a cheery sign of approaching spring. First of all, birth rates, along with all demographic statistics, show short-term fluctuations caused by many factors. For instance, the birth rate depends rather heavily on the number of women at reproductive age. In the United states the current low birth rates soon will be replace by higher rates as more post World War II "baby boom" children move into their reproductive years. In Japan, 1966, the Year of the Fire Horse, was a year of very low birth rates. There is widespread belief that girls born in the Year of the Fire Horse make poor wives, and Japanese couples try to avoid giving birth in that year because they are afraid of having daughters.

But, I repeat, it is the relationship between birth rate and death rate that is most critical. Indonesia, Laos, and Haiti all had birth rates around 46 per thousand in 1966. Costa Rica's birth rate was 41 per thousand. Good for Costa

Rica? Unfortunately, not very. Costa Rica's death rate was less than nine per thousand, while the other countries all had death rates above 20 per thousand. The population of Costa Rica in 1966 was doubling every 17 years, while the doubling times of Indonesia, Laos, and Haiti were all above 30 years. Ah, but, you say, it was good for Costa Rica—fewer people per thousand were dying each year. Fine for a few years perhaps, but what then? Some 50 per cent of the people in Costa Rica are under 15 years old. As they get older, they will need more and more food in a world with less and less. In 1983 they will have twice as many mouths to feed as they had in 1966, if the 1966 trend continues. Where will the food come from? Today the death rate in Costa Rica is low in part because they have a large number of physicians in proportion to their population.

How do you suppose those physicians will keep the death rate down when there's not enough food to keep the people alive?

One of the most ominous facts of the current situation is that roughly 40 per cent of the population of the underdeveloped world is made up of people *under 15 years old*. As that mass of young people moves into its reproductive years during the next decade, we're going to see the greatest baby boom of all time. Those youngsters are the reason for all the ominous predictions for the year 2000. They are the gunpowder for the population explosion.

How did we get into this bind? It all happened a long time ago, and the story involves the grades of natural selection, the development of culture, and man's swollen head. The essence of success in evolution is reproduction. Indeed, natural selection is simply defined as differential reproduction of genetic types. That is, if people with blue eyes have more children on average that those with brown eyes, natural selection is occurring. More genes for blue eyes will be passed on to the next generation than will genes for brown eyes. Should this continue, the population will have progressively larger and larger proportions of blue-eyed people. This differential reproduction of genetic types is the driving force of evolution for billions of years. Whatever types produced more offspring became the common types. Virtually all populations contain very many different genetic types (for reasons that need not concern us), and some are always outreproducing others. As I said, reproduction is the key to winning the evolutionary game. Any structure, physiological process, or pattern of behavior that leads to greater reproductive success will tend to be perpetuated. The entire process by which man developed involves thousands of millennia of our ancestors being more successful breeders than their relatives. Facet number one of our bind—the urge to reproduce has been fixed in us by billions of years of evolution.

Of course through all those years of evolution our ancestors were fighting a continual battle to keep the birth rate ahead of the death rate. That they were successful is attested to by our very existence, for, if the death rate had overtaken the birth rate for any substantial period of time, the evolutionary line leading to man would have gone extinct. Among our apelike ancestors, a few million years ago, it was still very difficult for a mother to rear her children successfully. Most of the offspring died before they reached reproductive age. The death rate was near the birth rate. Then another factor entered the picture—cultural evolution was added to biological evolution.

Culture can be loosely defined as the body of nongenetic information which people pass from generation to generation. It is the accumulated knowledge, that, in the old days, was passed on entirely by word of mouth, painting, and demonstration. Several thousand years ago the written word was added to the means of cultural transmission. Today culture is passed on in these ways, and also through television, computer tapes, motion pictures, records, blueprints, and other media. Culture is all the information man possessed except for that which is stored in chemical language of his genes.

The large size of the human brain evolved in response to the development of cultural information. A big brain is an advantage when dealing with such information. Big-brained individuals were able to deal more successfully with the culture of their group. They were thus more successful reproductively than their smaller-brained relatives. They passed on their genes for big brains to their numerous offspring. They also added to the accumulating store of cultural information, increasing slightly the premium placed on brain size in the next generation. A self-reinforcing selective trend developed—a trend toward increased brain size.[8]

But there was, quite literally, a rub. Babies had bigger and bigger heads. There were limits to how large a woman's pelvis could conveniently become. To make a long story short, the strategy of evolution was not to make a woman bellshaped and relatively immobile, but to accept the problem of having babies who were helpless for a long period while their brains grew after birth.[9] How could the mother defend and care for her infant during its unusually long period of helplessness? She couldn't, unless Papa hung around. The girls are still working on that problem, but an essential step was to get rid of the short, well-defined breeding season characteristic of most mammals. The year-round sexuality of the human female, the long period of infant dependence on the female, the long period of infant dependence on the female, the evolution of the family group, all are at the roots of our present problem. They are all essential ingredients in the vast social phenomenon that we call sex. Sex is not simple an act leading to the production of offspring. It is a varied and complex cultural phenomenon penetrating into all aspects of our lives—one involving our self-esteem, our choice of friends, cars, and leaders. It is tightly interwoven with our mythologies and history. Sex in man is necessary for the production of young, but it also envolved to ensure their successful rearing. Facet number two of our bind—our urge to reproduce is hopelessly entwined with most of our other urges.

Of course, in the early days the whole system did not prevent a very high mortality among the young, as well as among the older members of the group. Hunting and food-gathering is a risky business. Cavemen had to throw very impressive cave bears out of their caves before the men could move in. Witch doctors and shamans had a less than perfect record at treating wounds and curing disease. Life was short, if not sweet. Man's total population size doubtless increased slowly but steadily as human populations expanded out of the African cradle of our species.

Then about 8,000 years ago a major change occurred—the agricultural revolution. People began to give up hunting food and settle down to grow it. Suddenly some of the risk was removed from life. The chances of dying of starvation diminished greatly in some human groups. Other threats associated with the nomadic life were also reduced, perhaps balanced by new threats of disease and large-scale warfare associated with the development of cities. But the overall result was a more secure existence than before, and the human population grew more rapidly. Around 1800, when the standard of living in what are today the DCs was dramatically increasing due to industrialization, population growth really began to accelerate. The development of medical science was the straw that broke the camel's back. While lowering death rates in the DCs was due in part to other factors, there is no question that "instant death control", exported by the DCs, has been responsible for the drastic lowering of death rates in the UDCs. Medical science, with its efficient public health programs, has been able to depress the death rate with astonishing rapidity and at time drastically increase the birth rate; healthier people have more babies.

The power of exported death control can best be seen by an examination of the classic case of Ceylon's assault on malaria after World War II. Between 1933 and 1942 the death rate due directly to malaria was *reported* as almost two per cent per thousand. This rate, however, represented only a portion of the malaria deaths, as many were reported as being due to "pyrexia".[10] Indeed, in 1934-1935 a malaria epidemic may have been directly responsible for fully half of the deaths on the island. In addition, malaria, which infected a large portion of the population, made people susceptible to many other diseases. It thus contributed to the death rate indirectly as well as directly.

The introduction of DDT in 1946 brought rapid control over the mosquitoes which carry malaria. As a result, the death rate on the island was halved in less than a decade. The death rate in Ceylon in 1945 was 22. It dropped 34 per cent between 1946 and 1947 and moved down to ten in 1954. Since the sharp postwar drop it has continued to decline and now stands at eight. Although part of the drop is doubtless due to the killing of other insects which carry disease and to other public health measures, most of it can be accounted for by the control of malaria.

Victory over malaria, yellow fever, smallpox, cholera, and other infectious diseases has been responsible for similar plunges in death rate throughout most of the UDCs. In the decade 1940-1950 the death rate declined 46 per cent in Puerto Rico, 43 per cent in Formosa, and 23 per cent in Jamaica. In a sample of 18 undeveloped areas the average decline in death rate between 1945 and 1950 was 24 per cent.

It is, of course, socially very acceptable to reduce the death rate. Billions of years to evolution have given us all powerful will to live. Intervening in the birth rate goes against our evolutionary values. During all those centuries of our evolutionary past, the individuals who had the most children passed on their genetic endowment in greater quantities than those who reproduced less. Their genes dominate our heredity today. All our biological urges are for more reproduction, and they are all too often reinforced by our culture. In brief, death control goes with the grain, birth control against it.

In summary, the world's population will continue to grow as long as the birth rate exceeds the death rate; it's as simple as that. When it stops growing or starts to shrink, it will mean that either the birth rate has gone down or the death rate has gone up or a combination of the two. Basically, then, there are only two kinds of solutions to the population problem. One is a "birth rate solution", in which we find ways to lower the birth rate. The other is a "death rate solution", in which ways to raise the death rate—war, famine, pestilence—*find us*. The problem could have been avoided by *population control*, in which mankind consciously adjusted the birth rate so that a "death rate solution" did not have to occur.

Ehrlich, P. & A. Ehrlich, *The Population Bomb*, Ballantine, New York, 1969, p. 17-35.

Notes and References

1 Since this was written, 1968 figures have appeared, showing that the doubling time is now 35 years.

2 J.H. Fremlin, "How Many People Can the World Support?" *New Scientist*, October 29, 1964.

3 To understand this, simply consider what would happen if we held the population constant at three billion people by exporting all the surplus people. If this were done for 37 years (the time it now takes for one doubling) we would have exported three billion people— enough to populate a twin planet of the Earth to the same density. In two doubling times (74 years) we would reach a total human population for the solar system of 12 billion people, enough to populate the Earth and three similar planets to the density found on Earth today. Since the areas of the the planets and moons mentioned above are not three times that of the Earth, they can be populated to equal density in much less than two doubling times.

4 "Interstellar Migration and the Population Problem." *Heredity* 50: 68-70, 1959.

5 I.J. Cook, *New Scientist*, September 8, 1966.

6 The birth rate is more precisely the total number of births in a country during a year, divided by the total population at the midpoint of the year, multiplied by 1,000. Suppose that there were 80 births in Lower Slobbovia during 1967, and that the population of Lower Slobbovia was 2,000 on July 1, 1967. Then the birth rate would be:

Birth rate $= \dfrac{80 \text{ (total births in L. Slobbovia in 1967)}}{2,000 \text{ (total population, July 1, 1967)}} \times 1,000$

$= .04 \times 1,000 = 40$

Similarly if there were 40 deaths in Lower Slobbovia during 1967, the death rate would be:

Death rate $= \dfrac{40 \text{ (total deaths in L. Slobbovia in 1967)}}{2,000 \text{ (total population, July 1, 1967)}} \times 1,000$

$= .02 \times 1,000 = 20.$

Then the Lower Slobbovian birth rate would be 40 per thousand, and the death rate would be 20 per thousand. For every 1,000 Lower Slobbovians alive on July 1, 1967, 40 babies were born and 20 people died. Subtracting the death rate from the birth rate gives us the rate of natural increase of Lower Slobbovia for the year 1967. That is, 40-20 = 20; during 1967 the population grew at a rate of 20 people per thousand per year. Dividing that rate by ten expresses the increase as a per cent (the increase per hundred a year). The increase in 1967 in Lower Slobbivia was two per cent. Remember that this rate of increase ignores any movement of people into and out of Lower Slobbovia.

7 McGraw-Hill Book Company, Inc., New York, 1965.

8 Human brain size increased from an apelike capacity of about 500 cubic centimeters (cc) in *Australopithecus* to about 1,500 cc in moder *Homo sapiens*. among modern men small variations in brain size do not seem to be related to significant differences in the ability to use cultural information, and there is no particular reason to believe that our brain size will continue to increase. Further evolution may occur more readily in a direction of increased efficiency rather than increased size.

9 This is, of course, an oversimplified explanation. For more detail see Ehrlich and Holm, *The Process of Evolution*, McGraw-Hill Book Company, Inc., New York, 1963.

10 These data and those that follow on the decline of death rates are from Kingsley Davis's "The Amazing Decline of Mortality in Underdeveloped Areas", *The American Economic Review*, Vol. 46, pp. 305-318.

CHAPTER 9

The Strategy of Ecosystem Development

An Understanding of Ecological Succession Provides a Basis for Resolving Man's Conflict with Nature

E.P. ODUM

The principles of ecological succession bear importantly on the relationships between man and nature. The framework of successional theory needs to be examined as a basis for resolving man's present environmental crisis. Most ideas pertaining to the development of ecological systems are based on descriptive data obtained by observing changes in biotic communities over long periods, or on highly theoretical assumptions; very few of the generally accepted hypotheses have been tested experimentally. Some of the confusion, vagueness, and lack of experimental work in this area stems from the tendency of ecologists to regard "succession" as a single straightforward idea; in actual fact, it entails an interacting complex of processes, some of which counteract one another.

As viewed here, ecological succession involves the development of ecosystems; it has many parallels in the developmental biology of organisms, and also in the development of human society. The ecosystem, or ecological system, is considered to be a unit of biological organization made up of all of the organisms in a given area (that is, "community") interacting with the physical environment so that a flow of energy leads to characteristic trophic structure and material cycles within the system. It is the purpose of this article to summarize, in the form of a tabular model, components and stages of development at the ecosystem level as a means of emphasizing those aspects of ecological succession that can be accepted on the basis of present knowledge, those that require more study, and those that have special relevance to human ecology.

Definition of Succession

Ecological succession may be defined in terms of the following three parameters.[1] (i) It is an orderly process of community development that is reasonably directional and, therefore, predictable. (ii) It results from modification of the physical environment by the community; that is, succession is community-controlled even though the physical environment determines the pattern, the rate of change, and often sets limits as to how far development can go. (iii) It culminates in a stabilized ecosystem in which maximum biomass (or high information content) and symbiotic function between organisms are maintained per unit of available energy flow. In a word, the "strategy" of succession as a short-term process is basically the same as the "strategy" of long-term evolutionary development of the biosphere—namely, increased control of homeostasis with, the physical environment in the sense of achieving maximum protection from its perturbations. As I illustrate below, the strategy of "maximum protection" (that is trying to achieve maximum support of complex biomass structure) often conflicts with man's goal of "maximum production" (trying to obtain the highest possible yield). Recognition of the ecological basis for this conflict is, I believe, a first step in establishing rational land-use policies.

The earlier descriptive studies of succession on sand dunes, grasslands, forests, marine shores, or other sites, and more recent functional considerations, have led to the basic theory contained in the definition given above. H.T. Odum and Pinkerton,[2] building on Lotka's[3] "law of maximum energy in biological systems", were the first to point out that succession involves a fundamental shift in energy flows as increasing energy is relegated to maintenance. Margalef[4] has recently documented this bioenergetic basis for succession and has extended the concept.

Changes that occur in major structural and functional characteristics of a developing ecosystem are listed in Table 1. Twenty-four attributes of ecological systems are grouped, for convenience of discussion, under six headings. Trends are emphasized by contrasting the situation in early and late development. The degree of absolute change, the rate of change, and the time required to reach a steady state may vary not only with different climatic and physiographic situations but also with different ecosystem attributes in the same physical environment. Where good data are available, rate-of-change curves are usually convex, with changes occurring most rapidly at the beginning, but bimodal or cyclic patterns may also occur.

Bioenergetics of Ecosystem Development

Attributes 1 through 5 in Table 1 represent the bioenergetics of the ecosystem. In the early stages of ecological succession, or in "young nature", so to speak, the rate of primary production or total (gross) photosynthesis (P) exceeds the rate of community respiration (R), so that the P/R ratio is greater than 1. In the special case of organic pollution, the P/R ratio is typically less than 1. In both cases, however, the theory is that P/R approaches 1 as succession occurs. In other words, energy fixed tends to be balanced by the energy cost of maintenance (that is, total community respiration) in the mature or "climax" ecosystem. The *P/R* ratio, therefore, should be an excellent functional index of the relative maturity of the system.

So long as *P* exceeds *R*, organic matter and biomass (*B*) will accumulate in the system (Table 1, item 6), with the result that ratio *P/B* will tend to decrease or, conversely, the *B/P*, *B/R*, or *B/E* ratios (where *E* = *P* + *R*) will increase (Table 1, items 2 and 3). Theoretically, then, the amount of standingcrop biomass supported by the available energy flow (*E*) increases to a maximum in the mature or climax stages (Table 1, item 3). As a consequence, the net community production, or yield, in an annual cycle is large in young nature and small or zero in mature nature (Table 1, item 4).

Comparison of Succession in a Laboratory Microcosm and a Forest

One can readily observe bioenergetic changes by initiating succession in experimental laboratory microecosystems. Aquatic microecosystems, derived from various types of outdoor systems, such as ponds, have been cultured by Beyers[5], and certain of these mixed cultures are easily replicated and maintain themselves in the climax state indefinitely on defined media in a flask with only light input.[6] If samples from the climax system are inoculated into fresh media, succession occurs, the mature system developing in less than 100 days. In Fig. 1 the general pattern of a 100-day autotrophic succession in a microcosm based on data of Cooke[7] is compared with a hypothetical model of a 100-year forest succession as presented by Kira and Shidei.[8]

During the first 40 to 60 days in a typical microcosm experiment, daytime net production (*P*) exceeds nighttime respiration (*R*), so that biomass (*B*) accumulates in the system.[9] After an early "bloom" at about 30 days, both rates decline, and they become approximately equal at 60 to 80 days. the *B/P* ratio, in terms of grams of carbon supported per gram of daily carbon production, increases from less than 20 to more than 100 as the steady state is reached. Not only are autotrophic and heterotrophic metabolism balanced in the climax, but a large organic structure is supported by small daily production and respiratory rates.

While direct projection from the small laboratory microecosystem to open nature may not be entirely valid, there is evidence that the same basic trends that are seen in the laboratory are characteristic of succession on land and in large bodies of water. Seasonal successions also often follow the same pattern, an early seasonal bloom characterized by rapid growth of a few dominant species being followed by the development later in the season of high *B/P* ratios, increased diversity, and a relatively steady, if temporary, state in terms of *P* and *R*[4]. Open systems may not experience a decline, at maturity, in total or gross

FIGURE I Comparison of the energetics of succession in a forest and a laboratory microcosm.

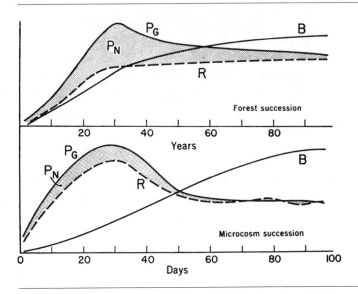

P_g, *gross production;* P_n, *net production, R, total community respiration;*
B, total biomass.

productivity, as the space-limited microcosms do, but the general pattern of bioenergetic change in the latter seems to mimic nature quite well.

These trends are not, as might at first seem to be the case, contrary to the classical limnological teaching which describes lakes as progressing in time from the less productive (oligotrophic) to the more productive (eutrophic) state. Table I, as already emphasized, refers to changes which are brought about by biological processes within the ecosystem in question. Eutrophication, whether natural or cultural, results when nutrients are imported into the lake from outside the lake—that is, from the watershed. This is equivalent to adding nutrients to the laboratory microecosystem or fertilizing a field; the system is pushed back, in successional terms, to a younger or "bloom" state. Recent studies on lake sediments,[10] as well as theoretical considerations,[11] have indicated that lakes can and do progress to a more oligotrophic condition when the nutrient input from the watershed slows or ceases. Thus, there is hope that the troublesome cultural eutrophication of our waters can be reversed if the inflow of nutrients from the watershed can be greatly reduced. Most of all, however, this situation emphasizes that it is the entire drainage or catchment basin, not just the lake or stream, that must be considered the ecosystem unit if we are to deal successfully with our water pollution problems. Ecosystematic study of entire landscape catchment units is a major goal of the American plan for the proposed International Biological Program. Despite the obvious logic of such a proposal, it is proving surprisingly difficult to get traditionbound scientists and granting agencies to look beyond their specialties toward the support of functional studies of large units of the landscape.

Food Chains and Food Webs

As the ecosystem develops, subtle changes in the network pattern of food chains may be expected. The manner in which organisms are linked together through food tends to be relatively simple and linear in the very early stages of succession, as a consequence of low diversity. Furthermore, heterotrophic utilization of net production occurs predominantly by way of grazing food chains—that is, plant-herbivore-carnivore sequences. In contrast, food chains become complex webs in mature stages, with the bulk of biological energy flow following detritus pathways (Table 1, item 5). In a mature forest, for example, less than 10 per cent of annual net production is consumed (that is, grazed) in the living state;[12] most is utilized as dead matter (detritus) through delayed and complex pathways involving as yet little understood animalmicroorganism interactions. The time involved in an uninterrupted succession allows for increasingly intimate associations and reciprocal adaptations between plants and animals, which lead to the development of many mechanisms that reduce grazing—such as the development of indigestible supporting tissues (cellulose, lignin, and so on), feedback control between plants and herbivores,[13] and increasing predatory pressure on herbivores.[14] Such mechanisms enable the biological community to maintain the large and complex organic structure that mitigates perturbations of the physical environment. Severe stress or rapid changes brought about by outside forces can, of course, rob the system of these protective mechanisms and allow irruptive, cancerous growths of certain species to occur, as man too often finds to his sorrow. An example of a stress-induced pest irruption occurred at Brookhaven National Laboratory, where oaks became vulnerable to aphids when translocation of sugars and amino acids was impaired by continuing gamma irradiation.[15]

Radionuclide tracers are providing a means of charting food chains in the intact outdoor ecosystem to a degree that will permit analysis within the concepts of network or matrix algebra. For example, we have recently been able to map, by use of a radiophosphorus tracer, the open, relatively linear food linkage between plants and insects in an early old-field successional stage.[16]

Diversity and Succession

Perhaps the most controversial of the successional trends pertain to the complex and much discussed subject of diversity[17]. It is important to distinguish between different kinds of diversity indices, since they may not follow parallel trends in the same gradient or developmental series. Four components of diversity are listed in Table 1, items 8 through 11.

The variety of species, expressed as a species-number ratio or a species-area ratio, tends to increase during the early stages of community development. A second component of species diversity is what has been called equitability, or evenness,[18] in the apportionment of individuals among the species. For example, two systems each containing 10 species and 100 individuals have the same diversity in terms of species-number ratio but could have widely different equitabilities depending on the apportionment of the 100 individuals among the 10 species—for example, 91-1-1-1-1-1-1-1-1-1 at one extreme or 10 individuals per species at the other. The Shannon formula,

$$-\sum \frac{ni}{N} \log_2 \frac{ni}{N}$$

where ni is the number of individuals in each species and N is the total number of individuals, is widely used as a diversity index because it combines the variety and equitability components one approximation. But, like all such lumping parameters, Shannon's formula may obscure the behavior of these two rather different aspects of diversity, for example, in our most recent field experiments, an acute stress from insecticide reduced the number of species of insects relative to the number of individuals but increased the evenness in the relative abundances of the surviving species.[19] Thus, in this case the "variety" and "evenness" components would tend to cancel each other in Shannon's formula. While an increase in the variety of species together with reduced dominance by any one species or small group of species (that is, increased evenness) can be accepted as a general probability during succession,[20] there are other community changes that may work against these trends. An increase in the size of organisms, an increase in the length and complexity of life histories, and an increase in interspecific competition that may result in competitive exclusion of species (Table 1, items 12-14) are trends that may reduce the number of species that can live in a given area. In the bloom stage of succession organisms tend to be small and to have simple life histories and rapid rates of reproduction. Changes in size appear to be a consequence of, or an adaptation to, a shift in nutrients from inorganic to organic (Table 1, item 7). In a mineral nutrient-rich environment, small size is of selective advantage, especially to autotrophs, because of the greater surface-to-volume ratio. As the ecosystem develops, however, inorganic nutrients tend to become more and more tied up in the biomass (that is, to become intrabiotic), so that the selective advantage shifts to larger organisms (either larger individuals of the same species or larger species, or both), which have greater storage capacities and more complex life histories, thus are adapted to exploiting seasonal or periodic releases of nutrients or other resources. The question of whether the seemingly direct relationship between organism size and stability is the result of positive feedback or is merely fortuitous remains unanswered.[21]

Thus, whether or not species diversity continues to increase during succession will depend on whether the increase in potential niches resulting from increased biomass, stratification (Table 1, item 9), and other consequences of biological organization exceeds the countereffects of increasing size and competition. No one has yet been able to catalogue all the species in any sizable area, much less follow total species diversity in a successional series. Data are so far available only for segments of the community (trees, birds, and so on). Margalef [4] postulates that diversity will tend to peak during the early or middle stages of succession and then decline in the climax. In a study of bird populations along a successional gradient we found a bimodal pattern;[22] the number of species increased during the early stages of old-field succession, declined during the early forest stages, and then increased again in the mature forest.

Species variety, equitability, and stratification are only three aspects of diversity which change during succession. Perhaps an even more important trend is an increase in the diversity of organic compounds, not only of those within the biomass but also of those excreted and secreted into the media (air, soil, water) as by-products of the increasing community metabolism. An increase in such "biochemical diversity" (Table 1, item 10) is illustrated by the increase in the variety of plant pigments along a successional gradient in aquatic situations, as described by Margalef .[4,23] Biochemical diversity within populations, or within systems as a whole, has not yet been systematically studied to the degree the subject of species diversity has been. Consequently, few generalizations can be made, except that it seems safe to say that, as succession progresses, organic extrametabolites probably serve increasingly important functions as regulators which stabilize the growth and composition of the ecosystem. Such metabolites may, in fact, be extremely important in preventing populations from overshooting the equilibrial density, thus in reducing oscillations as the system develops stability.

The cause-and-effect relationship between diversity and stability is not clear and needs to be investigated from many angles. If it can be shown that biotic diversity does indeed enhance physical stability in the ecosystem, or is the result of it, then we would have an important guide for conservation practice. Preservation of hedgerows, woodlots, noneconomic species, noneutrophicated waters, and other biotic variety in man's landscape could then be justified on scientific as well as esthetic grounds, even though such preservation often must result in some reduction in the production of food or other immediate consumer needs. In other words, is variety only the spice of life, or is it a necessity for the long life of the total ecosystem comprising man and nature?

Nutrient Cycling

An important trend in successional development is the closing or "tightening" of the biogeochemical cycling of major nutrients, such as nitrogen, phosphorus, and calcium (Table 1, items 15-17). Mature systems, as compared to developing ones, have a greater capacity to entrap and hold nutrients for cycling within the system. For example, Bormann and Likens[24] have estimated that only 8 kilograms per hectare out of a total pool of exchangeable calcium of 365 kilograms per hectare is lost per year in stream outflow from a North Temperate watershed covered with a mature forest. Of this, about 3 kilograms per hectare is replaced by rainfall, leaving only 5 kilograms to be obtained from weathering of the underlying rocks in order for the system to maintain mineral balance. Reducing the volume of the vegetation, or otherwise setting the succession back to a younger state, results in increased water yield by way of stream outflow,[25] but this greater outflow is accompanied by greater losses of nutrients, which may also produce downstream eutrophication. Unless there is a compensating increase in the rate of weathering, the exchangeable pool of nutrients suffers gradual depletion (not to mention possible effects on soil structure resulting from erosion). High fertility in "young systems" which have open nutrient cycles cannot be maintained without compensating inputs of new nutrients; examples of such practice are the continuous-flow culture of algae, or intensive agriculture where large amounts of fertilizer are imported into the system each year.

Because rates of leaching increase in a latitudinal gradient from the poles to the equator, the role of the biotic community in nutrient retention is especially important in the high-rainfall areas of the subtropical and tropical latitudes, including not only land areas but also estuaries. Theoretically, as one goes equatorward, a larger percentage of the available nutrient pool is tied up in the biomass and a correspondingly lower percentage is in the soil or sediment.

This theory, however, needs testing, since data to show such a geographical trend are incomplete. It is perhaps significant that conventional North Temperate row-type agriculture, which represents a very youthful type of ecosystem, is successful in the humid tropics only if carried out in a system of "shifting agriculture" in which the crops alternate with periods of natural vegetative redevelopment. Tree culture and the semiaquatic culture of rice provide much better nutrient retention and consequently have a longer life expectancy on a given site in these warmer latitudes.

TABLE I A tabular model of ecological succession: trends to be expected in the development of ecosystems.

Ecosystem attributes	Developmental stages	Mature stages
Community energetics		
1 Gross production/community respiration (P/R ratio)	Greater or less than 1	Approaches 1
2 Gross production/standing crop biomass (P/B ratio)	High	Low
3 Biomass supported/unit energy flow (B/E ratio)	Low	High
4 Net community production (yield)	High	Low
5 Food chains	Linear, predominantly grazing	Weblike, predominantly detritus
Community structure		
6 Total organic matter	Small	Large
7 Inorganic nutrients	Extrabiotic	Intrabiotic
8 Species diversity—variety component	Low	High
9 Species diversity—equitability component	Low	High
10 Biochemical diversity	Low	High
11 Stratification and spatial heterogeneity (pattern diversity)	Poorly organized	Well-organized
Life history		
12 Niche specialization	Broad	Narrow
13 Size of organism	Small	Large
14 Life cycles	Short, simple	Long, complex
Nutrient cycling		
15 Mineral cycles	Open	Closed
16 Nutrient exchange rate, between organisms and environment	Rapid	Slow
17 Role of detritus in nutrient regeneration	Unimportant	Important
Selection pressure		
18 Growth form	For rapid growth ("r-selection")	For feedback control ("K-selection")
19 Production	Quantity	Quality
Overall homeostasis		
20 Internal symbiosis	Undeveloped	Developed
21 Nutrient conservation	Poor	Good
22 Stability (resistance to external perturbations)	Poor	Good
23 Entropy	High	Low
24 Information	Low	High

Selection Pressure: Quantity versus Quality

MacArthur and Wilson[26] have reviewed stages of colonization of islands which provide direct parallels with stages in ecological succession on continents. Species with high rates of reproduction and growth, they find, are more likely to survive in the early uncrowded stages of island colonization. In contrast, selection pressure favors species with lower growth potential but better capabilities for competitive survival under the equilibrium density of late stages. Using the terminology of growth equations, where r is the intrinsic rate of increase and K is the upper asymptote or equilibrium population size, we may say that "r selection" predominates in early colonization, with "K selection" prevailing as more and more species and individuals attempt to colonize (Table 1, item 18). The same sort of thing is even seen within the species in certain "cyclic" northern insects in which "active" genetic strains found at low densities are replaced at high densities by "sluggish" strains that are adapted to crowding.[27]

Genetic changes involving the whole biota may be presumed to accompany the successional gradient, since, as described above, quantity production characterizes the young ecosystem while quality production and feedback control are the trademarks of the mature system (Table 1, item 19). Selection at the ecosystem level may be primarily interspecific, since species replacement is a characteristic of successional series or seres. However, in most well-studied seres there seem to be a few early successional species that are able to persist through to late stages. Whether genetic changes contribute to adaptation in such species has not been determined, so far as I know, but studies on population genetics of Drosophila suggest that changes in genetic composition could be important in population regulation.[28] Certainly, the human population, if it survives beyond its present rapid growth stage, is destined to be more and more affected by such selection pressures as adaptation to crowding becomes essential.

Overall Homeostasis

This brief review of ecosystem development emphasizes the complex nature of processes that interact. While one may well question whether all the trends described are characteristic of all types of ecosystems, there can be little doubt that the net result of community actions is symbiosis, nutrient conservation, stability, a decrease in entropy, and an increase in information (Table 1, items 20-24). The overall strategy is, as I stated at the beginning of this article, directed toward achieving as large and diverse an organic structure as is possible within the limits set by the available energy input and the prevailing physical conditions of existence (soil, water, climate, and so on). As studies of biotic communities become more functional and sophisticated, one is impressed with the importance of mutualism, parasitism, predation, commensalism, and other forms of symbiosis. Partnership between unrelated species is often noteworthy (for example, that between coral coelenterates and algae, or between mycorrhizae and trees). In many cases, at least, biotic control of grazing, population density, and nutrient cycling provide the chief positive-feedback mechanisms that contribute to stability in the mature system by preventing overshoots and destructive oscillations. The intriguing question is, Do mature ecosystems age, as organisms do? In other words, after a long period of relative stability or "adulthood", do ecosystems again develop unbalanced metabolism and become more vulnerable to diseases and other perturbations?

Relevance of Ecosystem Development Theory to Human Ecology

Figure 1 depicts a basic conflict between the strategies of man and of nature. The "bloom-type" relationship exhibited by the 30-day microcosm or the 30-year forest, illustrate man's present idea of how nature should be directed. For example, the goal of culture or intensive forestry, as now generally practiced, is to achieve high rates of production of readily harvestable products with little standing crop left to accumulate on the landscape—in other words, a high P/B efficiency. Nature's strategy, on the other hand, as seen in the outcome of the successional process, is directed toward the reverse efficiency—a high B/P ratio, as is depicted by the relationship at the right in Fig. 1. Man has generally been preoccupied with obtaining as much "production" from the landscape as possible, by developing and maintaining early successional types of ecosystems, usually monocultures. But, of course, man does not live by food and fiber alone; he also needs a balanced CO_2-O_2 atmosphere, the climatic buffer provided by oceans and masses of vegetation, and clean (that is, unproductive) water for cultural and industrial uses. Many essential life-cycle resources, not to mention recreational and esthetic needs, are best provided man by the less "productive" landscapes. In other words, the landscape is not just a supply depot but is also the oikos—the home—in which we must live. Until recently mankind has more or less taken for granted the gas-exchange, water-purification, nutrient-cycling, and other protective functions of self-maintaining ecosystems, chiefly because neither his numbers nor his environmental manipulations have been great enough to affect regional and global balances. Now, of course, it is painfully evident that such balances are being affected, often detrimentally. The "one problem, one solution approach" is no longer adequate and must be replaced by some form of ecosystem analysis that considers man as a part of, not apart from, the environment.

The most pleasant and certainly the safest landscape to live in is one containing a variety of crops, forests, lakes, streams, roadsides, marshes, seashores, and "waste places"—in other words, a mixture of communities of different ecological ages. As individuals we more or less instinctively surround our houses with protective, nonedible cover (trees, shrubs, grass) at the same time that we strive to coax extra bushels from our cornfield. We all consider the cornfield a "good thing", of course, but most of us would not want to live there, and it would certainly be suicidal to cover the whole land area of the biosphere with cornfields, since the boom and bust oscillation in such a situation would be severe.

The basic problem facing organized society today boils down to determining in some objective manner when we are getting "too much of a good thing". This is a completely new challenge to mankind because, up until now, he has had to be concerned largely with too little rather than too much. Thus, concrete is a "good thing", but not if half the world is covered with it. Insecticides are "good things", but not when used, as they now are, in an indiscriminate and wholesale manner. Likewise, water impoundments have proved to be very useful man-made additions to the landscape, but obviously we don't want the whole country inundated! Vast manmade lakes solve some problems, at least temporarily, but yield comparative little food or fiber, and, because of high evaporative losses, they may not even be the best device for storing water; it might better be stored in the watershed, or underground in aquafers. Also, the cost of building large dams is a drain on already overtaxed revenues. Although as individuals we readily recognize that we can have too many dams or other large-scale environmental changes, governments are so fragmented and lacking in systems-analysis capabilities that there is no effective mechanism whereby negative feedback signals can be received and acted on before there has been a serious overshoot. Thus, today there are governmental agencies, spurred on by popular and political enthusiasm for dams, that are putting on the drawing boards plans for damming every river and stream in North America!

TABLE 2 Contrasting characteristics of young and mature-type ecosystems.

Young	Mature
Production	Protection
Growth	Stability
Quantity	Quality

Society needs, and must find as quickly as possible, a way to deal with the landscape as a whole, so that manipulative skills (that is, technology) will not run too far ahead of our understanding of the impact of change. Recently a national ecological center outside of government and a coalition of governmental agencies have been proposed as two possible steps in the establishment of a political control mechanism for dealing with major environmental questions. The soil conservation movement in America is an excellent example of a program dedicated to the consideration of the whole farm or the whole watershed as an ecological unit. Soil conservation is well understood and supported by the public. However, soil conservation organizations have remained too exclusively farm-oriented, and have not yet risen to the challenge of the urban-rural landscape, where lie today's most serious problems. We do, then, have potential mechanisms in American society that could speak for the ecosystem as a whole, but none of them are really operational.[29]

The general relevance of ecosystem development theory to landscape planning can, perhaps, be emphasized by the "mini-model" of Table 2, which contrasts the characteristics of young and mature-type ecosystems in more general terms than those provided by Table 1. It is mathematically impossible to obtain a maximum for more than one thing at a time, so one cannot have both extremes at the same time and place. Since all six characteristics listed in Table 2 are desirable in the aggregate, two possible solutions to the dilemma immediately suggest themselves. We can compromise so as to provide moderate quality and moderate yield on all the landscape, or we can deliberately plan to compartmentalize the landscape so as to simultaneously maintain highly productive and predominantly protective types as separate units subject to different management strategies (strategies ranging, for example, from intensive cropping on the one hand to wilderness management on the other). If ecosystem development theory is valid and applicable to planning, then the so-called multiple-use strategy, about which we hear so much, will work only through one or both of these approaches, because, in most cases, the projected multiple uses conflict with one another. It is appropriate, then, to examine some examples of the compromise and the compartmental strategies.

Pulse Stability

A more or less regular but acute physical perturbation imposed from without can maintain an ecosystem at some intermediate point in the developmental sequence, resulting in, so to speak, a compromise between youth and maturity. What I would term "fluctuating water level ecosystems" are good examples. Estuaries, and intertidal zones in general, are maintained in an early, relatively fertile stage by the tides, which provide the energy for rapid nutrient cycling. Likewise, freshwater marshes, such as the Florida Everglades, are held at an early successional stage by the seasonal fluctuations in water levels. The dry-season drawdown speeds up aerobic decomposition of accumulated organic matter, releasing nutrients that, on reflooding, support a wet-season bloom in productivity. The life histories of many organisms are intimately coupled to this periodicity. The wood stork, for example, breeds when the water levels are falling and the small fish on which it feeds become concentrated and easy to catch in the drying pools. If the water level remains high during the usual dry season or fails to rise in the wet season, the stork will not nest.[30] Stabilizing water levels in the Everglades by means of dikes, locks, and impoundments, as is now advocated by some, would, in my opinion, destroy rather than preserve the Everglades as we now know them just as surely as complete drainage would. Without periodic drawdowns and fires, the shallow basins would fill up with organic matter and short-term economic and population pressures. Zoning the landscape would require a whole new order of thinking. Greater use of legal measures providing for tax relief, restrictions on use, scenic easements, and public ownership will be required if appreciable land and water areas are to be held in the "protective" categories. Several states (for example, New Jersey and California), where pollution and population pressure are beginning to hurt, have made a start in this direction by enacting "open space" legislation designed to get as much unoccupied land as possible into a "protective" status so that future uses can be planned on a rational and scientific basis. The United States as a whole is fortunate in that large areas of the country are in national forests, parks, wildlife refuges, and so on. The fact that such areas, as well as the bordering oceans, are not quickly exploitable gives us time for the accelerated ecological study and programming needed to determine what proportions of different types of landscape provide a safe balance between man and nature. The open oceans, for example, should forever be allowed to remain protective rather than productive territory, if Alfred Redfield's[35] assumptions are correct. Redfield views the oceans, the major part of the hydrosphere, as the biosphere's governor, which slows down and controls the rate of decomposition and nutrient regeneration, thereby creating and maintaining the highly aerobic terrestrial environment to which the higher forms of life, such as man, are adapted. Eutrophication of the ocean in a last-ditch effort to feed the populations of the land could well have an adverse effect on the oxygen reservoir in the atmosphere.

Until we can determine more precisely how far we may safely go in expanding intensive agriculture and urban sprawl at the expense of the protective landscape, it will be good insurance to hold inviolate as much of the latter as possible. Thus, the preservation of natural areas is not a peripheral luxury for society but a capital investment from which we expect to draw interest. Also, it may well be that restrictions in the use of land and water are our only practical means of avoiding overpopulation or too great an exploitation of resources, or both. Interestingly enough, restriction of land use is the analogue of a natural behavioral control mechanism known as "territoriality" by which many species of animals avoid crowding and social stress.[36]

FIGURE 2 Compartment model of the basic kinds of environment required by man, partitioned according to ecosystem development and life-cycle resource criteria.

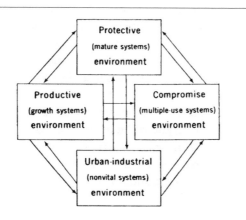

Since the legal and economic problems pertaining to zoning and compartmentalization are likely to be thorny, I urge law schools to establish departments, or institutes, of "landscape law" and to start training "landscape lawyers" who will be capable not only of clarifying existing procedures but also of drawing up new enabling legislation for consideration by state and national governing bodies. At present, society is concerned—and rightly so—with human rights, but environmental rights are equally vital. The "one man one vote" idea is important, but so also is a 'one man one hectare' proposition.

Education, as always, must play a role in increasing man's awareness of his dependence on the natural environment. Perhaps we need to start teaching the principles of ecosystem in the third grade. A grammar school primer on man and his environment could logically consist of four chapters, one for each of the four essential kinds of environment, shown diagrammatically in Fig. 2.

Of the many books and articles that are being written these days about man's environmental crisis, I would like to cite two that go beyond "crying out in alarm" to suggestions for bringing about a reorientation of the goals of society. Garrett Hardin, in a recent article in Science,[37] points out that, since the optimum population density is less than the maximum, there is no strictly technical solution to the problem of pollution caused by overpopulation; a solution, he suggests, can only be achieved through moral and legal means of "mutual coercion, mutually agreed upon by the majority of people". Earl F. Murphy, in a book entitled Governing Nature,[38] emphasizes that the regulatory approach alone is not enough to protect life-cycle resources, such as air and water, that cannot be allowed to deteriorate. He discusses permit systems, effluent charges, receptor levies, assessment, and cost-internalizing procedures as economic incentives for achieving Hardin's "mutually agreed upon coercion".

It goes without saying that the tabular model for ecosystem development which I have presented here has many parallels in the development of human society itself. In the pioneer society, as in the pioneer ecosystem, high birth rates, rapid growth, high economic profits, and exploitation of accessible and unused resources are advantageous, but, as the saturation level is approached, these drives must be shifted to considerations of symbiosis (that is, "civil rights", "law and order", "education", and "culture"), birth control, and the recycling of resources. A balance between youth and maturity in the socioenvironmental system is, therefore, the really basic goal that must be achieved if man as a species is to successfully pass through the present rapid-growth stage, to which he is clearly well adapted, to the ultimate equilibriumdensity stage, of which he as yet shows little understanding and to which he now shows little tendency to adapt.

Odum, E.P., The Strategy of Ecosystem Development; An Understanding of Ecological Succession Provides a Basis for Resolving Man's Conflict with Nature, in: *Science* 164, 1969, p. 262-270.

Notes and References

I   E. P. Odum, *Ecology* (Holt, Rinehart & Winston, New York, 1963), chap. 6.

2   H. T. Odum and R. C. Pinkerton, *Amer. Scientist* 43, 331 (1955).

3   A. J. Lotka. *Elements of Physical Biology* (Williams and Wilkins, Baltimore, 1925),

4   R. Margalef, *Advan. Frontiers Plant* Sci. 2, 137 (1963); *Amer. Naturalist* 97, 357 (1963).

5   R. J. Beyers, *Ecol. Monographs* 33, 281 (1963).

6   The systems so far used to test ecological principles have been derived from sewage and farm ponds and are cultured in halfstrength No. 36 Taub and Dollar medium [*Limnol. Oceanog.* 9. 61 (1964)]. They are closed to organic imput or output but are open to the atmosphere through the cotton plug in the neck of the flask. Typically, liter-sized microecosystems contain two or three species of nonflagellated algae and one to three species each of flagellated protozoans, ciliated protozoans, rotifers, nematodes, and ostracods; a system derived from a sewage pond contained at least three species of fungi and 13 bacterial isolates [R. Gordon, thesis, University of Georgia (1967)]. These cultures are thus a kind of minimum ecosystem containing those small species originally found in the ancestral pond that are able to function together as a self-contained unit under the restricted conditions of the laboratory flask and the controlled environment of a growth chamber [temperature, $65°$ to $75°F$ ($18°$ to $24°C$); photoperiod, 12 hours; illumination, 100 to 1000 footcandles].

7   G. D. Cooke, *BioScience* 17, 717 (1967).

8 T. Kira and T. Shidei, *Japan. J. Ecol.* 17, 70 (1967).

9 The metabolism of the microcosms was monitored by measuring diurnal pH changes, and the biomass (in terms of total organic matter and total carbon) was determined by periodic harvesting of replicate systems.

10 F. J. H. Mackereth, *Proc. Roy. Soc. London Ser. B* 161, 295 (1965); U. M. Cowgill and G. E. Hutchinson, *Proc. Intern. Limnol. Ass.* 15, 644 (1964); A, D. Harrison, *Trans, Roy. Soc. S. Africa* 36, 213 (1962).

11 R. Margalef, *Proc. Intern. Limnol. Ass.* 15, 169 (1964).

12 J. R. Bray, *Oikos* 12, 70 (1961).

13 D. Pimentel, *Amer. Naturalist* 95, 65 (1961).

14 R. T. Paine, *ibid.* 100, 65 (1966).

15 G. M. Woodwell, *Brookhaven Nat. Lab. Pub.* 924(T-381) (1965), pp. 1-15.

16 R. G. Wiezert, E. P. Odum, J. H. Schnell, *Ecology* 48, 75 (1967).

17 For selected general discussions of patterns of species diversity, see E. H. Simpson, *Nature* 163, 688 (1949); C. B. Williams, *J. Animal Ecol.* 22 14 (1953); G. E. Hutchinson, *Amer. Naturalist* 93, 145 (1959); R. Margalef, *Gen. Systems* 3, 36 (1958); R. MacArthur and J. MacArthur, *Ecology* 42 594 (1961); N. G. Hairston, *ibid.* 40, 404 (1959); B. C. Patten, *J. Marine Res. (Sears Found. Marine Res.)* 20, 57 (1960); E. G. Leigh, *Proc. Nat. Acad. Sci. US* 55, 777 (1965); E. R. Pianka, *Amer. Naturalist* 100, 33 (1966); E. C. Pielou, *J. Theoret. Biol.* 10, 370 (1966).

18 M. Lloyd and R. J. Ghelardi J, *Animal Ecol.* 33, 217 (1964); E. C. Pielou, *J. Theoret. Biol.* 13, 131 (1966).

19 G. W. Barrett, *Ecology* 49, 1019 (1969).

20 In our studies of natural succession following grain culture, both the species-to-numbers and the equitability indices increased for all trophic levels but especially for predators and parasites. Only 44 per cent of the species in the natural ecosystem were phytophagous, as compared to 77 per cent in the grain field.

21 J. T. Bonner, *Size and Cycle* (Princeton Univ. Press, Princeton, N.J., 1963); P, Frank, *Ecology* 49, 355 (1968).

22 D. W. Johnston and E. P. Odum, *Ecology* 37, 50 (1956).

23 R. Margalef, *Oceanog. Marine Biol. Annu. Rev.* 5, 257 (1967).

24 F. H. Bormann and G. E. Likens, *Science* 155, 424 (1967).

25 Increased water yield following reduction of vegetative cover has been frequently demonstrated in experimental watersheds throughout the world [see A. R. Hibbert, in *International Symposium on Forest hydrology* (Pergamon Press, New York, 1967), pp. 527-543]. Data on the long-term hydrologic budget (rainfall input relative to stream outflow) are available at many of these sites, but mineral budgets have got to be systematically studied. Again, this is a prime objective in the "ecosystem analysis" phase of the International Biological Program.

26 R. H. MacArthur and E. O. Wilson, *Theory of Island Biogeography* (Princeton Univer. Press, Princeton, N.J., 1967).

27 Examples are the tent caterpillar [see W. G. Wellington, *Can. J. Zool*, 35, 293 (1957)] and the larch budworm [see W. Baltensweiler, *Can. Entomologist* 96, 792 (1964)].

28 F. J. Ayala, *Science* 162, 1453 (1968).

29 Ira Rubinoff, in discussing the proposed sea level canal joining the Atlantic and Pacific oceans [*Science* 161, 857 (1968)], calls for a "control commission for environmental manipulation" with "broad powers of proving, disapproving, or modifying all major alterations of the marine or terrestrial environments. . . ".

30 See M. P. Kahl, *Ecol. Monographs* 34, 97 (1964).

31 The late Aldo Leopold remarked long ago [*Symposium on Hydrobiology* (Univ. of Wisconsin Press, Madison, 1941), p. 17] that man does not perceive oceanic behavior in systems unless he has built them himself. Let us hope it will not be necessary to build the entire biosphere before we recognize the worth of natural systems!

32 See C. F. Cooper, *Scf. Amer* 204, 150 (April 1961).

33 See "Proceedings Oyster Culture Workshop, Marine Fisheries Division, Georgia Game and Fish Commission, Brunswick" (1968), pp. 49-61.

34 See H. T. Odum, *in Symposium on Productivity and Mineral Cycling in Natural Ecosystems*, H. E. Young, Ed. (Univ. of Maine Press, Orono, 1967), p. 81; ——in *Pollution and Marine Ecology* (Wiley, New York, l967), p. 99; K. E. F. Watt, *Ecology and Resource Management* (McGraw-Hill, New York, 1968).

35 A. C. Redfield, *Amer. Scientist* 46, 205 (1958).

36 R. Ardrey, *The Territorial Imperative* (Atheneum, New York, 1967).

37 G. Hardin, *Science* 162, 1243 (1968).

38 E. F. Murphy, *Governing Nature* (Quadrangle Books, Chicago, 1967).

CHAPTER 10

The Historical Roots of Our Ecologic Crisis

L. WHITE, JR.

A conversation with Aldous Huxley not infrequently put one at the receiving end of an unforgettable monologue. About a year before his lamented death he was discoursing on a favorite topic: Man's unnatural treatment of nature and its sad results. To illustrate his point he told how, during the previous summer, he had returned to a little valley in England where he had spent many happy months as a child. Once it had been composed of delightful grassy glades; now it was becoming overgrown with unsightly brush because the rabbits that formerly kept such growth under control had largely succumbed to a disease, myxomatosis, that was deliberately introduced by the local farmers to reduce the rabbits' destruction of crops. Being something of a Philistine, I could be silent no longer, even in the interests of great rhetoric. I interrupted to point out that the rabbit itself had been brought as a domestic animal to England in 1176, presumably to improve the protein diet of the peasantry.

All forms of life modify their contexts. The most spectacular and benign instance is doubtless the coral polyp. By serving its own ends, it has created a vast undersea world favorable to thousands of other kinds of animals and plants. Ever since man became a numerous species he has affected his environment notably. The hypothesis that his fire-drive method of hunting created the world's great grasslands and helped to exterminate the monster mammals of the Pleistocene from much of the globe is plausible, if not proved. For 6 millennia at least, the banks of the lower Nile have been a human artifact rather than the swampy African jungle which nature, apart from man, would have made it. The Aswan Dam, flooding 5000 square miles, is only the latest stage in a long process. In many regions terracing or irrigation, overgrazing, the cutting of forests by Romans to build ships to fight Carthaginians or by Crusaders to solve the logistics problems of their expeditions, have profoundly changed some ecologies. Observation that the French landscape falls into two basic types, the open fields of the north and the *bocage* of the south and west, inspired Marc Bloch to undertake his classic study of medieval agricultural methods. Quite unintentionally, changes in human ways often affect nonhuman nature. It has been noted, for example, that the advent of the automobile eliminated huge flocks of sparrows that once fed on the horse manure littering every street.

The history of ecologic change is still so rudimentary that we know little about what really happened, or what the results were. The extinction of the European aurochs as late as 1627 would seem to have been a simple case of overenthusiastic hunting. On more intricate matters it often is impossible to find solid information. For a thousand years or more the Frisians and Hollanders have been pushing back the North Sea, and the process is culminating in our own time in the reclamation of the Zuider Zee. What, if any, species of animals, birds, fish, shore life, or plants have died out in the process? In their epic combat with Neptune have the Netherlanders overlooked ecological values in such a way that the quality of human life in the Netherlands has suffered? I cannot discover that the questions have ever been asked, much less answered.

People, then, have often been a dynamic element in their own environment, but in the present state of historical scholarship we usually do not know exactly when, where, or with what effects man-induced changes came. As we enter the last third of the 20th century, however, concern for the problem of ecologic backlash is mounting feverishly. Natural science, conceived as the effort to understand the nature of things, had flourished in several eras and among several peoples. Similarly there had been an age-old accumulation of technological skills, sometimes growing rapidly, sometimes slowly. But it was not until about four generations ago that Western Europe and North America arranged a marriage between science and technology, a union of the theoretical and the empirical approaches to our natural environment. The emergence in widespread practice of the Baconian creed that scientific knowledge means technological power over nature can scarcely be dated before about 1850, save in the chemical industries, where it is anticipated in the 18th century. Its acceptance as a normal pattern of action may mark the greatest event in human history since the invention of agriculture, and perhaps in nonhuman terrestrial history as well.

Almost at once the new situation forced the crystallization of the novel concept of ecology; indeed, the word *ecology* first appeared in the English language in 1873. Today, less than a century later, the impact of our race upon the environment has so increased in force that it has changed in essence. When the first cannons were fired, in the early 14th century, they affected ecology by sending workers scrambling to the forests and mountains for more potash, sulfur, iron ore, and charcoal, with some resulting erosion and deforestation. Hydrogen bombs are of a different order: a war fought with them might alter the genetics of all life on this planet. By 1285 London had a smog problem arising from the burning of soft coal, but our present combustion of fossil fuels threatens to change the chemistry of the globe's atmosphere as a whole, with consequences which we are only beginning to guess. With the population explosion, the carcinoma of planless urbanism, the now geological deposits of sewage and garbage, surely no creature other than man has ever managed to foul its nest in such short order.

There are many calls to action, but specific proposals, however worthy as individual items, seem too partial, palliative, negative: ban the bomb, tear down the billboards, give the Hindus contraceptives and tell them to eat their sacred cows. The simplest solution to any suspect change is, of course, to stop it, or, better yet, to revert to a romanticized past: make those ugly gasoline stations look like Anna Hathaway's cottage or (in the Far West) like ghost-town saloons. The "wilderness area" mentality invariably advocates deep-freezing an ecology, whether San Gimignano or the High Sierra, as it was before the first Kleenex was dropped. But neither atavism nor prettification will cope with the ecologic crisis of our time.

What shall we do? No one yet knows. Unless we think about fundamentals, our specific measures may produce new backlashes more serious than those they are designed to remedy.

As a beginning we should try to clarify our thinking by looking, in some historical depth, at the presuppositions that underlie modern technology and science. Science was traditionally aristocratic, speculative, intellectual in intent; technology was lowerclass, empirical, action-oriented. The quite sudden fusion of these two, towards the middle of the 19th century, is surely related to the slightly prior and contemporary democratic revolutions which, by reducing social barriers, tended to assert a functional unity of brain and hand. Our ecologic crisis is the product of an emerging, entirely novel, democratic culture. The issue is whether a democratized world can survive its own implications. Presumably we cannot unless we rethink our axioms.

The Western Traditions of Technology and Science

One thing is so certain that it seems stupid to verbalize it: both modern technology and modern science are distinctively *Occidental.* Our technology has absorbed elements from all over the world, notably from China; yet everywhere today, whether in Japan or in Nigeria, successful technology is Western. Our science is the heir to all the sciences of the past, especially perhaps to the work of the great Islamic scientists of the Middle Ages, who so often outdid the ancient Greeks in skill and perspicacity: al-Razi in medicine, for example; or ibn-al-Haytham in optics; or Omar Khayyam in mathematics. Indeed, not a few works of such geniuses seem to have vanished in the original Arabic and to survive only in medieval Latin translations that helped to lay the foundations for later Western developments. Today, around the globe, all significant science is Western in style and method, whatever the pigmentation or language of the scientists.

A second pair of facts is less well recognized because they result from quite recent historical scholarship. The leadership of the West, both in technology and in science, is far older than the so-called Scientific Revolution of the 17th century or the so-called Industrial Revolution of the 18th century. These terms are in fact outmoded and obscure the true nature of what they try to describe—significant stages in two long and separate developments. By AD 1000 at the latest—and perhaps, feebly, as much as 200 years earlier—the West began to apply water power to industrial processes other than milling grain. This was followed in the late 12th century by the harnessing of wind power. From simple beginnings, but with remarkable consistency of style, the West rapidly expanded its skills in the development of power machinery, labor-saving devices, and automation. Those who doubt should contemplate that most monumental achievement in the history of automation: the weight-driven mechanical clock, which appeared in two forms in the early 14th century. Not in craftsmanship but in basic technological capacity, the Latin West of the later Middle Ages far outstripped its elaborate, sophisticated, and esthetically magnificent sister cultures, Byzantium and Islam. In 1444 a great Greek ecclesiastic, Bessarion, who had gone to Italy, wrote a letter to a prince in Greece. He is amazed by the superiority of Western ships, arms, textiles, glass. But above all he is astonished by the spectacle of waterwheels sawing timbers and pumping the bellows of blast furnaces. Clearly, he had seen nothing of the sort in the Near East.

By the end of the 15th century the technological superiority of Europe was such that its small, mutually hostile nations could spill out over all the rest of the world, conquering, looting, and colonizing. The symbol of this technological superiority is the fact that Portugal, one of the weakest states of the Occident, was able to become, and to remain for a century, mistress of the East Indies. And we must remember that the technology of Vasco da Gama and Albuquerque was built by pure empiricism, drawing remarkably little support or inspiration from science.

In the present-day vernacular understanding, modern science is supposed to have begun in 1543, when both Copernicus and Vesalius published their great works. It is no derogation of their accomplishments, however, to point out that such structures as the *Fabrica* and the *De revolutionibus* do not appear overnight. The distinctive Western tradition of science, in fact, began in the late 11th century with a massive movement of translation of Arabic and Greek scientific works into Latin. A few notable books—Theophrastus, for example—escaped the West's avid new appetite for science, but within less than 200 years effectively the entire corpus of Greek and Muslim science was available in Latin, and was being eagerly read and criticized in the new European universities. Out of criticism arose new observation, speculation, and increasing distrust of ancient authorities. By the late 13th century Europe had seized global scientific leadership from the faltering hands of Islam. It would be as absurd to deny the profound originality of Newton, Galileo, or Copernicus as to deny that of the 14th century scholastic scientists like Buridan or Oresme on whose work they built. Before the 11th century, science scarcely existed in the Latin West, even in Roman times. From the 11th century onward, the scientific sector of Occidental culture has increased in a steady crescendo.

Since both our technological and our scientific movements got their start, acquired their character, and achieved world dominance in the Middle Ages, it would seem that we cannot understand their nature or their present impact upon ecology without examining fundamental medieval assumptions and developments.

Medieval View of Man and Nature

Until recently, agriculture has been the chief occupation even in "advanced" societies; hence, any change in methods of tillage has much importance. Early plows, drawn by two oxen, did not normally turn the sod but merely scratched it. Thus, crossplowing was needed and fields tended to be squarish. In the fairly light soils and semiarid climates of the Near East and Mediterranean, this worked well. But such a plow was inappropriate to the wet climate and often sticky soils of northern Europe. By the latter part of the 7th century after Christ, however, following obscure beginnings, certain northern peasants were using an entirely new kind of plow, equipped with a vertical knife to cut the line of the furrow, a horizontal share to slice under the sod, and a moldboard to turn it over. The friction of this plow with the soil was so great that it normally required not two but eight oxen. It attacked the land with such violence that cross-plowing was not needed, and fields tended to be shaped in long strips.

In the days of the scratch-plow, fields were distributed generally in units capable of supporting a single family. Subsistence farming was the presupposition. But no peasant owned eight oxen: to use the new and more efficient plow, peasants pooled their oxen to form large plow-teams, originally receiving (it would appear) plowed strips in proportion to their contribution. Thus, distribution of land was based no longer on the needs of a family but, rather, on the capacity of a power machine to till the earth. Man's relation to the soil was profoundly changed. Formerly man had been part of nature; now he was the exploiter of nature. Nowhere else in the world did farmers develop any analogous agricultural implement. Is it coincidence that modern technology, with its ruthlessness toward nature, has so largely been produced by descendants of these peasants of northern Europe?

This same exploitive attitude appears slightly before AD 830 in Western illustrated calendars. In older calendars the months were shown as passive personifications. The new Frankish calendars, which set the style for the Middle Ages, are very different: they show men coercing the world around them—plowing, harvesting, chopping trees, butchering pigs. Man and nature are two things, and man is master.

These novelties seem to be in harmony with larger intellectual patterns. What people do about their ecology depends on what they think about themselves in relation to things around them. Human ecology is deeply conditioned by beliefs about our nature and destiny—that is, by religion. To Western eyes this is very evident in, say, India or Ceylon. It is equally true of ourselves and of our medieval ancestors.

The victory of Christianity over paganism was the greatest psychic revolution in the history of our culture. It has become fashionable today to say that, for better or worse, we live in 'the post-Christian age'. Certainly the forms of our thinking and language have largely ceased to be Christian, but to my eye the substance often remains amazingly akin to that of the past. Our daily habits of action, for example, are dominated by an implicit faith in perpetual progress which was unknown either to GrecoRoman antiquity or to the Orient. It is rooted in, and is indefensible apart from, Judeo-Christian teleology. The fact that Communists share it merely helps to show what can be demonstrated on many other grounds: that Marxism, like Islam, is a Judeo-Christian heresy. We continue today to live, as we have lived for about 1700 years, very largely in a context of Christian axioms.

What did Christianity tell people about their relations with the environment?

While many of the world's mythologies provide stories of creation, Greco-Roman mythology was singularly incoherent in this respect. Like Aristotle, the intellectuals of the ancient West denied that the visible world had had a beginning. Indeed, the idea of a beginning was impossible in the framework of their cyclical notion of time. In sharp contrast, Christianity inherited from Judaism not only a concept of time as nonrepetitive and linear but also a striking story of creation. By gradual stages a loving and all-powerful God had created light and darkness, the heavenly bodies, the earth and all its plants, animals, birds, and fishes. Finally, God had created Adam and, as an afterthought, Eve to keep man from being lonely. Man named all the animals, thus establishing his dominance over them. God planned all of this explicitly for man's benefit and rule: no item in the physical creation had any purpose save to serve man's purposes. And, although man's body is made of clay, he is not simply part of nature: he is made in God's image.

Especially in its Western form, Christianity is the most anthropocentric religion the world has seen. As early as the 2nd century both Tertullian and Saint Irenaeus of Lyons were insisting that when God shaped Adam he was foreshadowing the image of the incarnate Christ, the Second Adam. Man shares, in great measure, God's transcendence of nature. Christianity, in absolute contrast to ancient paganism and Asia's religions (except, perhaps, Zoroastrianism), not only established a dualism of man and nature but also insisted that it is God's will that man exploit nature for his proper ends.

At the level of the common people this worked out in an interesting way. In Antiquity every tree, every spring, every stream, every hill had its own *genius loci*, its guardian spirit. These spirits were accessible to men, but were very unlike men; centaurs, fauns, and mermaids show their ambivalence. Before one cut a tree, mined a mountain, or dammed a brook, it was important to placate the spirit in charge of that particular situation, and to keep it placated. By destroying pagan animism, Christianity made it possible to exploit nature in a mood of indifference to the feelings of natural objects.

It is often said that for animism the Church substituted the cult of saints. True; but the cult of saints is functionally quite different from animism. The saint is not *in* natural objects; he may have special shrines, but his citizenship is in heaven. Moreover, a saint is entirely a man; he can be approached in human terms. In addition to saints, Christianity of course also had angels and demons inherited from Judaism and perhaps, at one remove, from Zoroastrianism. But these were all as mobile as the saints themselves. The spirits in natural objects, which formerly had protected nature from man, evaporated. Man's effective monopoly on spirit in this world was confirmed, and the old inhibitions to the exploitation of nature crumbled.

When one speaks in such sweeping terms, a note of caution is in order. Christianity is a complex faith, and its consequences differ in differing contexts. What I have said may well apply to the medieval West, where in fact technology made spectacular advances. But the Greek East, a highly civilized realm of equal Christian devotion, seems to have produced no marked technological innovation after the late 7th century, when Greek fire was invented. The key to the contrast may perhaps be found in a difference in the tonality of piety and thought which students of comparative theology find between the Greek and the Latin Churches. The Greeks believed that sin was intellectual blindness, and that salvation was found in illumination, orthodoxy—that is, clear thinking. The Latins, on the other hand, felt that sin was moral evil, and that salvation was to be found in right conduct. Eastern theology has been intellectualist. Western theology has been voluntarist. The Greek saint contemplates; the Western saint acts. The implications of Christianity for the conquest of nature would emerge more easily in the Western atmosphere.

The Christian dogma of creation, which is found in the first clause of all the Creeds, has another meaning for our comprehension of today's ecologic crisis. By revelation, God had given man the Bible, the Book of Scripture. But since God had made nature, nature also must reveal the divine mentality. The religious study of nature for the better understanding of God was known as natural theology. In the early Church, and always in the Greek East, nature was conceived primarily as a symbolic system through which God speaks to men: the ant is a sermon to sluggards; rising flames are the symbol of the soul's aspiration. This view of nature was essentially artistic rather than scientific. While Byzantium preserved and copied great numbers of ancient Greek scientific texts, science as we conceive it could scarcely flourish in such an ambience.

However, in the Latin West by the early 13th century natural theology was following a very different bent. It was ceasing to be the decoding of the physical symbols of God's communication with man and was becoming the effort to understand God's mind by discovering how his creation operates. The rainbow was no longer simply a symbol of hope first sent to Noah after the Deluge: Robert Grosseteste, Friar Roger Bacon, and Theodoric of Freiberg produced startlingly sophisticated work on the optics of the rainbow, but they did it as a venture in religious understanding. From the 13th century onward, up to and including Leibnitz and Newton, every major scientist, in effect, explained his motivations in religious terms. Indeed, if Galileo had not been so expert an amateur theologian he would have got into far less trouble: the professionals resented his intrusion. And Newton seems to have regarded himself more as a theologian than as a scientist. It was not until the late 18th century that the hypothesis of God became unnecessary to many scientists.

It is often hard for the historian to judge, when men explain why they are doing what they want to do, whether they are offering real reasons or merely culturally acceptable reasons. The consistency with which scientists during the long formative centuries of Western science said that the task and the reward of the scientist was "to think God's thoughts after him" leads one to believe that this was their real motivation. If so, then modern Western science was cast in a matrix of Christian theology. The dynamism of religious devotion, shaped by the Judeo-Christian dogma of creation, gave it impetus.

An Alternative Christian View

We would seem to be headed toward conclusions unpalatable to many Christians. Since both *science* and *technology* are blessed words in our contemporary vocabulary, some may be happy at the notions, first, that, viewed historically, modern science is an extrapolation of natural theology and, second, that modern technology is at least partly to be explained as an Occidental, voluntarist realization of the Christian dogma of man's transcendence of, and rightful mastery over, nature. But, as we now recognize, somewhat over a century ago science and technology—hitherto quite separate activities—joined to give mankind powers which, to judge by many of the ecologic effects, are out of control. If so, Christianity bears a huge burden of guilt.

I personally doubt that disastrous ecologic backlash can be avoided simply by applying to our problems more science and more technology. Our science and technology have grown out of Christian attitudes toward man's relation to nature which are almost universally held not only by Christians and neo-Christians but also by those who fondly regard themselves as post-Christians. Despite Copernicus, all the cosmos rotates around our little globe. Despite Darwin, we are *not*, in our hearts, part of the natural process. We are superior to nature, contemptuous of it, willing to use it for our slightest whim. The newly elected Governor of California, like myself a churchman but less troubled than I, spoke for the Christian tradition when he said (as is alleged), "when you've seen one redwood tree, you've seen them all". To a Christian a tree can be no more than a physical fact. The whole concept of the sacred grove is alien to Christianity and to the ethos of the West. For nearly 2 millennia Christian missionaries have been chopping down sacred groves, which are idolatrous because they assume spirit in nature.

What we do about ecology depends on our ideas of the man-nature relationship. More science and more technology are not going to get us out of the present ecologic crisis until we find a new religion, or rethink our old one. The beatniks, who are the basic revolutionaries of our time, show a sound instinct in their affinity for Zen Buddhism, which conceives of the man-nature relationship as very nearly the mirror image of the Christian view Zen, however, is as deeply conditioned by Asian history as Christianity is by the experience of the West, and I am dubious of its viability among us.

Possibly we should ponder the greatest radical in Christian history since Christ: Saint Francis of Assisi. The prime miracle of Saint Francis is the fact that he did not end at the stake, as many of his left-wing followers did. He was so clearly heretical that a General of the Franciscan Order, Saint Bonaventura, a great and perceptive Christian, tried to suppress the early accounts of Franciscanism. The key to an understanding of Francis is his belief in the virtue of humility—not merely for the individual but for man as a species. Francis tried to depose man from his monarchy over creation and set up a democracy of all God's creatures. With him the ant is no longer simply a homily for the lazy, flames a sign of the thrust of the soul toward union with God; now they are Brother Ant and Sister Fire, praising the Creator in their own ways as Brother Man does in his.

Later commentators have said that Francis preached to the birds as a rebuke to men who would not listen. The records do not read so: he urged the little birds to praise God, and in spiritual ecstasy they flapped their wings and chirped rejoicing. Legends of saints, especially the Irish saints, had long told of their dealings with animals but always, I believe, to show their human dominance over creatures. With Francis it is different. The land around Gubbio in the Apennines was being ravaged by a fierce wolf. Saint Francis, says the legend, talked to the wolf and persuaded him of the error of his ways. The wolf repented, died in the odor of sanctity, and was buried in consecrated ground.

What Sir Steven Ruciman calls "the Franciscan doctrine of the animal soul" was quickly stamped out. Quite possibly it was in part inspired, consciously or unconsciously, by the belief in reincarnation held by the Cathar heretics who at that time teemed in Italy and southern France, and who presumably had got it originally from India. It is significant that at just the same moment, about 1200, traces of metempsychosis are found also in western Judaism, in the Provençal *Cabbala*. But Francis held neither to transmigration of souls nor to pantheism. His view of nature and of man rested on a unique sort of pan-psychism of all things animate and inanimate, designed for the glorification of their transcendent Creator, who, in the ultimate gesture of cosmic humility, assumed flesh, lay helpless in a manger, and hung dying on a scaffold.

I am not suggesting that many contemporary Americans who are concerned about our ecologic crisis will be either able or willing to counsel with wolves or exhort birds. However, the present increasing disruption of the global environment is the product of a dynamic technology and science which were originating in the Western medieval world against which Saint Francis was rebelling in so original a way. Their growth cannot be understood historically apart from distinctive attitudes toward nature which are deeply grounded in Christian dogma. The fact that most people do not think of these attitudes as Christian is irrelevant. No new set of basic values has been accepted in our society to displace those of Christianity. Hence we shall continue to have a worsening ecologic crisis until we reject the Christian axiom that nature has no reason for existence save to serve man.

The greatest spiritual revolutionary in Western history, Saint Francis, proposed what he thought was an alternative Christian view of nature and man's relation to it: he tried to substitute the idea of the equality of all creatures, including man, for the idea of man's limitless rule of creation. He failed. Both our present science and our present technology are so tinctured with orthodox Christian arrogance toward nature that no solution for our ecologic crisis can be expected from them alone. Since the roots of our trouble are so largely religious, the remedy must also be essentially religious, whether we call it that or not. We must rethink and refeel our nature and destiny. The profoundly religious, but heretical, sense of the primitive Franciscans for the spiritual autonomy of all parts of nature may point a direction. I propose Francis as a patron saint for ecologists.

White, L. Jr., The Historical Roots of Our Ecologic Crisis, in: *Science*, Vol. 155, Number 3767, 10 March 1967, p. 1203-1207.

CHAPTER 11

Should Trees Have Standing?

Toward Legal Rights for Natural Objects

CHR.D. STONE

Toward Having Standing in its Own Right

It is not inevitable, nor is it wise, that natural objects should have no rights to seek redress in their own behalf. It is no answer to say that streams and forests cannot have standing because streams and forests cannot speak. Corporations cannot speak either; nor can states, estates, infants, incompetents, municipalities or universities. Lawyers speak for them, as they customarily do for the ordinary citizen with legal problems. One ought, I think, to handle the legal problems of natural objects as one does the problems of legal incompetents— human beings who have become vegetable. If a human being shows signs of becoming senile and has affairs that he is de jure incompetent to manage, those concerned with his well being make such a showing to the court, and someone is designated by the court with the authority to manage the incompetent's affairs. The guardian[1] (or "conservator"[2] or "committee"[3]—the terminology varies) then represents the incompetent in his legal affairs. Courts make similar appointments when a corporation has become "incompetent"—they appoint a trustee in bankruptcy or reorganization to oversee its affairs and speak for it in court when that becomes necessary.

On a parity of reasoning, we should have a system in which, when a friend of a natural object perceives it to be endangered, he can apply to a court for the creation of a guardianship.[4] Perhaps we already have the machinery to do so. California law, for example, defines an incompetent as "any person, whether insane or not, who by reason of old age, disease, weakness of mind, or other cause, is unable, unassisted, properly to manage and take care of himself or his property, and by reason thereof is likely to be deceived or imposed upon by artful or designing persons".[5] Of course, to urge a court that an endangered river is "a person" under this provision will call for lawyers as bold and imaginative as those who convinced the Supreme Court that a railroad corporation was a "person" under the fourteenth amendment, a constitutional provision theretofore generally thought of as designed to secure the rights of freedmen.[6] (As this article was going to press, Professor Byrn of Fordham petitioned the New York

Supreme Court to appoint him legal guardian for an unrelated foetus scheduled for abortion so as to enable him to bring a class action on behalf of all foetuses similarly situated in New York City's 18 municipal hospitals. Judge Holtzman granted the petition of guardianship.[7]) If such an argument based on present statutes should fail, special environmental legislation could be enacted along traditional guardianship lines. Such provisions could provide for guardianship both in the instance of public natural objects and also, perhaps with slightly different standards, in the instance of natural objects on "private" land.[8]

The potential "friends" that such a statutory scheme would require will hardly be lacking. The Sierra Club, Environmental Defense Fund, Friends of the Earth, Natural Resources Defense Counsel, and the Izaak Walton League are just some of the many groups which have manifested unflagging dedication to the environment and which are becoming increasingly capable of marshalling the requisite technical experts and lawyers. If, for example, the Environmental Defense Fund should have reason to believe that some company's strip mining operations might be irreparably destroying the ecological balance of large tracts of land, it could, under this procedure, apply to the court in which the lands were situated to be appointed guardian.[9] As guardian, it might be given rights of inspection (or visitation) to determine and bring to the court's attention a fuller finding on the land's condition. If there were indications that under the substantive law some redress might be available on the land's behalf, then the guardian would be entitled to raise the land's rights in the land's name, i.e., without having to make the roundabout and often unavailing demonstration, discussed below, that the "rights" of the club's members were being invaded. Guardians would also be looked to for a host of other protective tasks, e.g., monitoring effluents (and/or monitoring the monitors), and representing their "wards" at legislative and administrative hearings on such matters as the setting of state water quality standards. Procedures exist, and can be strengthened, to move a court for the removal and substitution of guardians, for conflicts of interest or for other reasons,[10] as well as for the termination of the guardianship.[11]

In point of fact, there is a movement in the law toward giving the environment the benefits of standing, although not in a manner as satisfactory as the guardianship approach. What I am referring to is the marked liberalization of traditional standing requirements in recent cases in which environmental action groups have challenged federal government action. *Scenic Hudson Preservation Conference v. FPC*[12] is a good example of this development. There, the Federal Power Commission had granted New York's Consolidated Edison a license to construct a hydroelectric project on the Hudson River at Storm King Mountain. The grant of license had been opposed by conservation interests on the grounds that the transmission lines would be unsightly, fish would be destroyed, and nature trails would be inundated. Two of these conservation groups, united under the name Scenic Hudson Preservation Conference, petitioned the Second Circuit to set aside the grant. Despite the claim that Scenic Hudson had no standing because it had not made the traditional claim "of any personal economic injury resulting from the Commission's actions",[13] the petitions were heard, and the case sent back to the Commission. On the standing point, the court noted that Section 313(b) of the Federal Power Act gave a right of instituting review to any party "aggrieved by an order issued by the Commission";[14] it thereupon read "aggrieved by" as not limited to those alleging the traditional personal economic injury, but as broad enough to include "those who by their activities and conduct have exhibited a special interest" in "the aesthetic, conservational, and recreational aspects of power development . . .".[15] A similar reasoning has swayed other circuits to allow proposed actions by the Federal Power Commission, the Department of Interior, and the Department of Health, Education and Welfare to be challenged by environmental action groups on the basis of, *e.g.*, recreational and esthetic interests of members, in lieu of direct economic injury.'[16] Only the Ninth Circuit has balked, and one of these cases, involving the Sierra Club's attempt to challenge a Walt Disney development in the Sequoia National Forest, is at the time of this writing awaiting decision by the United States Supreme Court.[17]

Even if the Supreme Court should reverse the Ninth Circuit in the Walt Disney-Sequoia National Forest matter, thereby encouraging the circuits to continue their trend toward liberalized standing in this area, there are significant reasons to press for the guardianship approach notwithstanding. For one thing, the cases of this sort have extended standing on the basis of interpretations of specific federal statutes—the Federal Power Commission Act,[18] the Administrative Procedure Act,[19] the Federal Insecticide, Fungicide and Rodenticide Act,[20] and others. Such a basis supports environmental suits only where acts of federal agencies are involved; and even there, perhaps, only when there is some special statutory language, such as "aggrieved by" in the Federal Power Act, on which the action groups can rely. Witness, for example, *Bass Angler Sportsman Society v. United States Steel Corp.*[21] There, plaintiffs sued 175 corporate defendants located throughout Alabama, relying on 33 USC 407 (1970), which provides:

> It shall not be lawful to throw, discharge, or deposit . . . any refuse matter . . .
> into any navigable water of the United States, or into any tributary of any navigable water from which the same shall float or be washed into such navigable water . . .[22]

Another section of the Act provides that one-half the fines shall be paid to the person or persons giving information which shall lead to a conviction.[23] Relying on this latter provision, the plaintiff designated his action a *qui tam* action[24] and sought to enforce the Act by injunction and fine. The District Court ruled that, in the absence of express language to the contrary, no one outside the Department of Justice had standing to sue under a criminal act and refused to reach the question of whether violations were occurring.[25]

Unlike the liberalized standing approach, the guardianship approach would secure an effective voice for the environment even where federal administrative action and public-lands and waters were not involved. It would also allay one of the fears courts—such as the Ninth Circuit—have about the extended standing concept: if any ad hoc group can spring up overnight, invoke some "right" as universally claimable as the esthetic and recreational interests of its members and thereby get into court, how can a flood of litigation be prevented?[26] If an ad hoc committee loses a suit brought *sub nom.* Committee to Preserve our Trees, what happens when its very same members reorganize two years later and sue *sub nom.* the Massapequa Sylvan Protection League? Is the new group bound by res judicata? Class action law may be capable of ameliorating some of the more obvious problems. But even so, court economy might be better served by simply designating the guardian de jure representative of the natural object, with rights of discretionary intervention by others, but with the understanding that the natural object is "bound" by an adverse judgment.[27] The guardian concept, too, would provide the endangered natural object with what the trustee in bankruptcy provides the endangered corporation: a continuous supervision over a period of time, with a consequent deeper understanding of a broad range of the ward's problems, not just the problems present in one particular piece of litigation. It would thus assure the courts that the plaintiff has the expertise and genuine adversity in pressing a claim which are the prerequisites of a true "case or controversy".

The guardianship approach, however, is apt to raise two objections, neither of which seems to me to have much force. The first is that a committee or guardian could not judge the needs of the river or forest in its charge; indeed, the very concept of "needs", it might be said, could be used here only in the most metaphorical way. The second objection is that such a system would not be much different from what we now have: is not the Department of Interior already such a guardian for public lands, and do not most states have legislation empowering their attorneys general to seek relief—in a sort of *parens patriae* way—for such injuries as a guardian might concern himself with?

As for the first objection, natural objects can communicate their wants (needs) to us, and in ways that are not terribly ambiguous. I am sure I can judge with more certainty and meaningfulness whether and when my lawn wants (needs) water, than the Attorney General can judge whether and when the United States wants (needs) to take an appeal from an adverse judgment by a lower court. The lawn tells me that it wants water by a certain dryness of the blades and soil—immediately obvious to the touch—the appearance of bald spots, yellowing, and a lack of springiness after being walked on; how does "the United States" communicate to the Attorney General? For similar reasons, the guardian-attorney for a smog-endangered stand of pines could venture with more confidence that his client wants the smog stopped, than the directors of a corporation can assert that "the corporation" wants dividends declared. We make decisions on behalf of, and in the purported interests of, others every day; these "others" are often creatures whose wants are far less verifiable, and even far more metaphysical in conception, than the wants of rivers, trees, and land.[28]

As for the second objection, one *can* indeed find evidence that the Department of Interior was conceived as a sort of guardian of the public lands.[29] But there are two points to keep in mind: First, insofar as the Department already is an adequate guardian it is only with respect to the federal public lands as per Article IV, section 3 of the Constitution.[30] Its guardianship includes neither local public lands nor private lands. Second, to judge from the environmentalist literature and from the cases environmental action groups have been bringing, the Department is itself one of the bogeys of the environmental movement (One thinks of the uneasy peace between the Indians and the Bureau of Indian Affairs.) Whether the various charges be right or wrong, one cannot help but observe that the Department has been charged with several institutional goals (never an easy burden), and is currently looked to for action by quite a variety of interest groups, only one of which is the environmentalists. In this context, a guardian outside the institution becomes especially valuable. Besides, what a person wants, fully to secure his rights, is the ability to retain independent counsel even when, and perhaps especially when, the government is acting "for him" in a beneficent way. I have no reason to doubt, for example, that the Social Security System is being managed "for me"; but I would not want to abdicate my right to challenge its actions as they affect me, should the need arise.[31] I would not ask more trust of national forests vis-à-vis the Department of Interior. The same considerations apply in the Instance of local agencies such as regional water pollution boards whose members' expertise in pollution matters is often all too credible.[32]

The objection regarding the availability of attorneys-general as protectors of the environment within the existing structure is somewhat the same. Their statutory powers are limited and sometimes unclear. As political creatures, they must exercise the discretion they have with an eye toward advancing and reconciling a broad variety of important social goals, from preserving morality to increasing their jurisdiction's tax base. The present state of our environment, and the history of cautious application and development of environmental protection laws long on the books,[33] testifies that the burdens of an attorney-general's broad responsibility have apparently not left much manpower for the protection of nature. (*Cf. Bass Anglers*, above.) No doubt, strengthening interest in the environment will increase the rest of public attorneys even where, as will often be the case, well-represented corporate polluters are the quarry. Indeed, the United States Attorney General has stepped up anti-pollution activity, and ought to be further encouraged in this direction.[34] The statutory powers of the attorneys-general should be enlarged, and they should be armed with criminal penalties made at least commensurate with the likely economic benefits of violating the law.[35] On the other hand, one cannot ignore the fact that there is increased pressure on public law-enforcement offices to give more attention to a host of other problems, from crime "on the streets" (why don't we say "in the rivers"?) to consumerism and school bussing. If the environment is not to get lost in the shuffle, we would do well, I think, to adopt the guardianship approach as an additional safeguard, conceptualizing major natural objects as holders of their own rights, raisable by the court-appointed guardian.

Stone, Chr.D., Should Trees Have Standing? Towards Legal Rights for Natural Objects, in: *Southern California Law Review,* 1972, p. 450-501.

Notes and references

1 See, e.g. *Cal. Prob. Code* §§ 1460-62 (West Supp. 1971).
2 Cal. Prob. Code §§ *1751* (West Supp. 1971) provides for the appointment of a "conservator".
3 In New York the Supreme Court and county courts outside New York City have jurisdiction to appoint a committee of the person and/or a committee of the property for a person "incompetent to manage himself or his affairs". *NY Mental Hygiene Law* § 100 (McKinney 1971).
4 This is a situation in which the ontological problems discussed in note 26 supra become acute. One can conceive a situation in which a guardian would be appointed by a county court with respect to a stream, bring a suit against alleged polluters, and lose. Suppose now that a federal court were to appoint a guardian with respect to the larger river system of which the stream were a part, and that the federally appointed guardian subsequently were to bring suit against the same defendants in state court, now on behalf of the river, rather than the stream. (Is it possible to bring a still subsequent suit, if the one above fails, on behalf of the entire hydrologic cycle, by a guardian appointed by an international court?)

While such problems are difficult, they are not impossible to solve. For one thing, pre-trial hearings and rights of intervention can go far toward their amelioration. Further, courts have been dealing with the matter of potentially inconsistent judgments for years, as when one state appear on the verge of handing down a divorce decree inconsistent with the judgment of another state's courts. *Kempson v. Kempson*, 58 N.J. Eg. 94, 43 A. 97 (Ch. Ct. 1899) Courts could, and of course would, retain some natural objects in the res nullius classification to help stave off the problem. Then, too, where (as is always the case) several "objects" are interrelated, several guardians could all be involved, with procedures for removal to the appropriate court—probably that of the guardian of the most encompassing "ward" to be acutely threatened. And in some cases subsequent suit by the guardian of more encompassing ward, not guilty of laches, might be appropriate. The problems are at least no more complex than the corresponding problems that the law has dealt with for years in the class action area.

5   *Cal. Prob. Code.* § 1460 (West Supp. 1971) The NY Mental Hygiene Law (McKinney 1971) provides for jurisdiction "over the custody of a person and his property if he is incompetent to manage himself or his affairs by reason of age, drunkenness, mental illness or other cause . . ."

6   *Santa Clara County v. Southern Pac. R. R.*, 118 US 394 (1886) Justice Black would have denied corporations the rights of "persons" under the fourteenth amendment. *See Connecticut Gen. Life Ins. Co. v. Johnson*, 303 US 77, 87 (1938) (Black, J. dissenting): "Corporations have neither race nor color."

7   *In re* Byrn, *LA Times*, Dec 5, 1971, § 1, at 16, col 1. A preliminary injunction was subsequently granted, and defendant's cross motion to vacate the guardianship was denied. Civ. 13113/71 (Sup. Ct. Queens Co., Jan. 4, 1972) (Smith, J.). Appeals are pending. Granting a guardianship in these circumstances would seem to be a more radical advance in the law than granting a guardianship over communal natural objects like lakes. In the former case there is a traditionally recognized guardian for the object—the mother—and her decision has been in favor of aborting the foetus.

8   The laws regarding the various communal resources had to develop along their own lines, not only because so many different persons' "right " to consumption and usage were continually and contemporaneously involved, but also because no one had to bear the costs of his consumption of public resources in the way in which the owner of resources on private land has to bear the costs of what he does. For example, if the landowner strips his land of trees, and puts nothing in their stead, he confronts the costs of what he has done in the form of reduced value of his land; but the river polluter's actions are costless, so far as he is concerned—except insofar as the legal system can somehow force him to internalize them. The result has been that the private landowner's power over natural objects on his land is far less restrained by law (as opposed to economics) than his power over the public resources that he can get his hands on. If this state of affairs is to be changed, the standard for interceding in the interests of natural objects on traditionally recognized "private" land might well parallel the rules that guide courts in the matter of people's children whose upbringing (or lack thereof) poses social threat. The courts can, for example, make a child "a dependent of the court" where the child's "home is an unfit place for him by reason of neglect, cruelty, or depravity of either of his parents . . ." *Cal. Welf. & Inst. Code* § 600(b) (West 1966). *See also id* at § 601: any child "who from any cause is in danger of leading an idle, dissolute, lewd, or immoral life [may be adjudged] a ward of the court".

9   *See* note 53 *supra*. The present way of handling such problems on "private" property is to try to enact legislation of general application under the police power, *see Pennsylvania Coal Co. v. Mahon*, 260 US 393 (1922), rather than to institute civil litigation which, though a piecemeal process, can be tailored to individual situations.

10 *Cal. Prob. Code* § 1580 (West Supp 1971) lists specific causes for which a guardian may, after notice and a hearing, be removed.

Despite these protections, the problem of overseeing the guardian is particularly acute where, as here, there are no immediately identifiable human beneficiaries whose self-interests will encourage them to keep a dose watch on the guardian To ameliorate this problem, a page might well be borrowed from the law of ordinary charitable trusts, which are commonly placed under the supervision of the Attorney General *See Cal. Corp. Code* §§ 9505, 10207 (West 1955).

11 See Cal. Prob. Code §§ 1472, 1590 (West 1956 and Supp. 1971).

12 354 F.2d 608 (2d Cir. 1965), *cert. denied,* Consolidated Edison Co. v. Scenic Hudson Preservation Conf, 384 US 941 (1966).

13 354 F.2d 608, 615 (2d Cir. 1965).

14 Act of Aug. 26, 1935, ch. 687, Title II, § 213, 49 Stat. 860 (codified in 16 USC § 8251(b) (1970).

15 354 F.2d 608, 616 (2d Cir. 1965) The court might have felt that because the New York New Jersey Trial Conference, one of the two conservation groups that organized Scenic Hudson, had some 17 miles of trailways in the area of Storm King Mountain, it therefore had sufficient economic interest to establish standing, Judge Hays' opinion does not seem to so rely, however.

16 Road Review League v. Boyd, 270 F. Supp. 650 (SDNY 1967) Plaintiffs who included the Town of Bedford and the Road Review League, a non-profit association concerned with community problems, brought an action to review and set aside a determination of the Federal Highway Administrator concerning the alignment of an interstate highway. Plaintiffs claimed that the proposed road would have an adverse effect upon local wildlife sanctuaries, pollute a local lake, and be inconsistent with local needs and planning. Plaintiffs relied upon the section of the Administrative Procedure Act, 5 USC § 702 (1970), which entitles persons "aggrieved by agency action within the meaning of a relevant statute" to obtain judicial review. The court held that plaintiffs had standing to obtain judicial review of proposed alignment of the road:

> I see no reason why the word "aggrieved" should have different meaning in the Administrative Procedure Act from the meaning given it under the Federal Power Act . . . The "relevant statute", i.e., the Federal Highway Act, contains language which seems even stronger than that of the Federal Power Act, as far as local and conservation interests are concerned.

Id. at 661.

In Citizens Comm. for the Hudson Valley v. Volpe, 425 F.2d 97 (2d Cir. 1970), plaintiffs were held to have standing to challenge the construction of a dike and causeway adjacent to the Hudson Valley. The Sierra Club and the Village of Tarrytown based their challenge upon the provision of the Rivers and Harbors Act of 1899. While the Rivers and Harbors Act does not provide for judicial review as does the Federal Power Act, the court stated that the plaintiffs were "aggrieved" under the Department of Transportation Act, the Hudson River Basin Compact Act, and a regulation under which the Corps of Engineers issued a permit, all of which contain broad provisions mentioning recreational and environmental resources and the need to preserve the same. Citing the Road Review League decision, the court held that as "aggrieved" parties under the Administrative Procedure Act, plaintiffs similarly had standing. Other decisions in which the court's grant of standing was based upon the Administrative Procedure Act include West Virginia Highlands Conservancy v. Island Creek Coal Co., 441 F.2d 231 (4th Cir. 1971); Environmental Defense Fund, Inc. v. Hardin, 428 F.2d 1093 (DC Cir. 1970); Allen v. Hickel, 424 F.2d 944 (DC Cir. 1970); Brooks v. Volpe, 329 F. Supp. 118 (WD Wash. 1971); Delaware v. Pennsylvania NY Cent. Transp. Co., 323 F. Supp. 487 (D. Del.

1971); Izaak Walton League of America v. St. Clair, 313 F. Supp. 1312 (D. Minn. 1970); Pennsylvania Environmental Council, Inc. v. Bartlett, 315 F. Supp. 238 (MD Pa. 1970).

17 Sierra Club v. Hickel, 433 F.2d 24 (9th Cir. 1970), *cert. granted sub. nom.* Sierra Club v. Morton, 401 US 907 (1971) (No. 70-34). The Sierra Club, a non-profit California corporation concerned with environmental protection, claimed that its interest in the conservation and sound management of natural parks would be adversely affected by an Interior permit allowing Walt Disney to construct the Mineral King Resort in Sequoia National Forest. The court held that because of the Sierra Club's failure to assert a direct legal interest, that organization lacked standing to sue. The court stated that the Sierra Club had claimed an interest only in the sense that the proposed course of action was displeasing to its members. The court purported to distinguish Scenic Hudson on the grounds that the plaintiff's claim of standing there was aided by the "aggrieved party" language of the Federal Power Act.

18 16 USC §§ 791(a) *et seq.* (1970). see note 59 and accompanying text *supra.*

19 5 USC §§ 551 *et seq.* (1970). Decisions relying upon 5 USC § 702 are listed in note 56 *supra.*

20 7 USC §§ 135 *et seq.* (1970). Section 135b(d) affords a right of judicial review to anyone "adversely affected" by an order under the Act. see Environmental Defense Fund, Inc. v. Hardin, 428 F.2d 1093, 1096 (DC Cir. 1970).

21 324 F. Supp. 412 (ND, MD & SD Ala. 1970), *aff'd mem., sub nom.* Bass Anglers Sportsman Soc'y of America, Inc. v. Koppen Co., 447 F.2d 1304 (5th Cir. 1971).

22 Section 13 of *Rivers and Harbors Appropriaton Act* of 1899

23 33 USC §§ 411 (1970) reads:

> Every person and every corporation that shall violate, or that shall knowingly aid, abet, authorize, or instigate a violation of the provisions of sections 407, 408 and 409 of the title shall . . . be punished by a fine . . . or by imprisonment . . . in the discretion of the court, one-half of aid fine to be paid to the person or persons giving information which shall lead to conviction.

24 This is from the latin, "who bring the action as well for the King as for himself", referring to an action brought by a citizen for the state as well as for himself.

25 These sections create a criminal liability. No civil action lies to enforce it; criminal statute can only be enforced by the government. A qui tam action lies only when expressly or impliedly authorized by statute to enforce a penalty by civil action, not a criminal fine.

324 F. Supp. 412, 415-16 (ND, MD & SD Ala. 1970). Other *qui tam* actions brought by the Bass Angler Sportsman Society have been similarly unsuccessful. See Bass Anglers Sportsman Soc'y of America v. Schoke Tannery, 329 F. Supp. 339 (ED Tenn 1971); Bass Anglers Sportsman's Soc'y of America v. United States Plywood-Champion Papers, Inc., 324 F. Supp. 302 (SD Tex. 1971).

26 Concern over an anticipated flood of litigation initiated by environmental organizations is evident in Judge Trask's opinion in Alameda Conservation Ass'n v California, 437 F.2d 1087 (9th Cir.), *cert. denied,* Leslie Salt Co. v. Alameda Conservation Ass'n, 402 US 908 (1971), where a non-profit corporation having as a primary purpose protection of the public's interest in San Francisco Bay was denied standing to seek an injunction prohibiting a land exchange that would allegedly destroy wildlife, fisheries and the Bay's unique flushing characteristics:

> Standing is not established by suit initiated by this association simply because it has as one of its purposes the protection of the "public interest" in the waters of the San Francisco Bay. However well intentioned the members may be, they may not by uniting create for themselves a super-administrative agency or a parens patriae official status with the capability of over-seeing and of challenging the action of the appointed and elected officials of

the state government. Although recent decisions have considerably broadened the concept of standing we do not find that they go this far [Citation.]

Were it otherwise the various clubs, political, economic and social now or yet to be organized, could wreak havoc with the administration of government, both federal and state. There are other forums where their voices and their views may be effectively presented, but to have standing to submit a "case or controversy" to a federal court, something more must be shown.

437 F.2d at 1090.

27 See note 49 *supra.*

28 Here, too, we are dogged by the ontological problem discussed in note 26 *supra.* It is easier to say that the smog-endangered stand of pines "wants" the smog stopped (assuming that to be a jurally significant entity) then it is to venture that the mountain, or planet earth, or the cosmos, is concerned about whether the pines stand or fall. The more encompassing the entity of concern, the less certain we can be in venturing judgments as to the "wants" of any particular substance, quality, or species within the universe. Does the cosmos care if we humans persist or not? "Heaven and earth . . . regard all things as significant, as though they were playthings made of straw." *Lao-Tzu, Tao Teh King* 13 (D. Goddard transl. 1919).

29 See Knight v. United States Land Ass'n, 142 US 161, 181 (1891).

30 Clause 2 gives Congress the power "to dispose of and make all needful Rules and Regulations respecting the Territory or other Property belonging to the United States".

31 See Flemming v. Nestor, 363 US 603 (1960).

32 See the LA Times editorial *Water: Public vs. Polluters* criticizing:

. . . the ridiculous built-in conflict of interest on Regional Water Quality Control Board. By law, five of the seven seats are given to spokesmen for industrial, governmental, agricultural or utility users. Only one representative of the public at large is authorized, along with a delegate from fish and game interests.

Feb. 12, 1969, Part II, at 8, cols. 1-2.

33 The Federal Refuse Act is over 70 years old. Refuse Act of 1899, 33 USC § 407 (1970).

34 See Hall, *Refuse Act of 1899 and the Permit Program,* 1 *Nat'l Res. Defense Council Newsletter* i (1971).

35 To be effective as a deterrent, the sanction ought to be high enough to bring about an internal reorganization of the corporate structure which minimizes the chances of future violations. Because the corporation is not necessarily a profit maximizing "rationally economic man", there is no reason to believe that setting the fine as high as—but no higher than—anticipated profits from the violation of the law, will bring the illegal behavior to an end.

CHAPTER 12

The Destruction of Nature in the Soviet Union

B. KOMAROV

Taking Stock
Or: Who Will Pay for the Tragedy?

The entire country, the entire area of the planet now colored red on the globe, covers 22 million square kilometers, as we know from our schoolbooks. Water, perpetual ice and snow, barren crags—in a word, uninhabitable land—occupy about one third, so that of the 22 million, 14 to 15 million square kilometers remain for us. That is the whole "pie". Now let's figure out who gets what slice.

In 1977:
- 175,000 to 220,000 square kilometers (17.5-22.0 million hectares) are covered by lands mutilated and made worthless by mining and peat works;
- 50,000 square kilometers (5 million hectares) go to dumps, tailings, slag, and sludge heaps from industrial enterprises, as well as municipal dumps;
- 120,000 square kilometers (12 million hectares) are buried under the waters of large reservoirs;
- 500,000 to 550,000 square kilometers (50-55 million hectares) are forest barrens and swamps left from logging and fires, which will not support even weeds;
- 630,000 square kilometers (63 million hectares) are severely eroded, salinated land (plowland and pastures and meadows), ravines, and sand dunes in place of former fields.[1]

The total: 1.45 million square kilometers of now sterile land, industrial wasteland or semiwasteland.

This represents about 10 per cent of the entire habitable territory of the USSR. And this tenth of our "pie", which we have devoured in little bites, was the most "succulent and tasty". The most suitable for use and, in many cases, the most fertile. We have picked the cherry from the cupcake. Hence we are parcelling out the second, third, and fourth shares as fast as possible. Especially since there are so many more of us, and our needs are growing at a very rapid pace.

If bourgeois Western Europe had treated its environment as we have, all the inhabitants of England, France, Italy, West Germany, Switzerland, and the Benelux countries would long ago have found themselves in a sterile desert. Their population is equal to our—240-250 million—but their total land area is approximately one tenth of ours—1.4 million square kilometers.

I. Laptev, one of the ideologues of our country's ecological policy and a functionary for the Central Committee of the CPSU, stated in one of his articles that a society's attitudes toward nature bear the mark of all the relations and attitudes that have evolved within society.[2] On this point we must agree with him.

"The depraved relations" that have evolved in our society do serious harm to our environment. The relationships between defects in the economic system and ecology are much more extensive than might appear at first glance.

They are:
- the chronic shortage of meat in the country—and the low productivity of collective farm stock raising—as well as the barbaric extermination of whales in the world's oceans and the coastal waters of the Far East;
- the same shortage of meat—and the organized poaching of birds on wintering grounds in the Kyzyl-Agach Preserve. Poaching that incidental is annihilating the little bustard, the flamingo, the red-breasted goose, the black partridge, and dozens of other species listed in the Red Book;
- the low productivity of collective farm stock raising—and the sharp decline of all species of deer and elk in Siberia and the Far East;
- a situation in which the Ministry of Agriculture is the sole proprietor of all the country's farmlands, while collective farm workers have become indifferent to the land—and hasty, ill-considered reclamation that is ruining a large portion of the land, forests, berrying, and hunting grounds;
- a situation in which a huge share of our cotton and paper goes for the manufacture of powder and other explosives (vast stockpiles of armaments for ourselves and for export)—and the excessive exploitation of all cotton-growing land in Central Asia as well as overcutting of forests. One might say that the successes of African and Asian national liberation movements, fighting with Soviet weapons, have been paid for with the destruction of many square kilometers of Siberian taiga;
- the recurrent need to purchase wheat and new technology abroad—and the devastation of rivers and valleys in Siberia thanks to gold mining at any price.

In turn, the degradation of the environment has an impact on economic development. The wheat harvest on fields exposed to factory smoke is reduced by 30 to 60 per cent. The forest has ceased to grow over broad areas. The pollution of rivers has reduced the value of meadows and pasture lands in their valleys to nothing.

The great "watershed" of 1929, as well as other "turning points" of the thirties and forties, the great construction campaigns of the five-year plans and the virgin lands, and the immediate strategic interests of the military, which shaped the fate of Baikal, are all factors that have affected out attitudes toward nature.

If today all pollution of the environment were suddenly to stop magically, the pollutants accumulated over the past years would continue to effect several subsequent generations as much as they do us.

If again by magic all the wretched legacy of "poor, backward Russia", the Second World War, "Stalin's Excesses", and the Hydrological Planning Service were to disappear, it would hardly reduce the plundering of nature. Not now nor for several more generations. The main reasons for the devastation of nature lie more in our present than in our past.

The phenomenal development of science is the greatest pride of modern times. Sciences has become a productive force, as we all know. What is probably less known is how much it produces in our society. The contribution of science to the annual growth of USSR national income is 4 per cent.[3]

Whereas in the past delays in introducing scientific discoveries retarded only the technical progress of production, now a slow development and introduction of ecological equipment and technology directly affect the state of the environment. The inability of our economy to realize its scientific potential is just as much a cause of the degradation of nature as the poor productivity of collective farm agriculture.

The figure 4 per cent will shake even the most optimistic faith in the power of modern science. It is powerful only under certain conditions.

How to change this situation?

How to interest industry in replenishing natural resources?

How to rectify inefficiency in agriculture?

How to see that legal measures effectively safeguard nature?

Wherever one trace the thread of ecology, it always leads deep into the economic and social structure of society.

Studies by all leading scientists say the same thing: largescale ecological problems cannot be resolved solely within the framework of ecology, solely within the framework of technology, or even within the framework of pure economics. They require simultaneous changes both in the economy and in the social and moral foundations of society.

We are approaching the issues of the possibility of reforms, of necessary changes that would, at least, slow the pace of the degradation, the impoverishment of nature and, ideally, would ensure the ecological well-being of our people in the future.

It goes without saying that the specific content of such reforms is an extremely complicated matter, a subject for the efforts of many government minds. We are speaking now only of the changes for such reforms. Do they exist? Who would venture such reforms and why?

The burden of ecological problems is shared far from equally by various groups and strata of the population. We all breathe roughly the same air and walk on the same earth; but with that, equality in our socialist society ends.

Sociological statistics say that about 5 to 6 per cent of our population lives wholly at the expenses of the state (this is not to be confuse with social security), i.e., in practical terms, beyond the economic laws of society. Regardless of the achievements of our agriculture and industry, 12 to 15 million of the top party and government leadership and members of their families have the best, the "cleanest" environment.

Another 8 to 9 per cent of the population—top military and scientific circles, the elite of the official art world, and the "new bourgeoisie" (dealers in commercial goods)—have enough money to or acquire everything they want. This includes suburban houses in the lap of nature. The other 85 to 87 per cent, the majority of the population living on wages, feel the direct effects of the "hitches" in production and supply. This majority lives in concrete cells in crowded houses indistinguishable from what our propaganda just fifteen years ago was calling "the jungles of capitalist cities". This majority willingly suffers from the noise, the smog of those cities where they can find better paid jobs, where they can find some products to buy, let alone worrying about the carcinogens, allergens, or substances causing chromosomal mutations present there.

More than 85 per cent of the population has no way to get real information about pollution; and if they think at all about ecology, it is usually only after they have contracted some sort of lung cancer and have been granted sufficient recuperation time in a hospital bed.

But the avant-garde of the people, as some of the representatives of the ruling elite call themselves, perceives the acuity of ecological problems only from the figures in various documents. The green fences around their suburban houses effectively screen them from the effects of both economic and ecological crises. Five to six per cent of our society has access to natural products (to the extent that this is now possible at all), special drinking water, and special swimming pools with filtered sea water (without oil and phenol).

Hence for us everything will always be ecologically all right. In fenced wild forests there will be enough "wild boars" to hunt; in lakes screened by underwater nets there will be fish for anglers. And no matter how Baikal is degraded, they will try to keep a few bays in virgin splendor, with pines on the high banks and the purest water, "where you can see every stone on the bottom". For them, for their guests, and sometimes for journalists and TV reporters there will be a clean Baikal and its *omul* salmon.

The bourgeois slogan—"Everyone pollutes, both producers and consumers, so everyone must pay for pollution"—is appropriate only for capitalists, the ideologue I. Laptev tells us.[4] The affluent classes in the West hardly suffer seriously from the ecological crises, although even Soviet authors acknowledge that their incomes have declined somewhat due to expenditures for environmental protection. But how much of a burden have our elite 5 to 6 per cent, to which Laptev belongs, assumed? He and his colleagues direct the whole of the country's industry, but have they paid even a bit? On the other hand, of course, Marshall Batitskii can no longer shoot polar bears from his helicopter. This is certainly a tremendous deprivation we cannot appreciate.

But we can understand that given such a distribution of the burden of environmental pollution among various strata of society, the chances for changing the situation are not great.

Government experts and referees know quite well about the true extent of the degradation of nature and how this threatens the country's future. They know it from studies by the Club of Rome and from our own forecasts, such as *Nature 1980* and *Nature 1990*, although the latter intentionally tone down the expected consequences. Semiofficially, ecological policy-makers explicate the government's present position toughly as follows:

Of course the state of the environment leaves much to be desired. Of course we destroy a lot and do so pointlessly. But it is no secret that investments in ecology are generally irretrievable and that this too is a question of strategy. As long as we do not invest very much, we can spend more on boosting the economy and on other targets. The more we are forced to invest, the less is left for economic development. In terms of land mass and many resources, our situation is better, hence we should use this strategic advantage over the United States.

Not only coal, oil and other minerals by also clean air, the soil, and water must now be counted among our natural and strategic resources. The vast expanses of unused lands, which could be used for dumping without spending millions to bury and eliminate wastes, are also a strategic reserve. The advantages of space were once acquired by the tsarist autocracy, but now they serve the triumph of the ideas of communism. We have more air into which we can spew the smoke of factories without risking suffocation; we have plenty of Baikal water that is difficult to pollute permanently; hence we can wait and accumulate funds while capitalism suffocates in its smoke—in both the literal and metaphorical sense, in the fumes of inflation. Right?

What part of the country can still be sacrificed so that the moneys saved on Baikal, the Sea of Azov, and on protecting the soil and water can be spent on "strategic and ideological power"??? How many dozens of countries can be devastated so that the last man in our civilization speaks Russian, and then just Marxist phrases? . . .

One would like to think that no one argues so fanatically.

The USSR's advantages in natural resources are substantial; one bough is actually thicker, but thicker does not always mean stronger. It does not mean that we can whittle away at it longer than the Americans of the West in general do theirs.

Estimates by economists say that ten years ago, when the ecological clouds had only begun to thicken, our society could compensate for each ruble of damage from pollution with fifty kopecks spent on environmental protection measures. Today compensation for each ruble of damages costs about 1.5 to 1.7 rubles. And after 1982-83 compensation for a ruble's damages will cost 3 to 3.5 rubles.

These statistics, and others, sit on the desks of the responsible leaders of the State Planning Committee and departments of the Central Committee; but as long as the objectives of spreading the USSR's political and ideological influence in the world are the prime elements in government policy, our attitude toward nature will not change. Among all the announced measures to protect the environment, there is one area where these steps have a good effect: it is propaganda, the showcase. The money invested in pilot-demonstration enterprises, in blustery articles that laud every such enterprise at each stage—in planning, at the start of construction, the end of building, opening, and so on—yields greater gains than all other protective measures. This money serves to placate society; it yields advantage in the form of enraptured neophytes and proselytes in Third World countries and in the West.

Our nature, our home is being destroyed by us. The roof in our home leaks, the beams under the floor have rotted, and a draft rattles the loose doors and windows. Capital repair is needed, but we are satisfies with the fact that in one respect—the showcase—everything has been done just as on a picture post-card; everything is fine. We should be concerned about the brick and the lumber, but we spend our money only for paint and brushes, while we tirelessly chatter about our unique qualities as proprietors and about what terrible landlords our neighbors are.

Our country's proprietors have made a simple calculation: if only one per cent of all the factories and plants in the country were clean on the outside and in technology, this would be more than enough for propaganda photographs, more than the most pedantic Western professor-pest could inspect.

The worse matters are with us in some are, then usually the more attention paid there to ideological issues, the fundamental advantages of socialism over capitalism, etc. This silly trait has now emerged in our approach to ecological problems. One need only to look on the shelves of the Lenin Library or the "public" libraries; the share of political and ideological questions among publications on nature conservation has increased sharply since 1974.[5]

Facts about "shortcomings" in matters of environmental protection in the

USSR somehow get into books and articles, but they would never see the light if beforehand or afterward their authors did not make fundamental declarations about the basic superiority of socialism to capitalism in all respects, including ecology, and that the USSR has everything needed to quickly resolve all problems that come up. Experiences editorial personnel say that any feeble impulse to see the ecological crisis as a global, worldwide phenomenon, without continual mention of "two natures—capitalist and socialist", are unconditionally squelched. For example, a chapter in the book *Ecology: Policy and Law*, by O.S. Kalbasov, in which some of the shortcomings of Soviet laws on environmental protections were analyzed in connection with the overall socialist structure of society, never saw the light, even though, judging from the other chapters, the author is an absolutely loyal Marxist.

Many trustworthy authors who attempt to speak about the common course of the ecological crisis throughout the world and about the need to regulate the economies of different countries and regions in accordance with the counsels of the Club of Rome and other serious institutions are accused of anti-Soviet tendencies of "Sakharovism", which, for the censors, are the same thing. Yet to any normal person it is obvious that broad cooperation in ecological problems is now more important than the development of trade or even cultural relations. "If you don't think about your future, you will not have one", Galsworthy once observed. More than ever before, ecology now shapes the future of countries and peoples. Thus after disarmament, cooperation in the solution of ecological problems has become the most important aspect of international détente.

The more thoroughly ecological policy is kneaded with the yeast of ideology, the less room remains for flexible economic decisions, for the exchange of valuable experience accumulated throughout the world.

The criticism of bourgeois theories and bourgeois practice in dealing with ecological problems is extremely superficial and opinionated and essentially cannot help solve similar problems in the USSR. What is most interesting in such books are the author's slips. For example, V. Bartov calls the attempts of the renowned economist Wassily Leontief to introduce planning into the operations of American monopolies evidence of the total decay of bourgeois economic thought.[6] Bartov, a person who knows at first hand the results of the planning system in the USSR, probably assumes that only a madman or a suicide would attempt to do something similar in the United States.

The ideological approach to genetics and cybernetics did tremendous harm to Soviet science, and ultimately to the country's economic development. Abuses of ideology in the realm of ecological problems bear a different character, although they threaten even greater disasters. The solution or aggravation of ecological problems touches the very foundations of existence of all of society are not merely isolated, although important, areas of its operation.

In the words of Barry Commoner, who is quoted by Soviet ecologists more often than any other author, "condemning as anathema any suggestion which reexamines basic economic values; . . . burying the issues revealed by logic in a morass of selfserving propaganda . . . failing fully to inform citizens of what they need to know . . ."—all this is an intolerable luxury for human society if it wants to overcome the ecological crisis and survive.[7]

Deliberately distorted, embellished information pervades our society on all levels. The effects of those few steps that have been taken to preserve the environment have been diminished by the fact that they are conceived and implemented on the basis of false theories and statistics. Not only does the ordinary citizen see a distorted picture of ecological problems, but data sent "to the very top" are also colored to a different extent and for completely different reasons.

"We don't need an institute to show that the socialist system runs badly. We know it ourselves." So, evidently, stated a top leader about the Institute of Concrete Sociological Research of the Academy of Sciences, which was trying to do an objective study of sociological problems. Ecological investigations, like the forecasts *Nature 1980* and *Nature 1990*, which provide accurate data on the course of events, "rounds off" the results in their conclusions, pass them through a prism of ideological terms, and dilute them with the rose water of optimism, since everyone knows that "upstairs", in the government, they don't want tot hear too much bad news.

Those governing the country can well imagine that if they fully unleashed citizens' personal initiative, if they slackened the reins of centralized planning, it would be possible to solve many problems of agriculture and industry. Many acute problems of ecology would be solved in the same way. However, people are not given full rein either creatively or legally. Their current position of servility, on all fours, suits our "best". Given such rules of the game, they always win however poorly they play. The price paid for our country's military and political power, for the esteem shown every representative of the Soviet government, for the respect with which every word uttered in the Kremlin is heard throughout the world, has been high. It includes the purity of unique Baikal and of dozens of rivers in the southern Ukraine and the Urals, the Sea of Azov, as well as the clarity of air in the Kuznets Basin and East Kazakhstan and the impoverishment of the soil in the Trans-Volga region and Kirghiz meadow lands. And ultimately it includes the health of the Soviet people, the health of generations to come.

Bertold Brecht wrote about times when a conversation among three people about trees represented a lie, hypocrisy, since behind it stood silence about atrocities, about people innocently murdered. We live in an era when in journalists' innocent conversations about saving the eagle, the bison, or the beaver in our country one hears the silence about the destruction of thousands of other animals. It is a time when reliable facts about the multiplication of the saiga or muskrat prove false, since behind them are silence about the thousands of deer slaughtered by bullets and rockets, the soft crackle of the last of the vanishing redbreasted geese, and the screams of cranes and bustards. The propagandistic clamor about million-ruble treatment systems, about new fish-breeding farms, is meant to drown the silence of wasted forests, moribund Azov, and degraded Baikal.

Even the most innocent conversation about trees is no longer just a lie; it is almost a crime if it drops from the blossoming branches to the bole of the trunk, the point at which in our country they cut millions of trees whose destructions is a violation of all nature laws.

One of the staunchest defenders of nature, V. Chivilikhin, said not long ago that he would write nothing about nature conservation. "Such grave words now weigh on my mind", he said. "No one will permit them anyway. Who needs them?"

The government does not want to hear bad news "about nature", but people need this news. It is not political motives, not the dictates of reason—it is the very air we breathe that forces us to understand: if we want to survive, we must know the truth. And tell it to others.

Komarov, B., *The Destruction of Nature in The Soviet Union,* translation from Russian, M.E. Sharpe, White Plains, New York, 1980, p. 130-140.

Notes and References

1   October [Oktiabr'], 1965, no. 4. A discussion of readers' letters on Chivilikin's essay about Baikal. His essay, entitled "The Radiant Orb of Siberia", was not only the first but also the most accurate statement about the Baikal tragedy.

2   N.M. Shavoronkov, academician, director of the Institute of Organic Chemistry of the Academy of Sciences, was concerned mainly with problems of chemical industry. Neither before nor since Baikal had he ever been concerned with anything even remotely resembling the problems of decomposition of organic matter, to say nothing of biochemical processes in rivers and lakes.

3   The preservation of Siberian Mountainscapes [Okhrana gornykh landshaftov Sibiri], Novosibirsk, 1973.

4   Ibid.

5   R. Shrilmark, *Taiga Fastnesses* [Taezhnye dali], Moscow, 1977.

6   Izvestia, March 4, 1976.

7   K. Mitriushkin in the magazine *Hunting and Hunting management,* [Okhota i okhotnich'e khoziaistvo], 1977, nos. 3 and 11; the book *Man and Nature* [Chelovek i priroda], Moscow, 1974, etc.

8   See K. Mitriushkin in the magazine *Hunting and Hunting Management,* 1977, no. 3.

9   Guy Biolat, *Marxism and the Environment* [Marksizm i okruzhaiushochaia sreda], Moscow, 1976, translation from the French edition, *Marxisme et environnement*, Paris, 1973. To support the assertion that only Marxism will provide the correct solution to the environmental problem, Biolat refers to the experience of the USSR, "where a national park of 30,000 square kilometers (!) has been created around Baikal, and the construction of new factories around the lake has been prohibited (!)".

10 Hundreds of books have been published about other aspects of the matter, especially the optimistic ones.

Part III

THE BESTSELLERS

Introduction to Part III: The Bestsellers

Most of the bestsellers were written in the first half on the 1970s, a period in which drastic changes in Western societies took place. In Part II, it was argued that in the 1960s critical movements dealing with societal questions in the field of nuclear weapons, women's liberation, environmental issues, the Vietnam War, and the North-South dialogue came up in Western Europe and in the USA. The impact of these movements was not limited to the 1960s. On the contrary, particularly in the first half of the 1970s, their societal influence became more manifest than ever before.

Growing awareness

This resulted in a growing awareness of environmental problems in society. However, this awareness was also based on experiences in everyday life. The increasing environmental problems resulting from a continuous growth in production was experienced by large sections of the population in Western countries. Air pollution expanded into many dwelling areas located in the neighbourhood of industrial complexes. During smoggy periods in the Netherlands and in the Ruhrgebiet in Germany, asthma sufferers had to be evacuated to less polluted areas. Swimming in lakes and rivers had always been a normal practice in many European countries. However, water pollution became so severe that swimming in some areas became dangerous.

As a result of pollution and intensive agricultural practices, many species of flowers and animals decreased dramatically in number. In spite of the continuous construction of motorways, roads were increasingly congested with cars. New industrial and residential areas were developed, crisscrossing rural areas and reducing their space. The enormous growth of the petrochemical industry increased the risk of calamities. In short, a situation arose in which people were daily experiencing the degradation of the environment. As a consequence, society and scientific circles paid more attention to environmental problems at the end of the 1960s and the first half of the 1970s.

The birth of environmental policy

Owing to the growing awareness of environmental problems, politicians became active in the field of environmental problems. Local, regional, and national governments started to work on environmental policy. In many countries the national government took the lead. In the beginning it was not clear what the range of environmental policy should be. There was a lot of uncertainty concerning the problems to be dealt with and about the best strategy for handling these problems. There were only a few public officers to work out such a policy and the national budgets to finance measures in this field were very limited. There was also a lack of well-defined knowledge concerning the use of policy instruments. The birth of environmental policy was more influenced by the need to react to the problems that had arisen than by the existing knowledge of how to handle new problems.

The character of the first environmental policy

Within this context, it was clear that the first environmental policy measures were based on trial and error. They were direct answers to perceived problems which in the early years, were regarded as relatively separate issues. The approach was sectoral, meaning that certain measures were taken to deal with water pollution, for example; while other measures were taken against air pollution, sometimes even by another department of the national (or federal) government. The whole range of early environmental measures was fragmented and, in essence, pragmatic. Another remarkable aspect of the early measures was the fact that laws and other legal regulations were considered the best answer to the problem: once a law was passed, the problem would be solved immediately. Early environmental policy was characterised by a strong belief in the effectiveness of legal measures, which resulted in a growth of policies in the 1970s based on legal regulations.

Environmental problems given low priority

In the second part of the 1970s, environmental problems were again marginalised. Instead, economic decline, increasing unemployment rates, and the growing financial deficit of the public sector dominated social as well as scientific debates. Before the decline in environmental concern in the second half of the 1970s, some important environmental publications appeared which influenced society very deeply. We call these the bestsellers, as most of them were sold in Western countries in very high numbers.

The publications which we discussed in Part 1 dealt mainly with the relationship between ecosystems and human society. Researchers from that period tried to prove that nature should be taken seriously and that we could not continue destroying nature and the environment. These types of messages were no longer necessary; people had had enough unpleasant experiences to know that environmental pollution was a serious problem. Environmental groups as such were a fairly new phenomenon in Western countries, which gave them significant opportunities to influence public opinion and the behaviour of politicians in the field of the protection of nature and the environment.

Environmental groups convinced people and political parties that the implementation of strict measures was the answer to environmental problems. Traditional political parties were afraid of these environmental groups, as they were convinced that the continued deterioration of nature and the environment would undoubtedly result in environmental political parties gaining voters. This is one reason why, particularly in this period, new laws protecting nature and the environment were passed in Parliament. The publications discussed in this section had a major influence on this policy formulation process.

Redefining the relationship

In the Revival period, ecologists, biologists, and economists took the lead by explaining that the relationship between the ecosystem and human society should be redefined. It was this relationship which was at the core of the societal debate. This influenced the scientific context considerably. In the 1970s, however, policy formulation and the implementation of strict norms were at the core of the societal and political debate. In such a climate, the disciplines themselves are not so relevant to the debate. It should be noted, however, that economists and public administrators played a more central role than previously. Indeed, generally speaking, economists and public administrators are more involved in policy formulation than biologists and ecologists.

Commoner: The Closing Circle

Commoner's book *The Closing Circle* (1971) is based on the experiences of progressive Americans at the end of the 1960s and the beginning of the 1970s. It follows the tradition of Kapp, Galbraith, and Heilbroner, who are quoted extensively. The book starts with the events of Earth Week in April 1970, during which many people actively protested environmental pollution and degradation. A link was established between the Vietnam War and the crisis in Western societies. Hence, environmental problems are analysed in this book as part of other societal problems.

In Commoner's view, pollution results from a free market economy, one based on profit, in which the goal of companies is to grow. As part of this competitive process, companies try to find new combinations resulting in new and improved products. This has produced a fundamental change in the use of natural resources, which gives the companies that use new technologies a marketing advantage. They can produce at lower cost levels, or put new products on the market, which will yield higher profits.

Generally speaking, it is the technological change in production during the last few decades which has introduced the use of materials with a higher level of toxicity, but also higher profits. The expansion of the economy is based, in this view, on the use of new raw materials which pollute the environment more than before. Commoner gives the examples of artificial fertilizers, plastics, heavy metals, etc. These new resources are mainly based on the use of fossil resources, which in principle are exhaustible. Furthermore, the increased use of these new resources affects the environment more than the industrial resources which are mainly made from organic materials and which do not have such a detrimental effect on nature and ecocycles.

Trying to reduce pollution is a real problem. This can only be done by using other, 'old-fashioned' raw materials, which pollute the environment less. Companies that are forced to use these materials have to lower wages or increase the price of their products, which has a negative effect on the purchasing power of people with low incomes. Nevertheless, development of other techniques and raw materials in production has to continue since any living thing that hopes to live on the earth must fit into the ecosphere or perish. The term Closing Circle is based on the necessity of closing the ecocycles again and using materials of an organic nature which are more compatible with the ecosystem itself. In broad outline, these are the environmental cycles which govern the behaviour of the three great global systems: the air, the water, and the soil. Full attention is given to theoretical elements as well as to empirical evidence. The Laws of Thermodynamics are given special attention and it is demonstrated that the production of goods and services has to be realised within these ecosystems. Additionally, the condition of the water, soil, air, and nuclear power at particular locations in the USA is discussed.

Commoner is not willing to blame overpopulation in the USA as a significant reason for the increase in pollution levels after the Second World War. Primary production of food and shelter is consistent with the increase in population, but pollution levels have increased much more than population levels. This increase in pollution is caused, in his view, by a sharp decrease in the use of organic raw materials with a low level of pollution. Furthermore, the types of products have changed. Thus, for example, there has been an increase in the use of cars and a decrease in the use of horses and trains. Production has gone in the wrong direction since more money can be earned by introducing new materials and goods which cause pollution.

Commoner's view is the absolute opposite of that held by Ehrlich and Hardin, who wrote in the same period. They argued that overpopulation was the most significant cause of pollution problems. Their general idea is that the main problems are caused by population density that is too high in relation to the natural resources. These authors do not or will not recognise the dramatic differences between people from the West and inhabitants of Third World countries. Commoner believes that the overall evidence is clear. The chief reason for the environmental crisis that has engulfed the United States in recent years is the sweeping transformation of productive technology since World War II. The economy has grown enough to give the United States population about the same amount of basic goods, per capita, as it did in 1946. However, productive technologies that have an intense impact on the environment have displaced less destructive ones. The environmental crisis is the inevitable result of this counter-ecological pattern of growth.

The basis of the crisis is the way in which technological development took place, ignoring the effects of these developments on society and societal development. Reductionism, which is rooted in technical science and ignores social questions, has to be avoided.

At the end of the 20th century, some of Commoner's criticisms need to be re-evaluated. For example, the use of polluting materials in technological processes has been (somewhat) reduced in modern industry, and material and energy efficiency have been improved during the last decade in several sectors of the economy. The first signs of a less polluting technology are visible and the concept of sustainable technology has taken hold.

Meadows: The Limits to Growth

Meadows' publication *The Limits to Growth* (1972) is undoubtedly the most influential work of this period and perhaps even of the last two decades. This is mainly due to its ambitious starting point which sought to establish the relationships between all relevant variables in the problem of environmental disruption. Previous publications such as those by Commoner, Ehrlich, and Hardin were based on the idea of one causal factor for environmental problems, as a result of which these authors did not pay enough attention to all the relevant combinations of the numerous variables. Jay Forrester designed the prototype model for this study, founded on earlier studies on *Industrial Dynamics, Urban Dynamics* and *World Dynamics*. This background gave the publication a high profile and was recognised by most scientists.

The relevant variables in the model are accelerating industrialisation, rapid population growth, widespread malnutrition, the depletion of non-renewable resources, and a deteriorating environment. These trends are interconnected in many ways. One of the characteristics of these variables is that their relevant changes are measured in decades or in centuries rather than in months or years. Such a model can provide information, for example, about the effect of a decrease in the availability of food on the other variables. It attempts to clarify the causes of these trends, their interrelationships and their implications as far as one hundred years in the future. The formal written model of the world constitutes a preliminary attempt to improve our mental models of long-term global problems by combining the large amount of information that is already available. New information-processing tools, such as systems analysis and the modern computer, are used.

The main conclusions of the report are:
1 If the current growth trends in world population, industrialisation, pollution, food production, and resource depletion continue unchanged, the limits to growth on this planet will be reached within the next one hundred years. The most probable result will be a rather sudden and uncontrollable decline in both population and industrial capacity.
2 It is possible to alter these growth trends and to establish a condition of ecological and economic stability that is sustainable in the distant future. The state of global equilibrium could be designed in such a way that the basic material needs of each person can be satisfied; additionally, each person has an equal opportunity to realise his or her individual human potential.
3 If policies are directed to the second option which is to reach a sustainable development of human society, the realisation of this option will be more easily effected the sooner these policies are implemented.

A model with these types of outcomes is bound to generate severe criticism. First, the authors were attacked from many sides for their assumptions about the level of fossil reserves and other natural resources. It was said that in many cases these reserves were significantly higher than was estimated in the report. The authors responded that this criticism was less relevant than it might seem at first sight. They argued that even relatively big mistakes in these estimations would only postpone the collapse by a few years if production continued to grow. They gave the example of a pond in which water plants are growing. They multiply, which means that every year one plant generates another plant. The question is how high the density of plants is just one year before the general catastrophe, which is a pond full of plants obstructing further growth of the system. The answer is that this will be the case only one year before the collapse. It is the exponential growth element which is responsible for these types of problems.

Secondly, other critics focused on the fixed construction of the model. If the model predicts a collapse based on a level of resources that is too low in relation to the increasing demand for these resources in the case of growing population and production levels, how can this situation be rectified by implementing new policies? The authors claimed that the model was not as fixed as many people thought. If certain parameters could be changed by the implementation of new policies, the total outcome would also change. Therefore, the implementation of new policies can be seen as mankind's reaction to the outcomes of the model which describes a situation of non-intervention.

Thirdly, many objected to the global character of the model; they argued that it is impossible to develop a model which can describe all the reactions of relevant variables in the whole world. One cannot compare Third World countries with the USA and Japan. This was probably the most serious criticism of the model, and one that could not be effectively dealt with. Later, a new model was developed which divided the globe into regions, making these types of models more realistic. Measarovic and Pestel divided the world into several regions and analyzed an elaborated model for each of the areas. In their publication *Mankind at the turning point* (1974), they came to the same general conclusions as Meadows, but with some more specific conclusions and recommendations based on the idea of 'organic growth'.

The Limits to Growth put the limited possibilities of the ecosystem in the production and consumption process at the centre of the political debate. This perspective found its way into many Western policies of that time.

Goldsmith: A Blueprint for Survival

A blueprint is a general sketch of the way things have to be done in the future. This is exactly the aim of the authors of the publication *A Blueprint for Survival* (1972). They take the results of other investigations as a starting point and then ask what the optimal construction of society should be. The goal of the blueprint is to steer society in a particular direction as indicated by environmental information.

In this publication, the relationship between the ecosystem and the system of production and consumption is the primary focus, as it is in many publications by biologists in the Revival period. Society has to be changed so it is in harmony with the ecosystem. In the opening lines of the publication, the authors argue that it is necessary to create a sustainable society fit for future generations as well.

The phenomenon of exponential growth in production is given particular attention in the publication. It is claimed that this mechanism will result in a sudden and sharp decrease in the use of natural resources and will therefore have a disastrous effect on society itself. The aim of the Blueprint is to convince governments, labour unions, and citizens of the need to change society itself. The authors formulate the principles of a stable society, i.e. one that can be sustained indefinitely while giving optimal satisfaction to its members.

These principles are:
- A minimum disruption of ecological processes;
- A maximum conservation of materials and energy;
- A stable population without elements of growth which will always put extra demands on the environment;
- A social system in which the individual can enjoy, rather than feel restricted by the first three conditions.

In order to achieve these principles, a controlled and well-orchestrated change on numerous fronts is required. This change can be brought about by implementing seven operations:

1 A control operation whereby environmental disruption will be reduced as much as possible. This reduction has to be realised by developing new techniques, new machines, and new combinations.

2 A period of stabilisation. In the concept of exponential production growth, a fixed stock of resources can be depleted fairly rapidly. Therefore, a stabilisation period is the only way to stop this process of exhaustion and disruption.

3 An asystemic substitution process, by which the elements of the most dangerous trends are analysed and replaced by technological substitutes, the effects of which are less negative and disruptive in the short term. In the long run, such a substitution process will reach its own limits and subsequently become ineffective.

4 Therefore a systemic substitution process is needed by which these technological substitutes are replaced by 'natural' or self-regulating ones. This means that they are replaced by those substitutes which either replicate or employ the normal processes of the ecosphere but without causing the traditional damage. Therefore, these substitutes can be seen as being sustainable over very long periods of time.

5 Alternative technologies which conserve energy and materials need to be developed, promoted and applied. They have to be designed for relatively 'closed' economic communities which cause only minimal disruption of ecological processes (e.g., intermediate technology).

6 A decentralisation of economic processes and policies at all levels is necessary. Communities have to be formed which are small enough to be reasonably self-regulating and self-supporting. By doing so, transport of materials and energy will be reduced. Furthermore, it becomes easier to control the flow of goods, waste, pollution, and exhaustion.

7 Education, in particular regarding these small communities, is necessary.

The authors are sufficiently aware of the 'political reality' to accept that many of these proposals will be considered impracticable. However, they believe that if a strategy for survival is to have any chance of success, the solutions must be formulated in the light of the problems and not from a timorous and superficial understanding of what may or may not be immediately feasible. If we base remedial action on political rather than ecological reality, then very reasonably, very practicably, and very surely, we will muddle our way to extinction.

A Blueprint for Survival was based on principles characteristic of that period. The call for a radical change in society was often articulated in progressive circles. People did not believe that progress would ever made without such a change. This is one reason why the blueprint involves so many principles and conditions which have to be fulfilled. Otherwise, according to the prevailing view, changes could not be realised under the existing social, economic, and political order. There was a need for radical social change and the Blueprint advocated an alternative social order based on small self-supporting communities. Criticisms were articulated about these small and closed communities necessary to optimise control of the economic processes by the policies of the people themselves. Basic democracy joins the need for environmental control. For many politicians this approach to modern problems was inspired by utopian socialism and was therefore considered unrealistic.

The importance of the Blueprint can be found in the way it demonstrated the necessity for a fundamental change in society. Economic processes have to be brought under the control of a societal policy which, in this view, is inevitable when a sustainable society is taken as a leading principle.

United Nations Conference on the Human Environment

During the 1960s and 1970s, international environmental problems were mainly seen in terms of the exhaustion of fossil resources. 'Normal' environmental problems were mainly defined as regional or national problems. The 'United Nations Conference on the Human Environment' (1972) was the first international platform where international environmental problems could be discussed. This resulted in a declaration. Additionally, the relevance of the Stockholm conference can be demonstrated by the fact that the significant international problem of acid rain (or air pollution as it was called then) was put on the international agenda.

Following World War II, when the industrial development of Europe was intensified, air pollution became increasingly significant. In this period, air pollution was seen as a threat to people living near industrial complexes. It became a public health issue. During smoggy periods, particularly in such areas as metropolitan London and the Dutch Rijnmond, higher than normal mortality rates were observed. Air pollution was therefore generally regarded as a health problem, a problem which could be 'solved' by building tall chimneys.

The construction of tall chimneys in the 1960s, and the mushrooming of automobile traffic and intensive farming in the last few decades has made air pollution an international problem. As long ago as the early 60s scientists were expressing their concern about the building of tall chimneys which would spread air pollution over the whole of Europe.

When air pollution was transported by the wind to the remote areas of Europe, it affected the ecosystems in those places not capable of coping with this threat. Indeed, the phenomenon of fish dying in the lakes of Scandinavia was a typical example which attracted a great deal of attention in the Scandinavian countries in the 1960s. In particular, Sweden suffered from this type of pollution. At the United Nations Conference in Stockholm in 1972, Sweden put this issue on the agenda. They argued that the damage to the Swedish lakes was being caused by the industrial emissions of Poland, Germany, England, and the Netherlands. Sweden argued that these emissions had to be reduced.

All these countries claimed, however, that there was no proof that they were responsible for the damage to Swedish lakes. They were not willing to reduce their emissions based on the claims made by Sweden. The Swedes, however, persisted as they were of the opinion that their natural resources were in danger. An agreement was made that international research would be done to provide more insight into these problems.

The first results were published by the OECD in 1977. It became clear that transboundary air pollution was very common. The deposition of acidifying substances from abroad was more than 50 per cent of the total deposition in countries such as Norway, Sweden, Austria, Finland and Switzerland. This implied that these problems could only be solved by international cooperation. However, most other countries were not willing to take measures at that time.

The first conference on a European policy for protecting clean air was convened in Geneva in 1979. Agreement was reached on general principles such as the dissemination of information and technology, the relevance of public health, and the necessity of reducing emissions. However, it was not possible to agree on specific measures. The participating nations were not able to agree on the type of measures, the scale and the division of the costs related to these measures. The crucial point was that every decision would benefit or injure one country more than others. It was not before 1985 that countries were able to agree upon a flat reduction of sulphur dioxide emissions by 30 per cent. It took 13 years of negotiating before the first measures in the field of international abatement policies were realised.

This brief description of the first stage of the implementation of international abatement policies on acid rain demonstrates how important the Stockholm conference of 1972 was. It was the first step on a long road towards a solution of international environmental problems. It made it clear that there is a continuous battle between national interests and international cooperation and that, at the same time, the existing international order was (and is) relatively powerless in attacking global problems.

Boulding: The Economics of the Coming Spaceship Earth

Toward a Steady-State Economy (1973) is a collection of articles by various influential authors, edited by Herman Daly. There are, for example, contributions by Georgescu-Roegen on entropy, by Paul Ehrlich on population growth, by Boulding on Spaceship Economy, by Schumacher on Buddhist Economics and by Cobb on Ethology. This book brought together relevant publications by well-known authors at an early stage of the societal debate on environmental issues.

The book can be seen as an attempt to formulate new economic paradigms necessary to describe and analyse modern environmental problems. The authors of these contributions are all aware of the shortcomings of traditional neoclassical economics and have tried to develop new ideas and approaches more appropriate to these new problems. Most of the authors concentrate on the level of production, consumption, or population. They are convinced that a limit should be set to the further expansion of economic development.

Boulding's article *The Economics of the Coming Spaceship Earth* (1973) is at the core of the debate for this period. The starting point is that the aims of a traditional economic approach will lead to a maximisation in the use of energy and raw materials. However, this process causes the environmental problems we are confronted with. Such a production process is based on a cowboy economy in which people are inclined to look for other resources in the Far West which will create new possibilities for mankind. This typical American approach has to be replaced by a spaceship economy in which people are aware that we only have a limited supply of energy, raw materials, and other resources. In such an economy, the aim is to reduce the throughput of the economy. For example, a low national product is, in this view, an expression of a low level of environmental disruption and not of a state of poverty.

The relevance of the Boulding paper can be found in the influence it has had on many environmental scientists and in particular environmental economists in a later period. They became aware of the need to develop new approaches and new paradigms more appropriate to modern environmental problems. At the end of the 1980s, the ideas articulated in this 1973 book became a crystallisation point leading to the establishment of the International Society for Ecological Economics, which held its first conference in Washington DC in 1990. In the course of the 1990s, the International Society for Ecological Economics became the leading organisation in the field of environmental economics. The leading principle of this organisation is that traditional economics cannot fully describe and analyse environmental problems. The relationship between the ecosystem and the system of production and consumption is the central point of all theoretical reasoning about environmental problems.

Tinbergen: Reshaping the International Order

Tinbergen's *Reshaping the International Order* (1977) does not deal with traditional environmental problems. It focuses on the question of how to distribute wealth and income in a world with limited resources. The author takes the outcome of the environmental discussion in a broader sense for granted; after that, he tries to analyse the distributional effects of these environmental conclusions. Consequently, this publication can be seen as a predecessor to *Our Common Future*, a report of the World Commission on Environment and Development in 1987. This publication claims that in a world with limited resources and limited possibilities for growth, underdeveloped countries should be allowed to take a more than proportional share in comparison to Western developed countries, which have already reached a certain level of wealth and income.

What we need, according to Tinbergen, is a New International Order based on the rights of all human beings to live a safe life, unhampered by fundamental shortages of essential resources. The report seeks to stimulate and contribute to the necessary further exchange of ideas between the many parties involved in the attempt to shape a fairer world. Attempts are made to formulate politically feasible first steps which the existing international community might steer in the direction of a more human and equitable international order.

Tinbergen does not believe that proposals for change should be limited to economic relations between nations. The world is too complex to be viewed in purely economic terms. The establishment of a New International Economic Order entails fundamental changes in political, social, cultural, and other aspects of society; changes which would bring about a New International Order.

In this publication, we do not find well-elaborated ideas about an institutional framework which could be constructed in due course. The document is based on ideas about the direction in which global economic development go in a finite world rather than presenting lists of instruments and institutions to be realised. The ideas are strongly influenced by the general principles of a fair world from social democracy.

One can conclude that the problems which Tinbergen describes and analyses are still in some way on the political agenda. On the one hand, there are some developing countries which have recently been able to build up a strong competitive industrial society. We should not overlook the fact, however, that many Third World countries are still suffering from malnutrition and a shortage of essential resources. This means that the New International Order is still rather far from being realised.

Lovelock: Gaia; a New Look at Life on Earth

The core of Lovelock's argument in his book *Gaia; a New Look at Life on Earth* (1979) is that the Earth and life on earth are regulated by life itself. He came to this conclusion while investigating the possibility of life on other planets such as Mars. He asked himself how, in general, we could detect the possibility of life. He then asked the same question about the planet Earth. He came to the conclusion that the presence and the level of the gasses most relevant for life on earth could not be explained by using traditional methods. Consequently, he concluded that life itself 'created' the optimal conditions necessary for life on earth. Most scientists, not being accustomed to this type of physical science, do not feel they are able to test these arguments.

Nevertheless, this approach attracted the attention of many scholars in a wide variety of disciplines. His approach has the advantage of sounding rather plausible and it will, in many cases, give the reader pleasant feelings of 'knowing' or 'believing' that environmental disruption and the deterioration of nature cannot be so serious that it will destroy the possibility of life on earth. This was particularly relevant in the 1970s, when many pessimists were arguing that life on earth could be destroyed by the ongoing environmental disruption.

The Gaia approach worked as a counterbalance to the negative ideas of the many environmentalists who strongly protested against environmental disruption. From this point of view, Lovelock's ideas can be evaluated as a positive instrument against the negative ideas which were rather common in that decade. On the other hand, it has to be argued that hardly any (traditional) scientific basis can be found for his ideas. Furthermore, it is not clear how these ideas can be put into practice. Would it be a good thing to implement strict norms in environmental policies, or does this not make sense, as Gaia is absolutely able to balance the attacks of modern industries and to counteract their negative effects?

Lovelock's ideas should be placed in the context of the debate at the end of the sixties and the seventies in which the attitude towards environmental pollution and how to deal with it was at the core of the societal debate. Later on in the eighties, environmental policies became much more accepted as 'normal' phenomena in society, which reduced the relevance of Lovelock's ideas.

Concluding remarks

The bestsellers were not only important from a quantitative point of view, namely the number of copies sold in bookshops, but also from a qualitative point of view. The ideas formulated in the books from this period became the hard core of scientific reasoning in environmental studies. These books were responsible for the scientific definition of environmental problems, for the way these problems were viewed, the methodology used in environmental studies, and for the close relationship between scientific analysis and political recommendations. The great advantage of the bestsellers was that they were written for a large audience, that they were published in paperback, and that they were promoted by the mass media. Most of these publications became a sort of bible for environmentalists. They used scientific reasoning and tried to convince non-environmentalists of the importance of considering actual environmental problems from an ecological point of view. Names such as Commoner, Meadows, Boulding, Tinbergen, and Lovelock became known all over the world, as did their scientific approach to global problems.

References
Baker, J.P.C. and W.A. Macfarlane, 1961. Fuel Selection and Utilization, World Health Organization, *Air Pollution*, pp. 345-363.
Daly, H.E. and J.B. Cobb Jr., 1990. *For the Common Good*, Green Print, London.
Mesarovic, M. and E. Pestel, 1974. *Mankind at the Turning Point*, New York.
Odèn, S., 1968. *Nederbördens och Luftens Försurning dess Orsaker, Förlopp och Verkan i Olika Miljöer*. Statens Naturvetenskopliga Forskningsråd. Ekologikomitteën, Bulletin 1.
OECD, 1977. *The OECD Programme on Long Range Transport of Air Pollutants; Measurements and Findings*, Paris, OECD.

CHAPTER 13

The Closing Circle

Nature, Man and Technology

B. COMMONER

The Economic Meaning of Ecology

A statistician, Daniel Fife, has recently made an interesting observation that helps to explain this paradoxical relationship between the profitability of a business and its tendency to destroy its own environmental base. His example is the whaling industry, which has been driving itself out of business by killing whales so fast as to ensure that they will soon become extinct. Fife refers to this kind of business operation as "irresponsible", in contrast with a "responsible" operation, which would only kill whales as fast as they can reproduce. He points out that even though the irresponsible business will eventually wipe itself out, it *may be profitable to do so*—at least for the entrepreneur, if not for society—if the extra profit derived from the irresponsible operation if high enough to yield a return on investment elsewhere that outweighs the ultimate effect of killing off the whaling business. To paraphrase Fife, the "irresponsible" entrepreneur finds it profitable to kill the goose that lays the golden eggs, so long as the goose lives long enough to provide him with sufficient eggs to pay for the purchase of a new goose. Ecological irresponsibility can pay—for the entrepreneur, but not for society as a whole.

The crucial link between pollution and profits appears to be modern technology, which is both the main source of recent increases in productivity—and therefore of profits—and of recent assaults on the environment. Driven by an inherent tendency to maximize profits, modern private enterprise has seized upon those massive technological innovations that promise to gratify this need, usually unaware that these same innovations are often also instruments of environmental destruction. Nor is this surprising, for, as shown earlier (Chapter 10), technologies tend to be designed at present as single-purpose instruments. Apparently, this purpose is, unfortunately, too often dominated by the desire to enhance productivity—and therefore profit.

Obviously, we need to know a great deal more about the connection between pollution and profits in private enterprise economies. Meanwhile, it would be prudent to give some thought to the meaning of the functional connection between pollution and profits, which is at least suggested by the present information.

The general proposition that emerges from these considerations is that environmental pollution is connected to the economics of the private enterprise system in two ways. First, pollution tends to become intensified by the displacement of older productive techniques by new, ecologically faulty, but more profitable technologies. Thus, in these cases, pollution is an unintended concomitant of the natural drive of the economic system to introduce new technologies that increase productivity. Second, the cost of environmental degradation are chiefly borne not by the producer, but by society as a whole, in the form of "externalities". A business enterprise that pollutes the environment is therefore being subsidized by society; to this extent, the enterprise, though free, is not wholly private.

If the course of environmental degradation is to be reserved, these relationships need to be changed. To begin with, the environmental costs must be met by introducing the needed changes in the processes of production. In a private enterprise system this means, necessarily, that costs, however they are ultimately met, must be introduced into the system through the producer's enterprise. The new, highly polluting technologies will, of course, be more seriously affected by these changes than the relatively low-impact technologies they have displaced. Thus, the added costs would have a larger effect on producer of detergents than on soap manufacturers, and on trucks more than railroads.

Now, the new, highly polluting technologies also represent a greater source of the over-all growth in productivity of the economic system than do the technologies they displace, as shown by their higher profits and rates of growth. However, when environmentally required changes in technology are imposed upon these highly productive enterprises, these activities do *not* thereby gain in productivity. This in contrast to the effect of introducing *conventional* new productive technology, which is always motivated by, and usually achieves, an increase in productivity. Thus, no matter how pollution costs are ultimately met, if they are imposed initially through the producer they will not contribute to the over-all growth of productivity. This has been pointed out by the economist G.F. Bloom, who finds, in a study of productivity:

> Pollution controls . . . will add millions of dollars to industry's costs of production and will pour additional purchasing power into the income stream without increasing productivity, as conventionally measured. Indeed, carrying on production without fouling the air and polluting the water may require an actual reduction in man-hour output [and therefore a reduction in productivity] in some industries.

The economics seem clear: the technology required for pollution controls, unlike ordinary technology, does not *add* to the value of the output of saleable goods. Hence, the extensive technological reform of agricultural and industrial production that is now demanded by the environmental crisis cannot contribute to the growth of productivity—to the continued expansion of the GNP. Bloom concludes that in part due to heightened environmental concern, the "prospect for increased productivity is therefore not bright". Since continued increase in productivity is closely linked to profitability, it is essential to the health of a private enterprise economy. Therefore there appears to be a basic conflict between pollution control and what is often regarded as a fundamental requirement of the private enterprise system—the continued maximization of productivity. Bloom sees a grave, underappreciated danger to the economic system in all this and concludes, pessimistically: "As far as productivity is concerned, there seems to be little awareness of the urgency of this problem . . . Business underestimated the power of consumerism; likewise, it discounted the drive against pollution."

Another difficulty is that in certain important ways the stress on the environment due to ecological faulty productive technology, so long as it is tolerated, seems to operate, for a time, to the advantage of the producer and to the disadvantage of the population as a whole. This situation arises out of certain time-dependent characteristics of ecological degradation and relates to a crucial feature of the private enterprise economic system—the competition between the entrepreneur's drive for maximal profit and the wage earner's interest in increased wages.

Thus, let us say, if through government regulation, the producer is prevented from passing the added costs of environmental control along to the consumer, he will need to find an alternative means of cutting general production costs in order to maintain profits. The obvious recourse is to reduce wages; this would, of course, exacerbate the conflict between entrepreneur and wage earner. On the other hand, if the added costs are met by raising prices, then the wage earner is confronted with a rising cost of living which will naturally lead him to demand higher wages; again the conflict is intensified. Moreover, increased prices would inevitably burden the poor most heavily. For example, given that present agricultural practice has heavily mortgaged soil and water ecosystems, their ecological reform would result in a very sharp increase in food prices. Inevitably, the poor would suffer most. Thus, the attempt to meet the real, social costs of environmental degradation, either through increased prices or reduced wages, would appear to intensify the long-standing competition between capital and labor over the division of the wealth produced by the private enterprise system and worsen the already intolerable incidence of poverty.

That the productive system as a whole "borrows" from the ecosystem and incurs the "debt to nature" represented by pollution is an immediate saving for the producer. At the same time, pollution often adds to the living costs of the population as a whole, most of whom are wage earners rather than entrepreneurs. Thus, when the workers in the vicinity of a power plant find their laundry bills increased because of soot emitted by its stacks, their wages are thereby reduced. In a sense, the workers' extra laundry bill subsidizes part of the cost of operating the power plant. In this hidden way, environmental deterioration erodes the wage earners' real wages.

However, some of these effects do not occur concurrently. For example, it may take from fifteen to twenty years of environmental pollution from, say, industrial plants along the shore of Lake Erie before the burden of waste reduces the water's oxygen content to zero, halts the self-purification process, and fouls the beaches—so that to continue to enjoy summer recreation, the plants' workers need to add to the cost of living the price of admission to a swimming pool. Similarly, chronic, lowlevel exposure to radiation, mercury, or DDT may shorten a wage earner's life without reducing his income or even incurring extra medical costs during his lifetime. In this case, the cost of pollution is not met by anyone for a long time; the bill is finally paid by exacting the wage earner's premature death, which—apart from the incalculable human anguish—can be reckoned in terms of some number of years of lost income. In this situation, then, during the "free" period, pollutants accumulate in the ecosystem or in a victim's body, but not all the resultant costs are immediately felt. Part of the value represented by the free abuse of the environment is then available to mitigate the economic conflict between capital and labor. The benefit *appears* to accrue to both parties and the conflict between them is reduced. Later, however, when the environmental bill is paid, it is met by labor more than by capital; the buffer is suddenly removed and the conflict between these two economic sectors is revealed in its full force.

Another way to look at this situation relates to the value of the capital created by the operation of the private enterprise system. In the creation of this capital, certain goods are regarded as freely and continuously available from nature: the fertility of the soil, oxygen, water—in general, nature, or the biological capital represented by the ecosphere. However, the environmental crisis tells us that these goods are no longer freely available, and that when they are treated as though they were, they are progressively degraded.

This suggests that we need to consider the true value of the conventional capital accumulated by the operation of the economic system. The effect of the operation of the system on the value of its *biological* capital needs to be taken into account in order to obtain a true estimate of the over-all wealth-producing capability of the system. The course of environmental deterioration shows that as conventional capital has accumulated, for example in the United States since 1946, the value of the biological capital has *declined*. Indeed, if the process continues, the biological capital may eventually be driven to the point of total destruction. Since the usefulness of conventional capital—the ecosystem—when the latter is destroyed, the usefulness of the former is also destroyed.

Thus despite its apparent prosperity, in reality the system is being driven into bankruptcy. Environmental degradation represents a crucial, potentially fatal, *hidden* factor in the operation of the economic system.

It should be evident, from nearly everything that has been said in this book, that no economic system can be regarded as stable if its operation strongly violates the principles of ecology. To what extent is this true of present economic systems?

In the case of the private enterprise system, this question has already been answered in part, for there does seem to be a tendency for that system to enhance productivity—and therefore profits—by means of technologies that also intensify environmental stress. A more theoretical basis for the incompatibility between the private enterprise system and the ecosystem relates to the matter of growth.

The total rate of exploitation of the earth's ecosystem has some upper limit, which reflects the intrinsic limit of the ecosystem's turnover rate. If this rate is exceeded, the system is eventually driven to collapse. This is firmly established by everything that we know about ecosystems. Hence it follows that there is an upper limit to the rate of exploitation of the biological capital on which any productive system depends. Since the rate of use of this biological capital cannot be exceeded without destroying it, it also follows that the actual rate of use of the *total* capital (i.e. biological capital plus conventional capital, or the means of production) is also limited. Thus there must be some limit to the growth of total capital, and the productive system *must* eventually reach a "no-growth" condition, at least with respect to the accumulation of capital goods designed to exploit the ecosystem, and the product which they yield.

In a private enterprise system, the no-growth condition means no further accumulation of capital. If, as seems to be the case, accumulation of capital, through profits, is the basic driving force of this system, it is difficult to see how it can continue to operate under conditions of no growth. At this point, it can be argued that some new form of growth can be introduced, such as increases in services. However, nearly all services represent the resultant of human labor expended through the agency of some form of capital goods. Any increase of services designed to achieve economic growth would have to be accomplished without increasing the amount of these service-orientated capital goods, if the ecological requirements are to be met.

The ecosystem poses another problem for the private enterprise system. Different ecological cycles vary considerably in their natural, intrinsic rates—which cannot be exceeded if breakdown is to be avoided. Thus, the natural turnover rate of the soil system is considerably lower than the intrinsic rate of an aquatic system (e.g. a fish farm). It follows, then, that if these different ecosystems are to be exploited concurrently by the private enterprise system without inducing ecological breakdown, they must operate at differential rates of economic return. However, the free operation of the private enterprise system tends to maximize rates of return from different enterprises. If a given enterprise yields a return less than that available from another one, investment funds will tend to be transferred to the latter. "Marginal" enterprises, i.e., operations that yield a profit significantly below that available elsewhere in the economic system, are eventually dropped. However, in ecological terms an enterprise which is based on an ecosystem with a relatively slow turnover rate is *necessarily* economically "marginal"—if it is to operate without degrading the environment. Such enterprises are of obvious *social* value, but, given the profit-maximizing tendency of the private enterprise system, are not likely to be operated for long. A corrective expedient is the provision of subsidies; but in some cases these may need to be large enough to amount to nationalization — a contradiction of private enterprise.

Commoner, B., *The Closing Circle; Nature, Man and Technology*, Knopf, New York, 1971, p. 267-276.

CHAPTER 14

The Limits to Growth

A Global Challenge; a Report for the Club of Rome Project on the Predicament of Mankind

D. MEADOWS

As the world system grows towards its ultimate limits, what will be its most likely behavior mode? What relationships now existent will change as the exponential growth curves level off? What will the world be like when growth comes to an end?

There are, of course, many possible answers to these questions. We will examine several alternatives, each dependent on a different set of assumptions about how human society will respond to problems arising from the various limits to growth.

Let us begin by assuming that there will be in the future no great changes in human values nor in the functioning of the global population-capital system as it has operated for the last one hundred years. The results of this assumption are shown in figure 35. We shall refer to this computer output as the "standard run" and use it for comparison with the runs based on other assumptions that follow. The horizontal scale in figure 35 shows time in years from 1900 to 2100. With the computer we have plotted the progress over time of eight quantities:

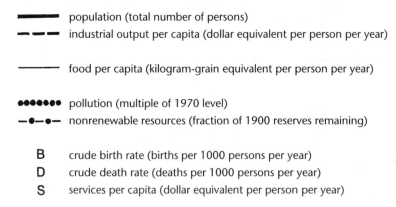

population (total number of persons)

industrial output per capita (dollar equivalent per person per year)

food per capita (kilogram-grain equivalent per person per year)

pollution (multiple of 1970 level)

nonrenewable resources (fraction of 1900 reserves remaining)

B   crude birth rate (births per 1000 persons per year)
D   crude death rate (deaths per 1000 persons per year)
S   services per capita (dollar equivalent per person per year)

Each of these variables is plotted on a different vertical scale. We have deliberately omitted the vertical scales and we have made the horizontal time scale somewhat vague because we want to emphasize the general behavior modes of these computer outputs, not the numerical values, which are only approximately known. The scales are, however, exactly equal in all the computer runs presented here, so results of different runs may be easily compared.

FIGURE 35   World Model Standard Run

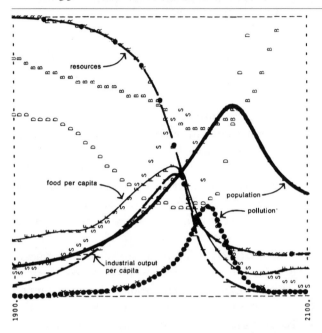

The "standard" world model run assumes no major change in the physical, economic, or social relationship that have historically governed the development of the world system. All variables plotted here follow historical values from 1900 to 1970. Food, industrial output, and population grow exponentially until the rapidly diminishing resources base forces a slowdown in industrial growth. Because of natural delays in the system, both population and pollution continue to increase for some time after the peak of industrialization. Population growth is finally halted by a rise in the death rate due to decreased food and medical services.

All levels in the model (population, capital, pollution, etc.) begin with 1900 values. From 1900 to 1970 the variables plotted in figure 35 (and numerous other variables included in the model but not plotted here) agree generally with their historical values to the extent that we know them. Population rises from 1.6 billion in 1900 to 3.5 billion in 1970. Although the birth rate declines gradually, the death rate falls more quickly, especially after 1940, and the rate of population growth increases. Industrial output, food, and services per capita increase exponentially. The resource base in 1970 is still about 95 per cent of its

1900 value, but it declines dramatically thereafter, as population and industrial output continue to grow.

The behavior more of the system shown in figure 35 is clearly that of over-shoot and collapse. In this run the collapse occurs because of nonrenewable resource depletion. The industrial capital stock grows to a level that requires an enormous input of resources. In the very process of that growth it depletes a large faction of the resources reserves available. As resource prices rise and mines are depleted, more and more capital must be used for obtaining resources, leaving less to be invested for future growth. Finally investment cannot keep up with depreciation, and the industrial base collapses, taking with it the service and agricultural systems, which have become dependent on industrial inputs (such as fertilizers, pesticides, hospital laboratories, computers, and especially energy for mechanization). For a short time the situation is especially serious because population, with the delays inherent in the age structure and the process of social adjustment, keeps rising. Population finally decreases when the death rate is driven upward by lack of food and health services.

The exact timing of these events is not meaningful, given the great aggregation and many uncertainties in the mode. It is significant, however, that growth is stopped well before the year 2100. We have tried in every doubtful case to make the most optimistic estimate of unknown quantities, and we have also ignored discontinuous events such as wars or epidemics, which might act to bring an end to growth even sooner than our model would indicate. In other words, the model is biased to allow growth to continue longer than it probably can continue in the real world. *We can thus say with some confidence that, under the assumption of no major change in the present system, population and industrial growth will certainly stop within the next century, at the latest.*

The system shown in figure 35 collapses because of a resource crisis. What if our estimate of the global stock of resources is wrong? In figure 35 we assumed that in 1970 there was a 250-year supply of all resources, at 1970 usage rates. The static reserve index column of the resource table in chapter II will verify that this assumption is indeed optimistic. But let us be even more optimistic and assume that new discoveries or advances in technology can *double* the amount of resources economically available. A computer run under that assumption is shown in figure 36.

The overall behavior mode in figure 36—growth and collapse—is very similar to that in the standard run. In this case the primary force that stops growth is a sudden increase in the level of pollution, caused by and overloading of the natural absorptive capacity of the environment. The death rate rises abruptly from pollution and from lack of food. At the same time resources are severely depleted, in spite of the doubled amount available, simply because a few more years of exponential growth in industry are sufficient to consume those extra resources.

FIGURE 36 World Model with Natural Resource Reserves Doubled

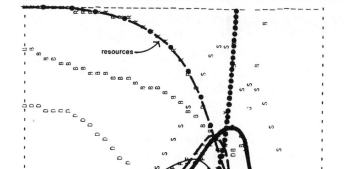

To test the model assumption about available resources, we doubled the resource reserves in 1900, keeping all other assumptions identical to those in the standard run. Now industrialization can reach a higher level since resources are not so quickly depleted. The larger industrial plant releases pollution at such a rate, however, that the environmental pollution absorption mechanisms become saturated. Pollution rises very rapidly, causing an immediate increase in the death rate and a decline in food production. At the end of the run resources are severely depleted in spite of the doubled amount initially available.

Is the future of the world system bound to be growth and then collapse into a dismal, depleted existence? Only if we make the initial assumption that our present way of doing things will not change. We have ample evidence of mankind's ingenuity and social flexibility. There are, of course, many likely changes in the system, some of which are already taking place. The Green Revolution is raising agricultural yields in nonindustrialized countries. Knowledge about modern methods of birth control is spreading rapidly. Let us use the world model as a tool to test the possible consequences of the new technologies that promise to raise the limits to growth.

Meadows, D., *The Limits to Growth, A Global Challenge; a Report for the Club of Rome Project on the Predicament of Mankind,* Universe Books, New York, 1972, p. 122-128.

CHAPTER 15

A Blueprint for Survival

E. GOLDSMITH

The Need for Change

The principal defect of the industrial way of life with its ethos of expansion is that it is not sustainable. Its termination within the lifetime of someone born today is inevitable—unless it continues to be sustained for a while longer by an entrenched minority at the cost of imposing great suffering on the rest of mankind. We can be certain, however, that sooner or later it will end (only the precise time and circumstances are in doubt) and that it will do so in one of two ways: either against our will, in a succession of famines, epidemics, social crisis and wars; or because we want it to—because we wish to create a society which will not impose hardship and cruelty upon our children—in a succession of thoughtful, humane and measured changes. We believe that a growing number of people are aware of this choice, and are more interested in our proposals for creating a sustainable society than in yet another recitation of the reasons why this should be done. We will therefore consider these reasons only briefly, reserving a fuller analyses for the four appendices which follow the *Blueprint* proper.

Radical change is both necessary and inevitable because the present increases in human numbers and *per capita* consumption, by disrupting ecosystems and depleting resources, are undermining the very foundations of survival. At present the world population of 3,600 million is increasing by 2 per cent per year (72 million), but this overall figure conceals crucially important differences between countries. The industrialized countries with one third of the world population have annual growth rates of between 0.5 and 1.0 per cent; the underdeveloped countries on the other hand, with two thirds of the world population, have annual growth rates of between 2 and 3 per cent, and from 40 to 45 per cent of their population is under 15. It is commonly overlooked that in countries with an unbalanced age structure of this kind the population will continue to increase for many years even after fertility has fallen to the replacement level. As the Population Council has pointed out: 'If replacement is achieved in the developed world by 2000 and in the developing world by 2040, then the world's population will stabilise at nearly 15.5 billion (15,500 million) about a century hence, or well over four times the present size.'

The *per capita* use of energy and raw material also shows a sharp division between the developed and the underdeveloped parts of the world. Both are increasing their use of these commodities, but consumption in the developed countries is so much higher that, even with their smaller share of the population, their consumption may well represent over 80 per cent of the world total. For the same reason, similar percentages increases are far more significant in the developed countries; to take one example, between 1957 and 1967 *per capita* steel consumption rose by 12 per cent in the US and by 41 per cent in India, but the actual increases (in kg per year) were from 568 to 634 and from 9.2 to 13 respectively. Nor is there any sign that an eventual end to economic growth is envisaged, and indeed industrial economies appear to break down if growth ceases or even slows, however high the absolute level of consumption. Even the US still aims at an annual growth of GNP of 4 per cent or more. Within this overall figure much higher growth rates occur for the use of particular resources, such as oil.

The combination of human numbers and *per capita* consumption has a considerable impact on the environment, in terms of both the resources we take from it and the pollutants we impose on it. A distinguished group of scientists, who came together for a Study of Critical Environmental Problems (SCEP) under the auspices of the Massachusetts Institute of Technology, state in their report the clear need for a means of measuring this impact, and have coined the term 'ecological demand', which they define as 'a summation of all man's demands on the environment, such as the extraction of resources and the return of wastes'. Gross Domestic Product (GDP), which is population multiplied by material standard of living, appears to provide most convenient measure of ecological demand, and according to the UN *Statistical Yearbook* this is increasing annually by 5 to 6 per cent, or doubling every 13.5 years. If this trend should continue, then in the time taken for world population to double (which is estimated to be by just after the year 2000) total ecological demand will have increased by a factor of six. SCEP estimate that 'such demand-producing activities as agriculture, and mining and industry have global annual rates of increase of 3.5 per cent and 7 per cent respectively. An integrated rate of increase is estimated to be between 5 and 6 per cent per year, in comparison with an annual rate of population increase of only 2 per cent.

It should go without saying that the world cannot accommodate this continued increase in ecological demand. *Indefinite* growth of whatever type cannot be sustained by *finite* resources. This is the nub of the environmental predicament. It is still less possible to maintain indefinite *exponential* growth—and unfortunately the growth of ecological demand is proceeding exponentially (i.e. it is increasing geometrically, by compound interest).

The implications of exponential growth are not generally appreciated and are well worth considering. As Professor Forrester explains it:

> . . . pure exponential growth possesses the characteristic of behaving according to a 'doubling time'. Each fixed time interval shows a doubling of the relevant system

variable. Exponential growth is treacherous and misleading. A system variable can continue through many doubling intervals without seeming to reach significant size. But then in one or two more doubling periods, still following the same law of exponential growth, it suddenly seems to become overwhelming.[1]

Thus, supposing world petroleum reserves stood at 2,100 billion barrels, and supposing our rate of consumption was increasing by 6.9 per cent per year, then, as can be seen from Figure 1, demand will exceed supply by the end of the century. What is significant, however, is not the speed at which such vast reserves can be depleted, but that as late as 1975 there will appear to be reserves fully ample enough to last for considerably longer. Such a situation can easily lull one into a false sense of security and the belief that a given growth rate can be sustained, if not indefinitely, at least for a good deal longer than is actually the case.[2]

FIGURE I World reserves of crude petroleum at exponential rate of consumption.

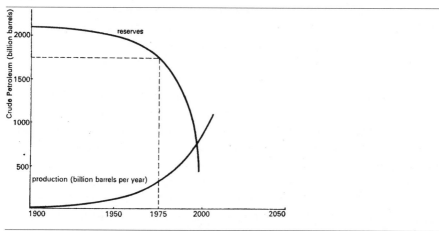

Note that in 1975, with no more than fifteen years left before demand exceeds supply, the total global reserve has been depleted by only 12.5 per cent.

The same basic logic applies to the availability of any resource including land, and it is largely because of this particular dynamic of exponential growth that the environmental predicament has come upon us so suddenly, and why its solution requires urgent and radical measures, many of which run counter to values which, in our industrial society, we have been taught to regard as fundamental.

If we allow the present growth rate to persist, total ecological demand will increase by a factor of 32 over the next 66 years—and there can be no serious person today willing to concede the possibility, or indeed the desirability, of our accommodating the pressures arising from such growth. For this can be done only at the cost of disrupting ecosystems and exhausting resources, which must lead to the failure of food supplies and the collapse of society. It is worth briefly considering each in turn.

Disruption of Ecosystems

We depend for our survival on the predictability of ecological processes. If they were at all arbitrary, we could not know when to reap or sow, and we would be at the mercy of environmental whim. We could learn nothing about the rest of nature, advance no hypotheses, suggest no 'laws'. Fortunately, ecological processes *are* predictable, and, although theirs is a relatively young discipline, ecologists have been able to formulate a number of important 'laws', one of which in particular relates to environmental predictability: namely, that all ecosystems tend toward stability, and further that the more diverse and complex the ecosystem the more stable it is; that is, the more species there are, and the more they interrelate, the more stable is their environment. By stability is meant the ability to return to the original position after any change, instead of being forced into a totally different pattern—and hence predictability.

Unfortunately, we behave as if we knew nothing of the environment and had no conception of its predictability, treating it instead with scant and brutal regard as if it were and idiosyncratic and extremely stupid slave. We seem never to have reflected on the fact that a tropical rain forest supports innumerable insect species and yet is never devastated by them; that its rampant luxuriance is not contingent on our overflying it once a month and bombarding it with insecticides, herbicides, fungicides, and what-have-you. And yet we tremble over our wheatfields and cabbage patches with a desperate battery of synthetic chemicals, in an absurd attempt to impede the operation of the immutable 'law' we have just mentioned—that all ecosystems tend towards stability, therefore diversity and complexity, therefore a growing number of different plant and animal species until a climax or optimal condition is achieved. If we were clever, we would recognize that successful long-term agriculture demands the achievement of an artificial climax, and imitation of the pre-existing ecosystem, so that the level of unwanted species could be controlled by those that did no harm to the crop-plants.

Instead we have put our money on pesticides, which although they have been effective have been so only to a limited and now diminishing extent: according to SCEP, the 34 per cent increase in world food production from 1951 to 1966 required increased investments in nitrogenous fertilizers of 146 per cent and in pesticides of 300 per cent. At the same time they have created a

number of serious problems, notably resistance—some 250 pest species are re-
sistant to one group of pesticides or another, while many others require increased applications to keep their populations within manageable proportions—and the promotion of formerly innocuous species to pest proportions,
because the predators that formerly kept them down have been destroyed. The
spread of DDT and other organochlorines in the environment has resulted in
alarming population declines among woodcock, grebes, various birds of prey
and seabirds, and in a number of fish species, principally the sea trout. SCEP
comments:

> The oceans are an ultimate accumulation site of DDT and its residues. As much as
> 25 per cent of the DDT compounds produced to date may have been transferred to
> the sea. The amount in the marine biota is estimated to be in the order of less than
> 0.1 per cent of total production and has already produced a demonstrable impact
> upon the marine environment . . . The decline in productivity of marine food fish
> and the accumulation of levels of DDT in their tissues which are unacceptable to
> man can only be accelerated by DDT's continued release to the environment.

There are half a million man-made chemicals in use today, yet we cannot predict the behaviour of properties of the greater part of them (either singly or in
combination) once they are released into the environment. We know, however,
that the combined effects of pollution and habitat destruction menace the survival of no fewer than 280 mammal, 350 bird, and 20,000 plant species. To
those who regret these losses but greet them with the comment that the survival of *Homo sapiens* is surely more important than that of an eagle or a prim-
rose, we repeat the *Homo sapiens* himself depends on the continued resilience
of those ecological networks of which eagles and primroses are integral parts.
We do not need to destroy the ecosphere utterly to bring catastrophe upon ourselves: all we have to do is to carry on as we are, clearing forests, 'reclaiming'
wetlands, and imposing sufficient quantities of pesticides, radioactive material,
plastics, sewage, and industrial wastes upon our air, water and land systems to
make them inhospitable to the species on which their continued stability and
integrity depend. Industrial man in the world today is like a bull in a china
stop, with the single difference that a bull with half the information about the
properties of china as we have about those of ecosystems would probably try
and adapt its behaviour to its environment rather than the reverse. By contrast,
Homo sapiens industrialis is determined that the china shop should adapt to him,
and has therefore set himself the goal of reducing it to rubble in the shortest
possible time.

Failure of Food Supplies
Increases in food production in the undeveloped world have barely kept

abreast of population growth. Such increases as there have been are due not to higher productivity but to the opening up of new land for cultivation. Unfortunately this will not be possible for much longer: all the good land in the world is now being farmed, and according to the FAO[3] at present rates of expansion none of the marginal land that is left will be unfarmed by 1985—indeed some of the land now under cultivation has been so exhausted that it will have to be returned to permanent pasture.

For this reason, FAO's programme to feed the world depends on a programme of intensification, at the heart of which are the new high-yield varieties of wheat and rice. These are highly responsive to inorganic fertilizers and quick-maturing, so that up to ten times present yields can be obtained from them. Unfortunately, they are highly vulnerable to disease, and therefore require increased protection by pesticides, and of course they demand massive inputs of fertilizers (up to 27 times present ones). Not only will these disrupt local ecosystems, thereby jeopardizing long-term productivity, but they force hard-pressed undeveloped nations to rely on the agro-chemical industries of the developed world.

Whatever their virtues and faults, the new genetic hybrids are not intended to solve the world food problem, but only to give us time to devise more permanent and realistic solutions. It is our view, however, that these hybrids are not the best means of doing this, since their use is likely to bring about a reduction in overall diversity, when the clear need is to develop an agriculture diverse enough to have long-term potential. We must beware of those 'experts' who appear to advocate the transformation of the ecosphere into nothing more than a food-factory for man. The concept of a world consisting solely of man and a few favoured food plants is so ludicrously impracticable as to be contemplated seriously only by those who find solace in their own wilful ignorance of the real world of biological diversity.

We in Britain must bear in mind that we depend on imports for half our food, and that we are unlikely to improve on this situation. The 150,000 acres which are lost from agriculture each year are about 70 per cent more productive than the average for all enclosed land[4], while are already beginning to experience diminishing returns from the use of inorganic fertilizers. In the period 1964-9, application of phosphates have gone up by 2 per cent, potash by 7 per cent, and nitrogen by 40 per cent[5], yet yields per acre of wheat, barley, lucerne and temporary grass have levelled off and are beginning to decline, while that of permanent grass has risen only slightly and may be levelling off.[6] As *per capita* food availability declines throughout the rest of the world, and it appears inevitable it will, we will find it progressively more difficult and expensive to meet our food requirements from abroad. The prospect of severe food shortages within the next thirty years is not so much a fantasy as that of continued abundance promised us by so many of our politicians.

FIGURE 2 Mineral resources: static and exponential reserves[7]

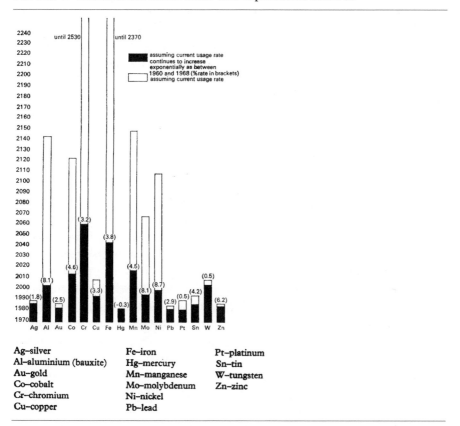

Ag–silver
Al–aluminium (bauxite)
Au–gold
Co–cobalt
Cr–chromium
Cu–copper

Fe–iron
Hg–mercury
Mn–manganese
Mo–molybdenum
Ni–nickel
Pb–lead

Pt–platinum
Sn–tin
W–tungsten
Zn–zinc

Exhaustion of Resources

As we have seen, continued exponential growth of consumption of materials and energy is impossible. Present reserves of all but a few metals will be exhausted within 50 years, if consumption rates continue to grow as they are (see Figure 2). Obviously there will be new discoveries and advances in mining technology, but these are likely to provide us with only a limited stay of execution. Synthetics and substitutes are likely to be of little help, since they must be made from materials which themselves are in short supply; while the hoped-for availability of unlimited energy would not be the answer, since the problem is the ratio of useful metal to waste matter (which would have to be disposed of without disrupting ecosystems), not the need for cheap power. Indeed, the availability of unlimited power holds more of a threat than a promise, since energy use is inevitably polluting, and in addition we would ultimately have to face the problem of disposing of an intractable amount of waste heat.

Collapse of Society

The developed nations consume such disproportionate amounts of protein, raw materials and fuels that unless they considerably reduce their consumption there is no hope of the undeveloped nations markedly improving their standards of living. This vast differential is a cause of much and growing discontent, made worse by our attempts at cultural uniformity on behalf of an expanding market economy. In the end, we are altering people's aspirations without providing the means for them to be satisfied. In the rush to industrialize we break up communities, so that the controls which formerly regulated behaviour are destroyed before alternatives can be provided. Urban drift is one result of this process, with a consequent rise in anti-social practices, crime, delinquency, and so on, which are so costly for society in terms of both of money and of well-being.

At the same time, we are sowing the seeds of massive unemployment by increasing the ratio of capital to labour so that the provision of each job becomes ever more expensive. In a world of fast-diminishing resources, we shall quickly come to the point when very great numbers of people will be thrown out of work, when the material compensations of urban life are either no longer available or prohibitively expensive, and consequently when whole sections of society will find good cause to express their considerable discontent in ways likely to be anything but pleasant for their fellows.

It is worth bearing in mind that the barriers between us and epidemics are not so strong as is commonly supposed. Not only is it increasingly difficult to control the vectors of disease, but it is more than probable that urban populations are being insidiously weakened by overall pollution levels, even when they are not high enough to be incriminated in any one illness. At the same time international mobility speeds the spread of disease. With this background, and at a time of widespread public demoralization, the collapse of vital social services such as power and sanitation could easily provoke a series of epidemics—and we cannot say with confidence that we would be able to cope with them.

At times of great distress and social chaos, it is more than probable that governments will fall into the hands of reckless and unscrupulous elements, who will not hesitate to threaten neighbouring governments with attack if they feel that they can wrest from them a larger share of the world's vanishing resources. Since a growing number of countries (an estimated 36 by 1980) will have nuclear power stations, and therefore sources of plutonium for nuclear warheads, the likelihood of a whole series of local (if not global) nuclear engagements is greatly increased.

Conclusion

A fuller discussion of ecosystems and their disruption, of social systems and their disruption, of population and food supply, and of resources and their depletion, can be found in Appendices A, B, C, D, respectively. There will be those who regard these accounts of the consequences of trying to accommodate present growth rates as fanciful. But the imaginative leap from the available scientific information to such predictions is negligible, compared with that required for those alternative predictions, laughably considered 'optimistic', of a world of 10,000 to 15,000 million people, all with the same material standard of living as the US, on a constant replica of this planet, the only moving parts being their machines and possibly themselves. Faced with inevitable change, we have to make decisions, and we must make these decisions *soberly* in the light of the best information, and not as if we were caricatures of the archetypal mad scientist.

By now it should be clear that the main problems of the environment do not arise from temporary and accidental malfunctions of existing economic and social systems. On the contrary, they are the warning signs of a profound incompatibility between deeply rooted beliefs in continuous growth and the dawning recognition of the earth as a space ship, limited in its resources and vulnerable to thoughtless mishandling. The nature of our response to these symptoms is crucial. If we refuse to recognize the cause of our trouble the result can only be increasing disillusion and growing strain upon the fragile institutions that maintain external peace and internal social cohesion. If, on the other hand, we can respond to this unprecedented challenge with informed and constructive action the rewards will be as great as the penalties for failure.

We are sufficiently aware of 'political reality' to appreciate that many of the proposals we will make in the next chapter will be considered impracticable. However, we believe that if a strategy for survival is to have any chance of success, the solutions must be formulated in the light of the problems and not from a timorous and superficial understanding of what may or may not be immediately feasible. If we plan remedial action with our eyes on political rather than ecological reality, then very reasonably, very practicably, and very surely, we will muddle our way to extinction.

A measure of political reality is that government has yet to acknowledge the impending crisis. This is to some extent because it has given itself no machinery for looking at energy, resources, food, environmental disruption and social disruption as a whole, as part of a general, global pattern, preferring instead to deal with its many aspects as if they were self-contained analytical units. Lord Rothchild's Central Policy Review Staff in the Cabinet Office, which is the only body in government which might remedy the situation, appears not to think it worthwhile: at the moment at least, they are undertaking 'no specific studies on the environment that would require an environmentalist or ecologist'. There is a strong element of positive feedback here, in that there can be no appreciation of our predicament unless we view it in totality, and yet government can see no cause to do so unless it can be shown that such a predicament exists.

Possibly because government sees the world in fragments and not as a totality, it is difficult to detect in its actions or words any coherent general policy, although both major political parties appear to be mesmerized by two dominating notions: that economic expansion is essential for survival and is the best possible index of progress and well-being; and that unless solutions can be devised that do not threaten this notion, then the problems should not be regarded as existing. Unfortunately, government has an increasingly powerful incentive for continued expansion in the tendency for economic growth to create the need for more economic growth. This it does in six ways:

Firstly, the introduction of technological devices, i.e. the growth of the technosphere, can only occur to the detriment of the ecosphere, which means that it leads to the destruction of natural controls which must then be replaced by further technological ones. It is in this way that pesticides and artificial fertilizers create the need for yet more pesticides and artificial fertilizers.

Secondly, for various reasons, industrial growth, particularly in its earlier phases, promotes population growth. Even in its later phases, this can still occur at a high rate (0.5 per cent in the UK). Jobs must constantly be created for the additional people—not just any job, but those that are judged acceptable in terms of current values. This basically means that the capital outlay per person employed must be maintained, otherwise the level of 'productivity' per man will fall, which is a determinant of both the 'viability' of economic enterprise and of the 'standard of living'.

Thirdly, no government can hope to survive widespread and protracted unemployment, and without changing the basis of our industrial society, the only way government can prevent it is by stimulating economic growth.

Fourthly, business enterprises, whether state-owned or privately owned, tend to become self-perpetuating, which means that they require surpluses for further investment. This favours continued growth.

Fifthly, the success of a government and its ability to obtain support is to a large extent assessed in terms of its ability to increase the 'standard of living' as measured by *per capita* gross national product (GNP).

Finally, confidence in the economy, which is basically a function of its ability to grow, must be maintained to ensure a healthy state of the stock market. Were confidence to fail, stock values would crash, drastically reducing the availability of capital for investment and hence further growth, which would lead to further unemployment. This would result in a further fall in stock-market values and hence give rise to a positive-feedback chain-reaction, which under the existing order might well lead to social collapse.

For all these reasons, we can expect our government (whether Conservative or Labour) to encourage further increases in GNP regardless of the consequences, which in any case tame 'experts' can be found to play down. It will curb growth only when public opinion demands such a move, in which case it will be politically expedient, and when a method is found for doing so without creating unemployment or excessive pressure on capital. We believe this is possible only within the framework of a fully integrated plan.

The emphasis must be on integration. If we develop relatively clean technologies but do not end economic growths then sooner or later we will find ourselves with as great a pollution problem as before but without the means of tackling it. If we stabilize our economies and husband our non-renewable resources without stabilizing our populations we will find we are no longer able to feed ourselves. As Forrester[1] and Meadows[8] convincingly make clear, daunting through an integrated programme may be, a piecemeal approach will cause more problems than it solves.

Our task is to create a society which is sustainable and which will give the fullest possible satisfaction to its members. Such a society by definition would depend not on expansion but on stability. This does not mean to say that it would be stagnant—indeed it could well afford more variety than does the state of uniformity at present being imposed by the pursuit of technological efficiency. We believe that the stable society, the achievement of which we shall discuss in the next chapter, as well as removing the sword of Damocles which hangs over the heads of future generations, is much more likely than the present one to bring the peace and fulfilment which hitherto have been regarded, sadly, as utopian.

Goldsmith, E. et al., A Blueprint for Survival, in: *The Ecologist*, 2, 1972, p. 1-22.

Notes and References

1   Jay Forrester, *World Dynamics*. Wright Allen Press, Cambridge, Mass, 1970.

2   It is perhaps worth bearing in mind that the actual rate of petroleum consumption *is* increasing by 6.9 per cent per year, and according to the optimistic estimate of W.P. Ryman, Deputy Exploration Manager of the Standard Oil Company of New Jersey, world petroleum reserves (including deposits yet to be discovered) are about 2,100 billion barrels.

3   FAO, *Provisional Indicative World Plan for Agriculture*. Rome, 1969.

4   Stated by the Ministry of Agriculture to the Select committee on Science and Technology, *Population of the United Kingdom*. London, HMSO, 1971.

5   Agricultural Advisory Council, *Modern Farming and the Soil*. London, HMSO, 1971.

6   Ministry of Agriculture (Statistics Division), *Output and Utilisation of Farm Produce in the United Kingdom, 1968-9*. London, HMSO, 1970.

7   The American Metal Market Co., *Metal Statistics*, 1970; P.T. Flawn, *Mineral Resources*, 1966; Dennis Meadows *et al.*, *The Limits to Growth*, 1972; United Nations, *The World Market for Iron Ore*, 1986; United Nations, *Statistical Yearbook*, 1969, 1970; US Bureau of Mines, *Commodity Data Summary*, 1971; *Yearbook of American Bureau of Metal Statistics*, 1970.
8   Dennis Meadows *et al.*, *The Limits to Growth*. London, Earth Island, 1972.

CHAPTER 16

Report of the Conference on the Human Environment

UNITED NATIONS

Declaration of the United Nations Conference on the Human Environment

The United Nations Conference on the Human Environment,
Having met at Stockholm from 5 to 16 June 1972,
Having considered the need for a common outlook and for common principles to inspire and guide the peoples of the world in the preservation and enhancement of the human environment,

I

Proclaims that:

1 Man is both creature and moulder of his environment, which gives him physical sustenance and affords him the opportunity for intellectual, moral, social and spiritual growth. In the long and tortuous evolution of the human race on this planet a stage has been reached when, through the rapid acceleration of science and technology, man has acquired the power to transform his environment in countless ways and on an unprecedented scale. Both aspects of man's environment, the natural and the man-made, are essential to his well-being and to the enjoyment of basic human rights—even the right to life itself.

2 The protection and improvement of the human environment is a major issue which affects the well-being of peoples and economic development throughout the world; it is the urgent desire of the peoples of the whole world and the duty of all governments.

3 Man has constantly to sum up experience and go on discovering, inventing, creating and advancing. In our time, man's capability to transform his surroundings, if used wisely, can bring to all peoples the benefits of development and the opportunity to enhance the quality of life. Wrongly or heedlessly applied, the same power can do incalculable harm to human beings and the human environment. We see around us growing evidence of man-made harm in many regions of the earth: dangerous levels of pollution in water, air, earth and living beings; major and undesirable disturbances to the ecological balance of the biosphere; destruction and depletion of irreplaceable resources; and gross deficiencies, harmful to the physical, mental and social health of man, in the man-made environment, particularly in the living and working environment.

4 In the developing countries most of the environmental problems are caused by under-development. Millions continue to live far below the minimum levels required for a decent human existence, deprived of adequate food and clothing, shelter and education, health and sanitation. Therefore, the developing countries must direct their efforts to development, bearing in mind their priorities and the need to safeguard and improve the environment. For the same purpose, the industrialized countries should make efforts to reduce the gap themselves and the developing countries. In the industrialized countries, environmental problems are generally related to industrialization and technological development.

5 The natural growth of population continuously presents problems for the preservation of the environment, and adequate policies and measures should be adopted, as appropriate, to face these problems. Of all things in the world, people are the most precious. It is the people that propel social progress, create social wealth, develop science and technology and, through their hard work, continuously transform the human environment. Along with social progress and the advance of production, science and technology, the capability of man to improve the environment increases with each passing day.

6 A point has been reached in history when we must shape our actions throughout the world with a more prudent care for their environmental consequences. Through ignorance or indifference we can do massive and irreversible harm to the earthly environment on which our life and well-being depend. Conversely, through fuller knowledge and wiser action, we can achieve for ourselves and our posterity a better life in an environment more in keeping with human needs and hopes. There are broad vistas for the enhancement of environmental quality and the creation of a good life. What is needed is an enthusiastic but calm state of mind and intense but orderly work. For the purpose of attaining freedom in the world of nature, man must use knowledge to build, in collaboration with nature, a better environment. To defend and improve the human environment for present and future generations has became an imperative goal for mankind—a goal to be pursued together with, and in harmony with, the established and fundamental goals of peace and of worldwide economic and social development.

7 To achieve this environmental goal will demand the acceptance of responsibility by citizens and communities and by enterprises and institutions at every level, all sharing equitably in common efforts. Individuals in all walks of life as well as organizations in many fields, by their values and the sum of their actions, will shape the world environment of the future. Local and national governments will bear the greatest burden for large-scale environmental policy and action within their jurisdictions. International co-operation is also needed in order to

raise resources to support the developing countries in carrying out their responsibilities in this field. A growing class of environmental problems, because they are regional or global in extent or because they affect the common international realm, will require extensive co-operation among nations and action by international organizations in the common interest. The Conference calls upon Governments and peoples to exert common efforts for the preservation and improvement of the human environment, for the benefit of all the people and for their posterity.

II

States the common conviction that:

Principle 1
Man has the fundamental right to freedom, equality and adequate conditions of life, in an environment of a quality that permits a life of dignity and well-being, and he bears a solemn responsibility to protect and improve the environment for present and future generations. In this respect, policies promoting or perpetuating *apartheid*, racial segregation, discrimination, colonial and other forms of oppression and foreign domination stand condemned and must be eliminated.

Principle 2
The natural resources of the earth, including the air, water, land, flora and fauna and especially representative samples of natural ecosystems, must be safeguarded for the benefit of present and future generations through careful planning or management, as appropriate.

Principle 3
The capacity of the earth to produce vital renewable resources must be maintained and, wherever practicable, restored or improved.

Principle 4
Man has a special responsibility to safeguard and wisely manage the heritage of wildlife and its habitat, which are now gravely imperilled by a combination of adverse factors. Nature conservation, including wildlife, must therefore receive importance in planning for economic development.

Principle 5
The non-renewable resources of the earth must be employed in such a way as to guard against the danger of their future exhaustion and to ensure that benefits from such employment are shared by all mankind.

Principle 6

The discharge of toxic substances or of other substances and the release of hear, in such quantities or concentrations as to exceed the capacity of the environment to render them harmless, must be halted in order to ensure that serious or irreversible damage is not inflict upon ecosystems. The just struggle of the people of all countries against pollution should be supported.

Principle 7

States shall take all possible steps to prevent pollution of the seas by substances that are liable to create hazard to human health, to harm living resources and marine life, to damage amenities or to interfere with other legitimate uses of the sea.

Principle 8

Economic and social development is essential for ensuring a favourable living and working environment for man and for creating conditions on earth that are necessary for the improvement of the quality of life.

Principle 9

Environmental deficiencies generated by the conditions of under-development and natural disasters pose grave problems and can best be remedied by accelerated development through the transfer of substantial quantities of financial and technological assistance as a supplement to the domestic effort of the developing countries and such timely assistance as may be required.

Principle 10

For the developing countries, stability of prices and adequate earnings for primary commodities and raw materials are essential to environmental management since economic factors as well as ecological processes must be taken into account.

Principle 11

The environmental policies of all States should enhance and not adversely affect the present or future development potential of developing countries, nor should they hamper the attainment of better living conditions for all, and appropriate steps should be taken by States and international organizations with a view to reaching agreement on meeting the possible national and international economic consequences resulting from the application of environmental measures.

Principle 12

Resources should be made available to preserve and improve the environment, taking into account the circumstances and particular requirements of developing countries and any costs which may emanate from their incorporating environmental safeguards into their development planning and the need for making available to them, upon their request, additional international technical and financial assistance for this purpose.

Principle 13
In order to achieve a more rational management of resources and thus to improve the environment, States should adopt an integrated and co-ordinated approach to their development planning so as to ensure that development is compatible with the need to protect and improve environment for the benefit of their population.

Principle 14
Rational planning constitutes an essential tool for reconciling any conflict between the needs of development and the need to protect and improve the environment.

Principle 15
Planning must be applied to human settlements and urbanization with a view to avoiding adverse effects on the environment and obtaining maximum social, economic and environmental benefits for all. In this respect, projects which are designed for colonialists and racist domination must be abandoned.

Principle 16
Demographic policies which are without prejudice to basic human rights and which are deemed appropriate by Governments concerned should be applied in those regions where the rate of population growth or excessive population concentrations are likely to have adverse effects on the environment of the human environment and impede development.

Principle 17
Appropriate national institutions must be entrusted with the task of planning, managing or controlling the environmental resources of States with a view to enhancing environmental quality.

Principle 18
Science and technology, as part of their contribution to economic and social development, must be applied to the identification, avoidance and control of environmental risks and the solution of environmental problems and for the common good of mankind.

Principle 19

Education in environmental matters, for the younger generation as well as adults, giving due consideration to the underprivileged, is essential in order to broaden the basis for an enlightened opinion and responsible conduct by individuals, enterprises and communities in protecting and improving the environment in its full human dimension. It is also essential that mass media of communications avoid contributing to the deterioration of the environment, but, on the contrary, disseminate information of an educational nature on the need to protect and improve the environment in order to enable man to develop in every respect.

Principle 20

Scientific research and development in the context of environmental problems, both national and multinational, must be promoted in all countries, especially the developing countries. In this connexion, the free flow of up-to-date scientific information and transfer of experience must be supported and assisted, to facilitate the solution of environmental problems; environmental technologies should be made available to developing countries on terms which would encourage their wide dissemination without constituting an economic burden on the developing countries.

Principle 21

States have, in accordance with the Charter of the United Nations and the principles of international law, the sovereign right to exploit their own resources pursuant to their own environmental policies, and the responsibility to ensure that activities within their jurisdiction or control do not cause damage to the environment of the other States or of area beyond the limits of national jurisdiction.

Principle 22

States shall co-operate to develop further the international law regarding liability and compensation for the victims of pollution and other environmental damage caused by activities within the jurisdiction or control of such States to areas beyond their jurisdiction.

Principle 23

Without prejudice to such criteria as may be agreed upon by the international community, or to standards which will have to be determined nationally, it will be essential in all cases to consider the systems of values prevailing in each country, and the extent of the applicability of standards which are valid for the most advanced countries but which may be inappropriate for the developing countries.

Principle 24

International matters concerning the protection and improvement of the environment should be handled in a co-operative spirit by all countries, big and small, on an equal footing. Co-operation through multilateral or bilateral arrangements or other appropriate means is essential to effectively control, prevent, reduce and eliminate adverse environmental effects resulting from activities conducted in all spheres, in such a way that due account is taken of the sovereignty and interest of all States.

Principle 25

States all ensure that international organizations play a co-ordinated, efficient and dynamic role for the protection and improvement of the environment.

Principle 26

Man and his environment must be spared the effects of nuclear weapons and all other means of mass destruction. States must strive to reach prompt agreement, in the relevant international organs, on the elimination and complete destruction of such weapons.

21st plenary meeting
16 June 1972

Report of the Conference on the Human Environment, Stockholm, 5-16 June 1972, United Nations, New York, 1973, p. 3-5.

CHAPTER 17

The Economics of the Coming Spaceship Earth

K.E. BOULDING

We are now in the middle of a long process of transition in the nature of the image which man has of himself and his environment. Primitive men, and to a large extent also men of the early civilizations, imagined themselves to be living on a virtually illimitable plane. There was almost always somewhere beyond the known limits of human habitation, and over a very large part of the time that man has been on earth, there has been something like a frontier. That is, there was always some place else to go when things got too difficult, either by reason of the deterioration of the natural environment or a deterioration of the social structure in places where people happened to live. The image of the frontier is probably one of the oldest images of mankind, and it is not surprising that we find it hard to get rid of.

Gradually, however, man has been accustoming himself to the notion of the spherical earth and a closed sphere of human activity. A few unusual spirits among the ancient Greeks perceived that the earth was a sphere. It was only with the circumnavigations and the geographical explorations of the fifteenth and sixteenth centuries, however, that the fact that the earth was a sphere became at all widely known and accepted. Even in the nineteenth century, the commonest map was Mercator's projection, which visualizes the earth as an illimitable cylinder, essentially a plane wrapped around the globe, and it was not until the Second World War and the development of the air gage that the global nature of the planet really entered the popular imagination. Even now we are very far from having made the moral, political, and psychological adjustments which are implied in the transition from the illimitable plane to the closed sphere.

Economists in particular, for the most part, have failed to come to grips with the ultimate consequences of the transition from the open to the closed earth. One hesitates to use the terms "open" and "closed" in this connection, as they have been used with so many different shades of meaning. Nevertheless, it is hard to find equivalents. The open system, indeed, has some similarities to the open system of von Bertalanffy,[1] in that it implies that some kind of structure is maintained in the midst of a throughput from inputs to outputs. In a closed system, the outputs of all parts of the system are linked to the inputs of other parts. There are no inputs from outside and no outputs to the outside; indeed, there is no outside at all. Closed systems, in fact, are very rare in human experience, in fact almost by definition unknowable, for if there are genuinely closed systems around us, we have no way of information into them or out of them; and hence if they are really closed, we would be quite unaware of their existence. We can only find out about a closed system if we participate in it. Some isolated primitive societies may have approximated to this, but even these had to take inputs from the environment and give outputs to it. All living organisms, including man himself, are open systems. They have to receive inputs in the shape of air, food, water, and give off outputs in the form of effluvia and excrement. Deprivation of input of air, even for a few minutes, is fatal. Deprivation of the ability to obtain any input or to dispose of any output is fatal in a relatively short time. All human societies have likewise been open systems. They receive inputs from the earth, the atmosphere, and the water, and they give outputs into these reservoirs; they also produce inputs internally in the shape of babies and outputs in the shape of corpses. Given a capacity to draw upon inputs and to get rid of outputs, an open system of this kind can persist indefinitely.

There are some systems—such as the biological phenotype, for instance the human body—which cannot maintain themselves indefinitely by inputs and outputs because of the phenomenon of aging. This process is very little understood. It occurs, evidently, because there are some outputs which cannot be replaced by any known input. There is not the same necessity for aging in organizations and in societies, although an analogues phenomenon may take place. The structure and composition of an organization or society, however, can be maintained by inputs of fresh personnel from birth and education as the existing personnel ages and eventually dies. Here we have an interesting example of a system which seems to maintain itself by the self-generation of inputs, and in this sense is moving towards closure. The input of people (that is, babies) is also an output of people (that is, parents).

Systems may be open or closed in respect to a number of classes of inputs and outputs. Three important classes are matter, energy, and information. The present world economy is open in regard to all three. We can think of the world economy or "econosphere" as a subset of the "world set", which is the set of all objects, people, organizations, and so on, which are interesting from the point of view of the system of exchange. This total stock of capital is clearly an open system in the sense that it has inputs and outputs, inputs being production which adds to the capital stock, outputs being consumption which subtracts from it. From a material point of view, we see objects passing from the non-economic into the economic set in the process of production, and we similarly see products passing out of the economic set as their value becomes zero. Thus we see the econosphere as a material process involving the discovery and mining of fossil fuels, ores, etc., and at the other end a process by which the effluents of the system are passed out into noneconomic reservoirs—for instance, the atmosphere and the oceans—which are not appropriated and do not enter into the exchange system.

From the point of view of the energy system, the econosphere involves inputs of available energy in the form, say, of water power, fossil fuels, or sunlight, which are necessary in order to create the material throughput and to move matter from the noneconomic set into the economic set or even out of it again; and energy itself is given off by the system in a less available form, mostly in the form of heat. These inputs of available energy must come either from the sun (the energy supplied by other stars being assumed to be negligible) or it may come from the earth itself, either through its internal heat or through its energy of rotation or other motions, which generate, for instance, the energy of the tides. Agriculture, a few solar machines, and water power use the current available energy income. In advanced societies this is supplemented very extensively by the use of fossil fuels, which represent, as it were, a capital stock of stored-up sunshine. Because of this capital stock of energy, we have been able to maintain an energy input into the system, particularly over the last two centuries, much larger than we would have been able to do with existing techniques if we had had to rely on the current input of available energy from the sun or the earth itself. This supplementary input, however, is by its very nature exhaustible.

The inputs and outputs of information are more subtle and harder to trace, but also represent an open system, related to, but not wholly dependent on, the transformations of matter and energy. By far the larger amount of information and knowledge is self-generated by the human-society, through a certain amount of information comes into the sociosphere in the form of light from the universe outside. The information that comes from the universe has certainly affected man's image of himself and of his environment, as we can easily visualize if we suppose that we lived on a planet with a total cloud-cover that kept out all information from the exterior universe. It is only in very recent times, of course, that the information coming in from the universe has been captured and coded into the form of a complex image of what the universe is like outside the earth; but even in primitive times, man's perception of the heavenly bodies has always profoundly affected within the planet, however, and particularly that generated by man himself, which forms by far the larger part of the information system. We can think of the stock of knowledge, or as Teilhard de Chardin called it, the "noosphere", and consider this as an open system, losing knowledge through aging and death and gaining it though birth and education and the ordinary experience of life.

From the human point of view, knowledge, or information, is by far the most important of the three systems. Matter only acquires significance and only enters the sociosphere or the econosphere insofar as it becomes an object of human knowledge. We can think of capital, indeed, as frozen knowledge or knowledge imposed on the material world in the form of improbable arrangements. A machine, for instance, originates in the mind of man, and both its construction and its use involve information processes imposed on the material world by man himself. The cumulation of knowledge, that is, the excess of its production over its consumption, is the key to human development of all kinds, especially to economic development. We can see this preeminence of knowledge very clearly in the experiences of countries where the material capital has been destroyed by a war, as in Japan and Germany. The knowledge of the people was not destroyed, and it did not take long, therefore, certainly not more than ten years, for most of the material capital to be reestablished again. In a country such as Indonesia, however, where the knowledge did not exist, the material capital did not come into being either. By "knowledge" here I mean, of course, the whole cognitive structure, which includes valuations and motivations as well as images of the factual world.

The concept of entropy, used in a somewhat loose sense, can be applied to all three of these open systems. In material systems, we can distinguish between entropic processes, which take concentrated materials and diffuse them through the oceans or over the earth's surface or into the atmosphere, and antientropic processes, which take diffuse materials and concentrate them. Material entropy can be taken as a measure of the uniformity of the distribution of elements and, more uncertainly, compounds and other structures on the earth's surface. There is, fortunately, no law of increasing material entropy, as there is in the corresponding case of energy, as it is quite possible to concentrate diffused materials if energy inputs are allowed. Thus the processes for fixation of nitrogen from the air, processes for the extraction of magnesium of other elements from the sea, and processes for the desalinization of sea water are antientropic in the material sense, though the reduction of material entropy has to be paid for by inputs of energy and also inputs of information, or at least a stock of information in the system. In regard to matter, therefore, a closed system is conceivable, that is, a system in which there is neither increase nor decrease in material entropy. In such a system all outputs from consumption would constantly be recycled to become inputs for production, as for instance, nitrogen in the nitrogen cycle of the natural ecosystem.

In the energy system there is, unfortunately, no escape from the grim second law of thermodynamics; and if there were no energy inputs into the earth, any evolutionary or developmental process would be impossible. The large energy inputs which we have obtained from fossil fuels are strictly temporary. Even the most optimistic predictions expect the easily available supply of fossil fuels to be exhausted in a mere matter of centuries at present rates of use. If the rest of the world were to rise to American standards of power consumption, and still more if world population continues to increase, the exhaustion of fossil fuels would be even more rapid. The development of nuclear energy has improved this picture, but has fundamentally altered it, at least in present technologies, for fissionable material is still relatively scarce. If we should achieve the economic use of energy through fusion, of course, a much larger source of energy materials would be available, which would expand the time horizons of supplementary energy input into an open social system by perhaps tens to hundreds of thousands of years. Failing this, however, the time is not very far distant, historically speaking, when man will once more have to retreat to his current energy input from the sun, even though with increased knowledge this could be used much more effectively than in the past. Up to now, certainly, we have not got very far with the technology of using current solar energy, but the possibility of substantial improvements in the future is certainly high. It may be, indeed, that the biological revolution which is just beginning will produce a solution to this problem, as we develop artificial organisms which are capable of much more efficient transformations of solar energy into easily available forms than any that we now have. As Richard Meier has suggested, we may run our machines in the future with methane-producing algae.[2]

The question of whether there is anything corresponding to entropy in the information system is a puzzling one, though of great interest. There are certainly many examples of social systems and cultures which have lost knowledge, especially in transition from one generation to the next, and in which the culture has therefore degenerated. One only has to look at the folk culture of Appalachian migrants to American cities to see a culture which started out as a fairly rich European folk culture in Elizabethan times and which seems to have lost skills, adaptability, folk tales, songs, and almost everything that goes up to make richness and complexity in a culture, in the course of about ten generations. The American Indians on reservations provide another example of such degradation of the information and knowledge system. On the other hand, over a great part of human history, the growth of knowledge in the earth as a whole seems to have been almost continuous, even though there have been times of relatively slow growth and times of rapid growth. As it is knowledge of certain kinds that produces the growth of knowledge in general, we have here a very subtle and complicated system, and it is hard to put one's finger on the particular elements in a culture which make knowledge grow more or less rapidly, or even which make it decline. One of the great puzzles in this connection, for instance, is why the takeoff into science, which represents an "acceleration", or an increase in the rate of growth of knowledge in European society in the sixteenth century, did not take place in China, which at that time (about 1600) was unquestionably ahead of Europe, and one would think even more ready for the breakthrough. This is perhaps the most crucial question in the theory of social development, yet we must confess that it is very little understood. Perhaps the most significant factor in this connection is the existence of "slack" in the culture, which permits a divergence from established patterns and activity which is not merely devoted to reproducing the existing society but is devoted to changing it. China was perhaps too well organized and had too little slack in its society to produce the kind of acceleration which we find in the somewhat poorer and less well organized but more diverse societies of Europe.

The closed earth of the future requires economic principles which are somewhat different from those of the open earth of the past. For the sake of picturesqueness, I am tempted to call the open economy the "cowboy economy", the cowboy being symbolic of the illimitable plains and also associated with reckless, exploitative, romantic, and violent behavior, which is characteristic of open societies. The closed economy of the future might similarly be called the "spaceman" economy, in which the earth has become a single spaceship, without unlimited reservoirs of anything, either for extraction or for pollution, and in which, therefore, man must find his place in a cyclical ecological system which is capable of continuous reproduction of material form even though it cannot escape having inputs of energy. The difference between the two types of economy becomes most apparent in the attitude towards consumption. In the cowboy economy, consumption is regarded as a good thing and production likewise; and the success of the economy is measured by the amount of the throughput from the "factors of production", a part of which, at any rate, is extracted from the reservoirs of raw materials and noneconomic objects, and another part of which is output into the reservoirs of pollution. If there are infinite reservoirs from which material can be obtained and into which effluvia can be deposited, then the throughput is at least a plausible measure of the success of the economy. The Gross National Product is a rough measure of this total throughput. It should be possible, however, to distinguish that part of the GNP which is derived from exhaustible and that which is derived from reproducible resources, as well as that part of consumption which represents effluvia and that which represents input into the productive system again. Nobody, as far as I know, has ever attempted to break down the GNP in this way, although it would be an interesting and extremely important exercise, which is unfortunately beyond the scope of this paper.

By contrast, in spaceman economy, throughput is by no means a desideratum, and is indeed to be regarded as something to be minimized rather than maximized. The essential measure of the success of the economy is nor production and consumption at all, but the nature, extent, quality, and complexity of the total capital stock, including in this the state of the human bodies and minds included in the system. In the spaceman economy, what we are primarily concerned with is stock maintenance, and any technological change which results in the maintenance of a given total stock with a lessened throughput (that is, less production and consumption) is clearly a gain. This idea that both production and consumption are bad things rather than good things is very strange to economists, who have been obsessed with the income-flow concepts to the exclusion, almost, of capital-stock concepts.

There are actually some very tricky and unsolved problems involved in the questions as to whether human welfare or well-being is to be regarded as a stock or a flow. Something of both these elements seems actually to be involved in it, and as far as I know these two dimensions of human satisfaction. Is it, for instance, eating that is a good thing, or is it being well fed? Does economic welfare involve having nice clothes, fine houses, good equipment, and so on, or is to be measured by the depreciation and the wearing out of these things? I am inclined myself to regard the stock concept as most fundamental, that is, to think of being well fed as more important than eating, and to think even of so-called services as essentially involving the restoration of a depleting psychic capital. Thus I have argued that we go to a concert in order to restore a psychic condition which might be called "just having gone to a concert", which, once established, tends to depreciate. When it depreciates beyond a certain point, we go to another concert in order to restore it. If it depreciates rapidly, we go to a lot of concerts; if it depreciates slowly, we go to a few. On this view, similarly, we eat primarily to restore bodily homeostasis, that is, to maintain a condition of being well fed, and so on. On this view, there is nothing desirable in consumption at all. The less consumption we can maintain a given state with, the better off we are. If we had clothes that did not wear out, houses that did not depreciate, and even if we could maintain our bodily condition without eating, we would clearly be much better off.

It is this last consideration, perhaps, which makes one pause. Would we, for instance, really want an operation that would enable us to restore all our bodily tissues by intravenous feeding while we slept? Is there not, that is to say, a certain virtue in throughput itself, in activity itself, in production and consumption itself, in raising food and in eating it? It would certainly be rash to exclude this possibility. Further interesting problems are raised by the demand for variety. We certainly do not want a constant state to be maintained; we want fluctuations in the state. Otherwise there would be no demand for variety in food, for variety in scene, as in travel, for variety in social contact, and so on. The demand for variety can, of course, be costly, and sometimes it seems to be too costly to be tolerated or at least legitimated, as in the case of marital partners, where the maintenance of a homeostatic state in the family is usually regarded as much more desirable than the variety and excessive throughput of the libertine. There are problems here which the economies profession has neglected with astonishing singlemindedness. My own attempts to call attention to some of them, for instance, in two articles, as far as I can judge, produced no response whatsoever: and economists continue to think and act as if production, consumption, throughput, and the GNP were the sufficient and adequate measure of economic success.

It may be said, of course, why worry about all this when the spaceman economy is still a good way off (at least beyond the lifetimes of any now living), so let us eat, drink, spend, extract and pollute, and be as merry as we can, and let posterity worry about the spaceship earth. It is always a little hard to find a convincing answer to the man who says, "What has posterity ever done for me?" and the conservationist has always had to fall back on rather vague ethical principles postulating identity of the individual with some human community or society which extends not only back into the past but forward into the future. Unless the individual identifies with some community of this kind, conservation is obviously "irrational". Why should we not maximize the welfare of this generation at the cost of posterity? "*Après nous, le déluge*" has been the motto of not insignificant numbers of human societies. The only answer to this, as far as I can see, is to point out that the welfare of the individual depends on the extent to which he can identify himself with others, and that the most satisfactory individual identity is that which identifies not only with a community in space but also with a community extending over time from the past into the future. If this kind of identity is recognized as desirable, then posterity has a voice, even if it does not have a vote; and in a sense, if its voice can influence votes, it has votes too. This whole problem is linked up with the much larger one of the determinants of the morale, legitimacy, and "nerve" of a society, and there is a great deal of historical evidence to suggest that a society which loses its identity with posterity and which loses its positive image of the future loses also its capacity to deal with present problems, and soon falls apart.[4]

Even if we concede that posterity is relevant to our present problems, we still face the question of time-discounting. It is a well-known phenomenon that individuals discount the future, even in their own lives. The very existence of a positive rate of interest may be taken as at least strong supporting evidence of this hypothesis. If we discount our own future, it is certainly not unreasonable to discount posterity's future even more, even if we do give posterity a vote. If we discount this at five per cent per annum, posterity's vote or dollar halves every fourteen years as we look into the future, and after even a mere hundred years it is pretty small—only about one-and-a-half cents on the dollar. If we add another five per cent for uncertainty, even the vote of our grandchildren reduces almost to insignificance. We can argue, of course, that the ethical thing to do is not to discount the future at all, that time-discounting is mainly the result is myopia and perspective, and hence is an illusion which the moral man should not tolerate. It is a very popular illusion, however, and one that must certainly be taken into consideration in the formulation of policies. It explains, perhaps, why conservationist policies almost have to be sold under some other excuse which seems more urgent, and why, indeed, necessities which are visualized as urgent, such as defense, always seem to hold priority over those which involve the future.

All these considerations add some credence to the point of view which says that we should not worry about the spaceman economy at all, and that we should just go on increasing the GNP and indeed the Gross World Product, or GWP, in the expectation that the problems of the future can be left to the future, that when scarcities arise, wether this is of raw materials or of pollutable reservoirs, the needs of the then present will determine the solutions of the then present, and there is no use giving ourselves ulcers by worrying about problems that we really do not have to solve. There is even high ethical authority for this point of view in the New Testament, which advocates that we should take no thought for tomorrow and let the dead bury their dead. There has always been something rather refreshing in the view that we should live like the birds, and perhaps posterity is for the birds in more senses than one; so perhaps we should all call it a day and go out and pollute something cheerfully. As an old taker of thought for the morrow, however, I cannot quite accept this solution: and I would argue, furthermore, that tomorrow is not only very close, but in many respects it is already here. The shadow of the future spaceship, indeed, is already falling over our spendthrift merriment. Oddly enough, it seems to be in pollution rather than in exhaustion that the problem is first becoming salient. Los Angeles has run out of air, Lake Erie has become a cesspool, the oceans are getting full of lead and DDT, and the atmosphere may become man's major problem in another generation, at the rate at which we are filling it up with gunk. It is, of course, true that at least on a microscale, things have been worse at times in the past. The cities of today, with all their foul air and polluted waterways, are probably not as bad as the filthy cities of the pretechnical age. Nevertheless, that fouling of the nest which has been typical of man's activity in the past on a local scale now seems to be extending to the whole world society: and one certainly cannot view with equanimity the present rate of pollution of any of the natural reservoirs, whether the atmosphere, the lakes, or even the oceans.

I would argue strongly also that our obsession with production and consumption to the exclusion of the "state" aspects of human welfare distorts the process of technological change in a most undesirable way. We are all familiar, of course, with the wastes involved in planned obsolescence, in competitive advertising, and in poor quality of consumer goods. These problems may not be so important as the "view with alarm" school indicates, and indeed the evidence at many points if conflicting. New materials especially seem to edge towards the side of improved durability, such as, for instance, neolite soles for footwear, nylon socks, wash and wear shirts, and so on. The case of household equipment and automobiles is a little less clear. Housing and building construction generally almost certainly has declined in durability since the Middle Ages, but this decline also reflects a change in tastes towards flexibility and fashion and a need for novelty, so that it is not easy to access. What is clear is that no serious attempt has been made to assess the impact over the whole of economic life of changes in durability, that is, in the ratio of capital in the widest possible sense to income. I suspect that we have underestimated, even in our spendthrift society, the gains from increase durability, and that this might very well be one of the places where the price system needs correction through government-sponsored research and development. The problems which the spaceship earth is going to present, therefore, are not all in the future by any means, and a strong case can be made for paying much more attention to them in the present than we now do.

It may be complained that the consideration I have been putting forth relate only to the very long run, and they do not much concern our immediate problems. There may be some justice in this criticism, and my main excuse is that other writers have dealt adequately with the more immediate problems of deterioration in the quality of the environment. It is true, for instance, that many of the immediate problems of the pollution of the atmosphere of bodies of water arise because of the failure of the price system, and many of them could be solved by corrective taxation. If people had to pay the losses due to the nuisances which they create, a good deal more resources would go into the prevention of nuisances. These arguments involving external economies and diseconomies are familiar to economists and there is no need to recapitulate them. The law of torts is quite inadequate to provide for the correction of the price system which is required, simply because where damages are widespread and their incidence on any particular person is small, the ordinary remedies of the civil law are quite inadequate and inappropriate. There needs, therefore, to be special legislation to cover these cases, and though such legislation seems hard to get in practice, mainly because of the widespread and small personal incidence of the injuries, the technical problems involved are not insuperable. If we were to adopt in principle a law for tax penalties for social damages, with an apparatus for making assessments under it, a very large proportion of current pollution and deterioration of the environment would be prevented. There are tricky problems of equity involved, particularly where old established nuisances create a kind of "right by purchase" to perpetuate themselves, but these are problems again which a few rather arbitrary decisions can bring some kind of solution.

The problems which I have been raising in this paper are of larger scale and perhaps much harder to solve than the more practical and immediate problems of the above paragraph. Our success in dealing with the larger problems, however, is not unrelated to the development of skill in the solution of more immediate and perhaps less difficult problems. One can hope, therefore, that as a succession of mounting crises, especially in pollution, arouse public opinion and mobilize support for the solution of the immediate problems, a learning process will be set in motion which will eventually lead to an appreciation of and perhaps solutions for the larger ones. My neglect of the immediate problems, therefore, is in no way intended to deny their importance, for unless we make at least a beginning on a process for solving the immediate problems we will not have much chance of solving the larger ones. On the other hand, it may also be true that a long-run vision, as it were, of the deep crisis which faces mankind may predispose people to taking more interest in the immediate problems and to devote more effort for their solution. This may sound like a rather modest optimism, but perhaps a modest optimism is better than no optimism at all.

Boulding, K.E., The Economics of the Coming Spaceship Earth, in: Daly, H.E. (ed.), *Toward a Steady-state Economy,* W.H. Freeman and Company, San Francisco, 1973, p. 121-132.

Notes and References

1   Ludwig von Bertalanffy, *Problems of Life* (New York: John Wiley and Sons, 1952).

2   Richard L. Meier, *Science and Economic Development* (New York: John Wiley and Sons, 1956).

3   Kenneth E. Boulding, *"The Consumption Concept in Economic Theory"*, American Economic Review, 35:2 (May 1945), pp. 1-14: and *"Income or Welfare?"*, Review of Economic Studies, 17 (1949, 50), pp. 77-86.

4   Fred L. Polak, *The Image of the Future*, Vols. I and II, translated by Elise Boulding (New York: Sythoff, Leyden and Oceana, 1961).

CHAPTER 18

Reshaping the International Order

A Report to the Club of Rome

J. TINBERGEN (COORD.) & A.J. DOLMAN (ED.),

From World Disorder to International Order

'. . . the international economic system is not as free as is often claimed and our choice is not one between a free system based on free enterprise and a fully centrally planned economy. The real choice we have to make is between sticking to our present system, which is largely guided and manipulated for the benefit of the rich countries, and opting for a system directed towards finding solutions to the problems of an equitable division of income and property, of scarcity of natural resources and of despoliation of the environment.'

Joop den Uyl, Netherlands Premier. 1975

The Industrialized World: From Cornucopia to Pandora's Box

The call for a new international economic order was made in a period of economic turmoil without precedent in the post-war world. The industrialized countries were experiencing economic dislocations unknown since the agonies of the Great Depression of the 1930s. The international system, which they had largely created and which had appeared to serve them well, was in serious disequilibrium.

Behind them was a period of unparalleled economic growth. The planetary product, for which they were responsible for by far the largest part, had trebled in the twenty years between 1950 and 1970, a period in which most of the world's industrial capacity was created. This growth had brought material prosperity for most of their citizens, a more equitable income distribution within their societies, and achievements in many fields of science.

The rich countries had created an enormously powerful industrial machine. Fed by stimulated demand, it was, in the Western world, fired by abundant and cheap supplies of oil. At just over one dollar a barrel, oil supplies stimulated growth in energy consumption at between 6 and 11 per cent a year. The very cheapness of the supplies ensured rapid growth. It also encouraged extravagance and waste.

A colonial history had also helped bestow upon many countries in the Western world access to cheap supplies of other of the Third World's raw materials. Of the nine major minerals (excluding oil) required to sustain an industrial economy, the industrialized market economy countries consumed nearly 70 per cent of the world's output. The Third World, economically tied to the industrial machine, was compelled to sell for the price determined by international market mechanisms which worked to the advantage of the industrialized importing countries.

The two most powerful nations were able to construct a mighty military capability and devise weapons of incredible destructiveness in order to protect their competing social systems. This military capability not only required the industrial machine to sustain it, it also made it possible for the machine to grow still further. It was, moreover, a capability which threatened the lives of every single man, woman and child.

TABLE I Gross national product and population, 1973 (percentages)

	GNP	Population
North America	30.0	6.1
Europe (excluding U.S.S.R.)	31.8	13.2
U.S.S.R.	10.7	6.5
Asia (including Middle East and excluding Japan)	10.2	52.7
Japan	8.3	2.8
Central and South America[a]	5.2	7.9
Africa	2.4	10.2
Oceania	1.5	0.6
Total	100.0	100.0
Developed Market Economies[b]	65.7	17.9
Centrally Planned Economies[c]	20.2	32.0
Developing Countries	14.2	50.1
Total	100.0	100.0

a includes Mexico.

b Australia, Austria, Belgium, Canada, Denmark, Finland, France Fed. Rep. of Germany, Iceland, Ireland, Italy, Japan, Luxembourg, Netherlands, New Zealand, Norway, Portugal, Puerto Rico, South Africa, Sweden, Switzerland, United Kingdom, United States.

c Albania, Bulgaria, People's Rep. of China, Cuba, Czechoslovakia, Dem Rep. of Germany, Hungary, Dem. Rep. of Korea, Poland, Romania, U.S.S.R., Dem. Rep. of Vietnam.

Note In 1973, world GNP was $ 4.8 trillion; population was 3.8 billion.

Source: Based on World Bank Atlas, 1975: Population, per Capita Product, and Growth Rates (Washington, D.C.: World Bank Group, 1975).

By the early 1970s it had become clear that the cornucopia of economic growth was turning into a Pandora's box. The main props upon which the economic system was resting began to crumble, for the industrialized countries, in uncomfortably quick succession. The world monetary system, agreed upon by the Western powers at Bretton Woods towards the end of the Second World War, had all but collapsed by 1971. Despite the fact that this laid the basis for gigantic financial disruptions, world wide inflation, trade dislocations and, for some countries, enormous balance of payments difficulties, the Western powers found it hard to cooperate on international monetary reforms.

The disruptions contributed to wild movements in the price of most primary products which, because of their unstable markets, were already prone to serious fluctuation. Prices of industrial products had increasingly risen, partly as a consequence of increased demand and partly because of wage claims far surpassing increases in labour productivity. These developments not only jeopardized the growth prospects of the industrialized countries but also resulted in continuous increases in the import bills of most Third World countries.

The Development Strategy for the Second United Nations Development Decade, solemnly adopted by the UN General Assembly in 1970, and reviewed in 1973, had hardly been taken seriously by the larger industrialized countries and a general feeling of frustration prevailed among the countries of the Third World. A sudden and historically important change took place, however, when in 1973 the Organization of Petroleum Exporting Countries (OPEC) took the initiative to use their power and raised the price of crude oil—which in real terms had actually declined between 1950 and 1970—to about four times the previous level. This development, facilitated by a temporary and perhaps unexpected coincidence of interests between Western oil companies and OPEC nations, caused the industrialized countries considerable distress and brought vailed threats of military retaliation. It did, however, result in a temporary two per cent transfer of the GNP of the industrialized countries to the OPEC nations; it also contributed towards accelerating the recession in economic activity which had started in 1972.

The world situation was further aggravated by adverse weather conditions which brought disastrous crop failures in many parts of the world, and by the concerted action of the main producers of basic staple foods which enabled them to increase their prices in 1974 to a level some three to four times higher than in 1970. This placed enormous pressure on world reserve stocks of food grains—which dwindled to virtually nothing in 1972 as well as in 1975—and on the importers of staple foods, especially the poorest countries.

The full impact of all these developments were felt in the industrialized world in 1974 and 1975 when the recession assumed proportions larger than any experienced since the Second World War. Economists struggled to explain 'stagflation', the unique combination of high inflation and industrial recession;

FIGURE I Dependence of selected developed countries on imported energy, 1973 (percentages)

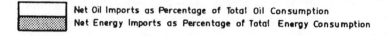

Net Oil Imports as Percentage of Total Oil Consumption
Net Energy Imports as Percentage of Total Energy Consumption

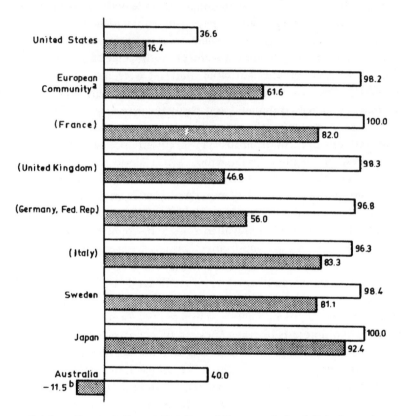

a Belgium, Denmark, France, Fed. Rep. of Germany, Ireland, Italy, Luxembourg, Netherlands, and United Kingdom.
b Net energy exporter

Source: Committee for Economic Development, Research and Policy Committee, *International Economic Consequences of High-Priced Energy: A Statement on National Policy,* New York, September 1975, p. 77.

Keynesian economics, which had helped to steer courses away from impending crises in the past, this time seemed perilously inadequate. Industrial production in many industrialized countries fell in 1975 for the first time since the Second World War. In September of that year, seventeen million people flooded the employment offices of the richest countries. Nor was it the industrial machine that alone was misfiring; some sectors of Western agriculture, which only a few years earlier had appeared strong, were in difficulties.

To many it became increasingly apparent that the economic crisis which plagued the Western world was more than a temporary phenomenon, a pocket of economic turbulence along the route towards even greater riches. They view it as a crisis in the very international structures and mechanisms which the Western world itself has largely created. If the inexorable workings of market forces have helped create the problems, it is clear that they, if left to their own devices, will not be able to solve them. Some economists even warn us that the much heralded 'recovery' might prove little more than a brief respite, a lull before an even greater economic storm, not long off, even more damaging than the last.

Prosperity has brought anxiety to the Western world, a gnawing fear that the food times might well be over, even though aspirations for still greater material benefits remain. If the Western world is to come to grips with this, with its growing list of problems—social as well as economic—it must, in its own long-term interests, seek to create new international structures based upon global cooperation.

The Third World: From Deference to Defiance

If the period since the Second World War was an era of growth in the industrialized world, it was an age of political liberation in the Third World. In little more than one decade, some one third of the world's then population was freed from foreign suzerainty. But the poor nations found that with fewer resources, less know-how, and limited opportunities to utilize that which they had, they were in fact less free than the rich countries. They discovered that political liberation does not necessarily bring economic liberation and that the two are inseparable: that without political independence it is impossible to achieve economic independence; and without economic power, a nation's political independence is incomplete and insecure.

This economic dependence is rooted in the main institutions of the international system created largely by the industrialized countries to deal essentially with their own problems at a time in which the voices of the world's poor were unheard in international fora. The poor nations have been forced to question the basic premises of an international system which leads to ever widening disparities between the richest and poorest nations and to a persistent denial of equality of opportunity. They contend that the 'free' market is in fact not 'free' but works to the advantage of industrialized nations, who have used it to construct a protective wall around their affluence and life-styles. And even if it were 'free', it would still work to the advantage of the industrialized nations because of their enormous political and economic strength. As at the national level, the market mechanism tends to mock poverty, or simply ignores it, since the poor hardly have the purchasing power to influence market decisions. This is even more true at the international level, since there is no world government and none of the established mechanisms which exist within countries which create pressures for the redistribution of income and opportunity.

Inevitably and rightly, the Third World is demanding change in an international system which, it contends, systematically discriminates against its interests and is characterized by institutional distortions which, according to some estimates, cost the poor nations in the order of $50—$100 billion a year.[1] It is insisting on fundamental structural change; not remedial tinkering at international institutions but a new world order which will redress past patterns of hopeless dependency and provide real opportunities to more equitable share in global growth.

Meeting in Dakar in 1975 the poor nations declared that to achieve their 'full and complete economic emancipation', it was necessary 'to recover and control their natural resources and wealth, and the means of economic development.' They agreed that there was 'an urgent need for the developing countries to change their traditional approach to negotiations with the developed countries, hitherto consisting of the presentation of a list of requests to the developed countries and an appeal to their political goodwill which reality was seldom forthcoming.'

It must be made clear that the Third World is not demanding massive redistribution of the past income and wealth of the rich nations. It is not seeking charity from the prosperous nor equality in income. It is asking for equality of opportunity and insisting on the right to share in future growth. The basic objectives of the emerging 'trade union' of poor nations is to negotiate a 'new deal' with the rich nations on the basis of reasonable demands through the instrument of collective bargaining and participation. In attempting to secure greater equality of opportunity, they are simply insisting on the right to sit as equals around the bargaining tables of the world.

The Centrally Planned Nations: We Do Not Live in a Hothouse

The economic and financial crisis besetting the industrialized world sent their ripples into Eastern Europe. The effect of OPEC's increase in oil prices, for example, has been felt by all the oil-importing countries of CMEA, particularly Hungary, Czechoslovakia, the German Democratic Republic and Bulgaria.

Since most obtain their oil from the Soviet Union, the shock was delayed; it was not until 1973 that CMEA substantially increased the price of oil in line with the world market. Poland and Romania were the least affected: Poland derives over 80 per cent of its energy from its coal resources while Romania is a producer of crude oil. Even Romania, however, failed to escape the economic dislocations. Its refining capacity is greater than its production and the difference had to be made up by imports from the Middle East.[2]

Most Eastern European nations ran balance of payment deficits with the West in 1974 and 1975 and were seriously affected by world inflation. An unusual editorial in a Hungarian newspaper, under the eloquent heading: 'We do not live in a hothouse' observed: 'Some time ago we believe that we are not affected by what is happening on the world capitalist market. Inflation could be stopped, as an unwelcome guest, at our borders and here in our country we could live and work under the same conditions as before. That 'hothouse' atmosphere has cost us 20 billion forints'.[3]

The situation is somewhat different with the Soviet Union, a global economic power. The Soviets produce about 8 million barrels of oil daily (nearly the same as Saudi Arabia), but export only one-fifth. The increase in gold prices has also advantaged the Soviet Union as has the general increase in the price of raw materials. Soviet exports to the West amounted, for oil alone, to $3 billion in 1974 and were expected to increase to $3.5 billion in 1975.[4]

Despite this, Western European nations ran a $1.6 billion trading surplus with Eastern Europe in 1973 (a figure slightly higher in 1974).[5] Soviet-American trade in 1975 resulted in a $1.5 billion surplus to the advantage of the US mainly as a result of the Soviet Union's substantial grain imports. The Soviet-American grain agreement points to a continued US trading surplus.

The main cause of Eastern Europe's trading deficit lies in the asymmetrical structure of East-West trade. The share of manufactured goods in Eastern European exports is relatively small (less than 20 per cent in the late sixties and early seventies), whereas the machinery imported from the West accounts for over 40 per cent of total West-East intercourse. The highly unfavourable structure of trade is compounded by the privileged position of Western currencies in financial transactions. As Soviet economist Pichugin has pointed out: '. . . such a situation can become a brake on the further expansion of exports from industrial capitalist countries to socialist nations, in as much as the latter's purchasing capacities with regard to capitalist countries are determined in the ultimate end by the size of their export receipts from those countries'.[6]

The centrally planned nations are disadvantaged on international markets with the manufactured goods of their infant industries, barely competitive both in prices and technological performance with the products of the industrially advanced nations. Moreover, the debts incurred by Eastern European countries with Western banks have reached an all-time high and run into billions of dollars. These observations lead inescapably to the conclusion that we are dealing here with a *specific pattern generated by a disparity in development* and that difficulties in East-West economic relations are in some cases more the result of different levels of development than of differences in social and economic systems.[7] Indeed development is not an exclusive North-South or West-South problem affecting the developing continents only. Europe has a *'development gap' of its own*, in a milder version to be sure, but nevertheless one that makes for an important factor in correctly understanding the vital stake of Eastern Europe in the restructuring of international economic relations.

As things stand now, in spite of the tremendous efforts of the Eastern European nations to industrialize and to develop a modern economy, their main economic indicators still lag behind those of their Western counterparts. Whereas per capita GNP in Eastern Europe ranges *grosso modo* from $1,000 to $3,000 annually, the same indicator in Western Europe goes as high as $2,500 to $6,000.[8] This is an imbalance with serious economic, political and ideological consequences.

To really understand the peculiarity of this phenomenon it must be realized that it has a much longer ancestry than that currently accorded to it by authors who deal with the East-West conflict as though it were purely and exclusively an ideological antagonism that started with the Russian Revolution and took shape in the aftermath of the Second World War with the extension of the revolution in Eastern Europe. Such an approach may at best help explain the origins of the Cold War, but not its underlying economic background. This goes back to the very inception of the modern international state system, when the two major convergent processes, namely the vigorous expansion of capitalism and the formation of nation-States in Europe, gave the Western part of the continent a strong edge over its Eastern part.

It was the renaissance with its blend of antiquity and feudalism that at once produced the many breakthroughs in science and the historical turning point from which Europe outdistanced all other continents. And since the Renaissance was a Western European phenomenon, *par excellence*, both the early start of the Absolutist state as the maker of modern nations and the capitalist expansionist thrust established there the centre of the new international system.

Such were the historical conditions that allowed the Western nations to fully benefit from the industrial revolution. Eastern European peoples (most of them still struggling for nationhood) remained in predominantly agrarian economies with strong feudal structures which survived until the twentieth century. And since the socialist revolution started in backward Russia and subsequently expanded into underdeveloped or less developed countries, they were all faced with the enormous task of industrialization at a pace as rapid as possible, a task so overriding that the whole social, economic and political fabric of those new societies bear its imprint.

It would logically follow that Eastern Europe has a vital interest in the creation of a new international order that would eliminate existing imbalances among nations. Indeed, all centrally planned nations have supported the drive initiated to that end by the Group of 77; Romania even recently joined the Group.[9]

The Need for a New International Order

The inequities in the international system are of tremendous significance. They have given rise to essentially two worlds and the disparities between them are growing. One is the world of the rich, the other the world of the poor, united by its heritage of common suffering. A poverty curtain divides the worlds materially and philosophically. One world is literate, the other largely illiterate; one industrial and urban, the other predominantly agrarian and rural; one consumption oriented, the other striving for survival. In the rich world, there is concern about the quality of life, in the poor world about life itself which is threatened by disease, hunger and malnutrition. In the rich world there is concern about the conservation of non-renewable resources and learned books written about how the world should be kept in a stationary state. In the poor world there is anxiety, not about the depletion of resources, but about their exploitation and distribution for the benefit of all mankind rather than a few privileged nations. While the rich world is concerned about the impact of its pollutive activities on life-support systems, the poor world is concerned by the pollution of poverty, because its problems arise not out of an excess of development and technology but out of the lack of development and technology and inadequate control over natural phenomena.

TABLE 2 Growing income disparities: per capita income ($ US) in selected regions, 1913 and 1957.

Region	1913		1957	
	Population (millions)	Income/capita ($ US)	Population (millions)	Income/capita ($ US)
North America	105	917	188	1,868
N.W. Europe	184	454	211	790
S.E. Asia	323	65	518	67
China	370	50	640	61
World population	1,463		2,373	

Source: L.J. Zimmerman: *Arme en Rijke Landen*, The Hague, 1959, p. 29, 31.

FIGURE 2 Countries according to population

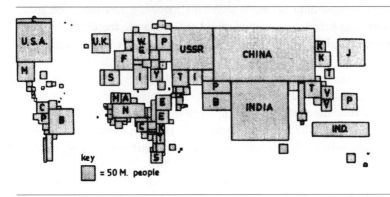

FIGURE 3 Countries according to GNP

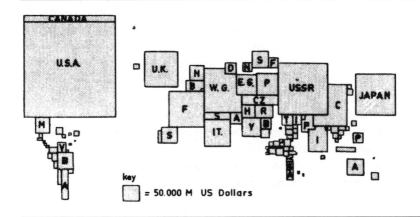

FIGURE 4 Countries according to energy consumption

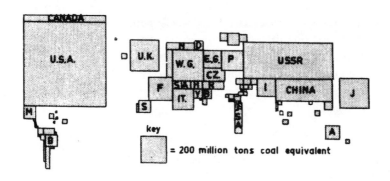

Source: Oosthoek-Times Wereldatlas, Times Newspapers Ltd/Kluwer, 1973, p.21, 24.

We have today about two-thirds of mankind living—if it can be called living—
on less than 30 cents a day. We have today a situation where there are about
one billion illiterate people around the world, although the world has both the
means and technology to spread education. We have nearly 70 per cent of the
children in the Third World suffering from malnutrition, although the world
has the resources to feed them. We have maldistribution of the world's resources on a scale where the industrialized countries are consuming about
twenty times more of the resources per capita than the poor countries. We have
a situation where, in the Third World, millions of people toil under a broiling
sun from morning till dusk for miserable rewards and premature death without
ever discovering the reasons why.

Both the rich and poor worlds have pressing, unparalleled problems. They
are not separate; they cannot be solved independently. Mankind's predicament
is rooted in its past, in the economic and social structures that have emerged
within and between nations. The present crisis, in the world economy and in
the relations between nations, is a crisis of international structures. What both
worlds must come to grips with is basically a sick system which cannot be
healed by expeditious economic first aid. Marginal changes will not be sufficient. What is required are fundamental institutional reforms, based upon a
recognition of a common interest and mutual concern, in an increasingly interdependent world. What is required is a new international order in which all
benefit from change.

Whatever agreements are eventually negotiated, they must balance the interests of the rich and poor nations. All nations have to carefully weigh the costs of
disruption against the costs of accommodation and to consider the fact that
any conceivable cost of a new deal will be a very small proportion of their future growth in an orderly cooperative framework.

Not only is there an overwhelming need for change, but there are increased
opportunities for organizing change also. Mankind's history is the story of a
continuous process of change, of evolution in the face of threats and dangers. It
is this process which provides the dynamism in the struggle to continually improve living conditions and to increase control over nature. The process, always subject to autonomous human interference, is occasionally subject to
'historical discontinuities', breaks or 'mutations' in trends which provide increased opportunities for taking new initiatives in redirecting the evolutionary
process. Four such discontinuities have recently occurred and are of very great
importance internationally.

The process of social change is subject to a variety of *quantitative development patterns*. These, familiar in many sciences, have a time dimension and may apply to any system of interaction. *Cycles* ('l'histoire se répète') are a familiar component of such patterns. Well-known examples include cycles in climate and business. Other components may be one-sided rather than cyclical movements. These can be either downward or upward, either accelerated ('*runaway*' or 'escalating') or decelerated, moving towards some *saturation* level ('trees don't grow into heaven'). In some cases saturation might constitute an optimum and therefore comprise a desirable target. In other cases, it may move to a level deemed too low, suggesting, among other possible states, *stagnation*. Of these various prototypes of evolution only a movement towards saturation or harmony is generally considered desirable. In contrast, cycles or stagnation often mean waste or annoyance and runaway movements are prone to result in disaster.

All the movement types described can be suddenly interrupted by great 'mutations' or unexpected interferences. In the field of technological development, such proved to be the case with, for example, the discovery of the steam engine, electricity and nuclear power. Social mutations have sometimes taken the form of *revolutions*, the French, the Russian, the Cuban, the Chinese and the Algerian revolutions being important examples. In terms of the process of change, the events comprised a moment of great 'historical discontinuity' for the countries concerned and were of such momentous importance that they changed the pattern of world relations. Following the revolutions, decades of evolutionary change were required to attain the aims pursued; the communist phase in Marxist terms, or social equity in the Western European democracies were by no means achieved overnight and have not been fully achieved yet.

Firstly, one of the world's superpowers was obliged against its will to leave a part of the world where it had intervened in support of a non-representative regime. That this immensely powerful nation was unable to use its full military capability is of enormous importance. Secondly, there are the demands of the Third World, strengthened by the OPEC actions, a development which will be increasingly felt in international fora in the coming decade. Thirdly, there is the increasing incapacity of the rich and privileged nations, the industrialized world, to come to grips with the economic imbalances in the present international system and to deal creatively with the collapse of their own invention; this combined with growing alienation and frustration and the threats to basic human values and human environment engendered by pressures to consume. And fourthly, there is a growing recognition of global interdependencies and of the fact that no nation, however powerful it may believe itself to be, can realistically pursue its policies in isolation.

The demands for a new international order must be placed within this historical process. On one level of reasoning, it is a natural evolution of the philosophy already accepted at the national level: that the government must actively intervene on behalf of the poorest segments of their populations ('the bottom 40 per cent') who will otherwise be bypassed by economic development. In a fast shrinking planet, it was inevitable that this 'new' philosophy would not stop at national borders; and, since there is no world government, the poor nations are bringing this concern to its closest substitute, the United Nations. At another level, the demand for structural change is, as we have seen, a natural second state in the evolution of Third World countries; a movement from political to economic equality.

There is of course nothing new in the existence of rich and poor. History has known nothing else and has in part been shaped by the struggles between them. But the rich and poor have in the past mainly existed within individual societies. What is relatively new is the enormous differences among societies. It is moreover a visible difference; the rich cannot conceal their wealth in a 'global village'. The glaring differences are perceived by the poor thanks, perhaps paradoxically, to the rich world's technological dexterity. And their perception of these differences will, in a shrinking world, exert growing stress on already frail international institutions.

Mankind's future depends upon it coming to terms with these differences, with developing a new understanding and awareness, based upon interdependence and mutual interest of working and living together. Recent discontinuities in the process of change have placed mankind on the threshold of new choices. In choosing among them, it will have to accept the harsh fact that, perhaps contrary to previous times, it has just one future or no future at all.

Tinbergen, J. (coord.) & A.J. Dolman (ed.), *Reshaping the International Order; A Report to the Club of Rome,* Hutchinson, 1977, p. 11-24.

Notes and References

1   The word billion when used in this report represents one thousand million, being equivalent to 'milliard' used in Continental Europe. Trillion refers to one million million. All dollars are US dollars.

2   While in the past Romania paid $21 per ton for such imports, in 1974 the price increased to $126, a loss which had to be offset by export of petroleum products. This, combined with increases in the price of other imported raw materials, left Romania in 1974 with a balance of payments deficit with centrally planned nations of over $200 million.

3   Magyar Nemzet, Boedapest, 1974.

4   'Oil and Gold-Price Increases Russian Windfall Profits', *The New York Times,* january 11, 1975.

5   *Economic Bulletin for Europe,* no. 1, Volume 26, 1974.

6 B. Pichugin: 'East West Economic Cooperation', *International Affairs*, Moscow, no. 8, 1975.

7 See Gunnar Adler-Karlsson: 'Economic and Trade Policies', *International Institute for Peace Research*, Vienna, no.3, september 1972, pp. 36-37.

8 These figures are based on official exchange rates and do not reflect differences in purchasing power.

9 In an official document submitted to the UN, the Romanian Government takes a clear-cut position in favour of the new order:

'In a situation of growing economic interdependence of states in which no country can isolate itself from world economic development, it is clearly necessary that all states, regardles of their social system, the extent of their territories or their economic potential, should make an active contribution to the solution of the major economic problems currently facing the world'.

CHAPTER 19

Gaia

A New Look at Life on Earth

J.E. LOVELOCK

Introductory

As I write, two Viking spacecraft are circling our fellow planet Mars, awaiting landfall instructions from the Earth. Their mission is to search for life, or evidence of life, now or long ago. This book also is about a search for life, and the quest for Gaia is an attempt to find the largest living creature on Earth. Our journey may reveal no more than the almost infinite variety of living forms which have proliferated over the Earth's surface under the transparent case of the air and which constitute the biosphere. But if Gaia does exist, then we may find ourselves and all other living things to be parts and partners of a vast being who in her entirety has the power to maintain our planet as a fit and comfortable habitat for life.

The quest for Gaia began more than fifteen years ago, when NASA (the National Aeronautics and Space Administration of the USA) first made plans to look for life on Mars. It is therefore right and proper that this book should open with a tribute to the fantastic Martian voyage of those two mechanical Norsemen.

In the early nineteen-sixties I often visited the Jet Propulsion Laboratories of the California Institute of Technology in Pasadena, as consultant to a team, later to be led by that most able of space biologists Norman Horowitz, whose main objective was to devise ways and means of detecting life on Mars and other planets. Although my particular brief was to advise on some comparatively simple problems of instrument design, as one whose childhood was illuminated by the writings of Jules Verne and Olaf Stapledon I was delighted to have the chance of discussing at first hand the plans for investigating Mars.

At that time, the planning of experiments was mostly based on the assumption that evidence for life on Mars would be much the same as for life on Earth. Thus one proposed series of experiments involved dispatching what was, in effect, an automated microbiological laboratory to sample the Martian soil and judge its suitability to support bacteria, fungi, or other micro-organisms. Additional soil experiments were designed to test for chemicals whose presence would indicate life at work: proteins, amino-acids, and particularly optically active substances with the capacity that organic matter has to twist a beam of polarized light in a counter-clockwise direction.

After a year or so, and perhaps because I was not directly involved, the euphoria arising from my association with this enthralling problem began to subside, and I found myself asking some rather down-to-earth questions, such as, 'How can we be sure that the Martian way of life, if any, will reveal itself to tests based on Earth's life style?' To say nothing of more difficult questions, such as 'What is life, and how should it be recognized?'

Some of my still sanguine colleagues at the Jet Propulsion Laboratories mistook my growing scepticism for cynical disillusion and quite properly asked, 'Well, what would you do instead?' At that time I could only reply vaguely, 'I'd look for an entropy reduction, since this must be a general characteristic of all forms of life.' Understandably, this reply was taken to be at the best unpractical and at worst plain obfuscation, for few physical concepts can have caused as much confusion and misunderstanding as has that of entropy.

It is almost a synonym for disorder and yet, as a measure of the rate of dissipation of a system's thermal energy, it can be precisely expressed in mathematical terms. It has been the bane of generations of students and is direfully associated in many minds with decline and decay, since its expression in the Second Law of Thermodynamics (indicating that all energy will eventually dissipate into heat universally distributed and will no longer be available for the performance of useful work) implies the predestined and inevitable run-down and death of the Universe.

Although my tentative suggestion had been rejected, the idea of looking for a reduction or reversal of entropy as a sign of life had implanted itself in my mind. It grew and waxed fruitful until, with the help of many colleagues, Dian Hitchcock, Sidney Epton, Peter Simmonds, and especially Lynn Margulis, it evolved into the hypothesis which is the subject of this book.

Back home in the quiet countryside of Wiltshire, after my visits to the Jet Propulsion Laboratories, I had time to do more thinking and reading about the real character of life and how one might recognize it anywhere and in any guise. I expected to discover somewhere in the scientific literature a comprehensive definition of life as a physical process, on which one could base the design of life-detection experiments, but I was surprised to find how little had been written about the nature of life itself. The present interest in ecology and the application of systems analysis to biology had barely begun and there was still in those days the dusty academic air of the classroom about the life sciences. Data galore had been accumulated on every conceivable aspect of living species, from their outermost to their innermost parts, but in the whole vast encyclopaedia of facts the crux of the matter, life itself, was almost totally ignored. At best, the literature read like a collection of expert reports, as if a group of scientists from another world had taken a television receiver home with them and had reported on it. The chemist said it was made of wood, glass, and metal. The physicist said it radiated heat and light. The engineer said the supporting wheels were too small and in the wrong place for it to run smoothly on a flat surface. But nobody said what it was.

This seeming conspiracy of silence may have been due in part to the division of science into separate disciplines, with each specialist assuming that someone else has done the job. Some biologists may believe that the process of life is adequately described by some mathematical theorem of physics or cybernetics, and some physicists may assume that it is factually described in the recondite writings on molecular biology which one day he will find time to read. But the most probable cause of our closed minds on the subject is that we already have a very rapid, highly efficient life-recognition programme in our inherited set of instincts, our 'read-only' memory as it might be called in computer technology. Our recognition of living things, both animal and vegetable, is instant and automatic, and our fellow-creatures in the animal world appear to have the same facility. This powerful and effective but unconscious process of recognition no doubt originally evolved as a survival factor. Anything living may be edible, lethal, friendly, aggressive, or a potential mate, all questions of prime significance for our welfare and continued existence. However, our automatic recognition system appears to have paralysed our capacity for conscious thought about a definition of life. For why should we need to define what is obvious and unmistakable in all its manifestations, thanks to our built-in programme? Perhaps for that very reason, it is an automatic process operating without conscious understanding, like the autopilot of an aircraft.

Even the new science of cybernetics has not tackled the problem, although it is concerned with the mode of operation of all manner of systems from the simplicity of a valve-operated water tank to the complex visual control process which enables your eyes to scan this page. Much, indeed, has already been said and written about the cybernetics of artificial intelligence, but the question of defining real life in cybernetic terms remains unanswered and is seldom discussed.

During the present century a few physicists have tried to define life. Bernal, Schroedinger, and Wigner all came to the same general conclusion, that life is a member of the class of phenomena which are open or continuous systems able to decrease their internal entropy at the expense of substances or free energy taken in from the environment and subsequently rejected in a degraded form. This definition is not only difficult to grasp but is far too general to apply to the specific detection of life. A rough paraphrase might be that life is one of those processes which are found whenever there is an abundant flow of energy. It is characterized by a tendency to shape or form itself as it consumes, but to do so it must always excrete low-grade products to the surroundings.

We can now see that this definition would apply equally well to eddies in a flowing stream, to hurricanes, to flames, or even to refrigerators and many other man-made contrivances. A flame assumes a characteristic shape as it burns, and needs an adequate supply of fuel and air to keep going, and we are now only too well aware that the pleasant warmth and dancing flames of an open fire have to be paid for in the excretion of waste heat and pollutant gases. Entropy is reduced locally by the flame formation, but the overall total of entropy is increased during the fuel consumption.

Yet even if too broad and vague, this classification of life at least points us in the right direction. It suggests, for example, that there is a boundary, or interface, between the 'factory' area where the flow of energy or raw materials is put to work and entropy is consequently reduced, and the surrounding environment which receives the discarded waste products. It also suggests that life-like processes require a flux of energy above some minimal value in order to get going and keep going. The nineteenth-century physicist Reynolds observed that turbulent eddies in gases and liquids could only form if the rate of flow was above some critical value in relation to local conditions. The Reynolds dimensionless number can be calculated from simple knowledge of a fluid's properties and its local flow boundaries. Similarly, for life to begin, not only the quantity but also the quality, or potential, of the energy flow must be sufficient. If, for example, the sun's surface temperature were 500 degrees instead of 5,000 degrees Centigrade and the Earth were correspondingly closer, so that we received the same amount of warmth, there would be little difference in climate, but life would never have got going. Life needs energy potent enough to sever chemical bonds; mere warmth is not enough.

It might be a step forward if we could establish dimensionless numbers like the Reynolds scale to characterize the energy conditions of a planet. Then those enjoying, with the Earth, a flux of free solar energy above these critical values would predictably have life whilst those low on the scale, like the cold outer planets, would not.

The design of a universal life-detection experiment based on entropy reduction seemed at this time to be a somewhat unpromising exercise. However, assuming that life on any planet would be bound to use the fluid media—oceans, atmosphere, or both—as conveyor-belts for raw materials and waste products, it occurred to me that some of the activity associated with concentrated entropy reduction within a living system might spill over into the conveyor-belt regions and alter their composition. The atmosphere of a life-bearing planet would thus become recognizably different from that of a dead planet.

Mars has no oceans. If life had established itself there, it would have had to make use of the atmosphere or stagnate. Mars therefore seemed a suitable planet for a life-detection exercise based on chemical analysis of the atmosphere. Moreover, this could be carried out regardless of the choice of landing site. Most life-detection experiments are effective only within a suitable target area. Even on Earth, local search techniques would be unlikely to yield much positive evidence of life if the landfall occurred on the Antarctic ice sheet or the Sahara desert or in the middle of a salt lake.

While I was thinking on these lines, Dian Hitchcock visited the Jet Propulsion Laboratories. Her task was to compare and evaluate the logic and information-potential of the many suggestions for detecting life on Mars. The notion of life detection by atmospheric analysis appealed to her, and we began developing the idea together. Using our own planet as a model, we examined the extent to which simple knowledge of the chemical composition of the Earth's atmosphere, when coupled with such readily accessible information as the degree of solar radiation and the presence of oceans as well as land masses on the Earth's surface, could provide evidence for life.

Our results convinced us that the only feasible explanation of the Earth's highly improbable atmosphere was that it was being manipulated on a day-to-day basis from the surface, and that the manipulator was life itself. The significant decrease in entropy—or, as a chemist would put it, the persistent state of disequilibrium among the atmospheric gases—was on its own clear proof of life's activity. Take, for example, the simultaneous presence of methane and oxygen in our atmosphere. In sunlight, these two gases react chemically to give carbon dioxide and water vapour. The rate of this reaction is such that to sustain the amount of methane always present in the air, at least 1,000 million tons of this gas must be introduced into the atmosphere yearly. In addition, there must be some means of replacing the oxygen used up in oxidizing methane and this requires a production of at least twice as much oxygen as methane. The quantities of both of these gases required to keep the earth's extraordinary atmospheric mixture constant was improbable on an abiological basis by at least 100 orders of magnitude.

Here, in one comparatively simple test, was convincing evidence for life on Earth, evidence moreover which could be picked up by an infra-red telescope sited as far away as Mars. The same argument applies to other atmospheric gases, especially to the ensemble of reactive gases constituting the atmosphere as a whole. The presence of nitrous oxide and of ammonia is as anomalous as that of methane in our oxidizing atmosphere. Even nitrogen in gaseous form is out of place, for with the Earth's abundant and neutral oceans, we should expect to find this element in the chemically stable form of the nitrate ion dissolved in the sea.

Our findings and conclusions were, of course, very much out of step with conventional geochemical wisdom in the mid-sixties. With some exceptions, notably Rubey, Hutchinson, Bates, and Nicolet, most geochemists regarded the atmosphere as an end-product of planetary outgassing and held that subsequent reactions by abiological processes had determined its present state. Oxygen, for example, was thought to come solely from the breakdown of water vapour and the escape of hydrogen into space, leaving an excess of oxygen behind. Life merely borrowed gases from the atmosphere and returned them unchanged. Our contrasting view required an atmosphere which was a dynamic extension of the biosphere itself. It was not easy to find a journal prepared to publish so radical a notion but, after several rejections, we found an editor, Carl Sagan, prepared to publish it in his journal, *Icarus.*

Nevertheless, considered solely as a life-detection experiment, atmospheric analysis was, if anything, too successful. Even then, enough was known about the Martian atmosphere to suggest that it consisted mostly of carbon dioxide and showed no signs of the exotic chemistry characteristic of Earth's atmosphere. The implication that Mars was probably a lifeless planet was unwelcome news to our sponsors in space research. To make matters worse, in September 1965 the US Congress decided to abandon the first Martian exploration programme, then called Voyager. For the next year or so, ideas about looking for life on other planets were to be discouraged.

Space exploration has always served as a convenient whipping-boy to those needing money for some worthy cause, yet it is far less expensive than many a stuck-in-the-mud, down-to-earth technological failure. Unfortunately, the apologists for space science always seem over-impressed by engineering trivia and make far too much of non-stick frying pans and perfect ball-bearings. To my mind, the outstanding spin-off from space research is not new technology. The real bonus has been that for the first time in human history we have had a chance to look at the Earth from space, and the information gained from seeing from the outside our azuregreen planet in all its global beauty has given rise to a whole new set of questions and answers. Similarly, thinking about life on Mars gave some of us a fresh standpoint from which to consider life on Earth and led us to formulate a new, or perhaps revive a very ancient, concept of the relationship between the Earth and its biosphere.

By great good fortune, so far as I was concerned, the nadir of the space programme coincided with an invitation from Shell Research Limited for me to consider the possible global consequences of air pollution from such causes as the ever-increasing rate of combustion of fossil fuels. This was in 1966, three years before the formation of Friends of the Earth and similar pressure-groups brought pollution problems to the forefront of the public mind.

Like artists, independent scientists need sponsors but this rarely involves a possessive relationship. Freedom of thought is the rule. This should hardly need saying, but nowadays many otherwise intelligent individuals are conditioned to believe that all research work supported by a multi-national corporation must be suspect by origin. Others are just as convinced that similar work coming from an institution in a Communist country will have been subject to Marxist theoretical constraint and will therefore be diminished. The ideas and opinions expressed in this book are inevitably influenced to some degree by the society in which I live and work, and especially by close contact with numerous scientific colleagues in the West. So far as I know, these mild pressures are the only ones which have been exerted on me.

The link between my involvement in problems of global air pollution and my previous work on life detection by atmospheric analysis was, of course, the idea that the atmosphere might be an extension of the biosphere. It seemed to me that any attempt to understand the consequences of air pollution would be incomplete and probably ineffectual if the possibility of a response or an adaptation by the biosphere was overlooked. The effects of poison on a man are greatly modified by his capacity to metabolize or excrete it; and the effect of loading a biospherically controlled atmosphere with the products of fossil fuel combustion might be very different from the effect on a passive inorganic atmosphere. Adaptive changes might take place which would lessen the perturbations due, for instance, to the accumulation of carbon dioxide. Or the perturbations might trigger some compensatory change, perhaps in the climate, which would be good for the biosphere as a whole but bad for man as a species.

Working in a new intellectual environment, I was able to forget Mars and to concentrate on the Earth and the nature of its atmosphere. The result of this more single-minded approach was the development of the hypothesis that the entire range of living matter on Earth, from whales to viruses, and from oaks to algae, could be regarded as constituting a single living entity, capable of manipulating the Earth's atmosphere to suit its overall needs and endowed with faculties and powers far beyond those of its constituent parts.

It is a long way from a plausible life-detection experiment to the hypothesis that the Earth's atmosphere is actively maintained and regulated by life on the surface, that is, by the biosphere. Much of this book deals with more recent evidence in support of this view. In 1967 the reasons for making the hypothetical stride were briefly these:

Life first appeared on the Earth about 3,500 million years ago. From that time until now, the presence of fossils shows that the Earth's climate has changed very little. Yet the output of heat from the sun, the surface properties of the Earth, and the composition of the atmosphere have almost certainly varied greatly over the same period.

The chemical composition of the atmosphere bears no relation to the expectations of steady-state chemical equilibrium. The presence of methane, nitrous oxide, and even nitrogen in our present oxidizing atmosphere represents violation of the rules of chemistry to be measured in tens of orders of magnitude. Disequilibria on this scale suggest that the atmosphere is not merely a biological product, but more probably a biological construction: not living, but like a cat's fur, a bird's feathers, or the paper of a wasp's nest, an extension of a living system designed to maintain a chosen environment. Thus the atmospheric concentration of gases such as oxygen and ammonia is found to be kept at an optimum value from which even small departures could have disastrous consequences for life.

The climate and the chemical properties of the Earth now and throughout its history seem always to have been optimal for life. For this to have happened by chance is as unlikely as to survive unscathed a drive blindfold through rush-hour traffic.

By now a planet-sized entity, albeit hypothetical, had been born, with properties which could not be predicted from the sum of its parts. It needed a name. Fortunately the author William Golding was a fellow-villager. Without hesitation he recommended that this creature be called Gaia, after the Greek Earth goddess also known as Ge, from which root the sciences of geography and geology derive their names. In spite of my ignorance of the classics, the suitability of this choice was obvious. It was a real four-lettered word and would thus forestall the creation of barbarous acronyms, such as Biocybernetic Universal System Tendency/Homoeostasis. I felt also that in the days of Ancient Greece the concept itself was probably a familiar aspect of life, even if not formally expressed. Scientists are usually condemned to lead urban lives, but I find that country people still living close to the earth often seem puzzled that anyone should need to make a formal proposition of anything as obvious as the Gaia hypothesis. For them it is true and always has been.

I first put forward the Gaia hypothesis at a scientific meeting about the origins of life on Earth which took place in Princeton, New Jersey, in 1969. Perhaps it was poorly presented. It certainly did not appeal to anyone except Lars Gunnar Sillen, the Swedish chemist now sadly dead, and Lynn Margulis, of Boston University, who had the task of editing our various contributions. A year later in Boston Lynn and I met again and began a most rewarding collaboration which, with her deep knowledge and insight as a life scientist, was to go far in adding substance to the wraith of Gaia, and which still happily continues.

We have since defined Gaia as a complex entity involving the Earth's biosphere, atmosphere, oceans, and soil; the totality constituting a feedback or cybernetic system which seeks an optimal physical and chemical environment for life on this planet. The maintenance of relatively constant conditions by active control may be conveniently described by the term 'homoeostasis'.

Gaia has remained a hypothesis but, like other useful hypotheses, she has already proved her theoretical value, if not her existence, by giving rise to experimental questions and answers which were profitable exercises in themselves. If, for example, the atmosphere is, among other things, a device for conveying raw materials to and from the biosphere, it would be reasonable to assume the presence of carrier compounds for elements essential in all biological systems, for example, iodine and sulphur. It was rewarding to find evidence that both were conveyed from the oceans, where they are abundant, through the air to the land surface, where they are in short supply. The carrier compounds, methyl iodide and dimethyl sulphide respectively, are directly produced by marine life. Scientific curiosity being unquenchable, the presence of these interesting compounds in the atmosphere would no doubt have been discovered in the end and their importance discussed without the stimulus of the Gaia hypothesis. But they were actively sought as a result of the hypothesis and their presence was consistent with it.

If Gaia exists, the relationship between her and man, a dominant animal species in the complex living system, and the possibly shifting balance of power between them, are questions of obvious importance. I have discussed them in later chapters, but this book is written primarily to stimulate and entertain. The Gaia hypothesis is for those who like to walk or simply stand and stare, to wonder about the Earth and the life it bears, and to speculate about the consequences of our own presence here. It is an alternative to that pessimistic view which sees nature as a primitive force to be subdued and conquered. It is also an alternative to that equally depressing picture of our planet as a demented spaceship, forever travelling, driverless and purposeless, around an inner circle of the sun.

Lovelock, J.E., *Gaia; A New Look at Life on Earth*, Oxford University Press, Oxford, 1979, p. 1-12.

Part IV

THE EIGHTIES

Introduction to Part IV: The Eighties

The 1980s was an important decade for the field of environmental studies. In this period many books were published that have proven very influential, even today. This may seem strange, as the '80s are not on record as being the best years for the environment. There were many environmental catastrophes and scandals, as well as Reaganism and Thatcherism; policies developed by a president and a prime minister who became known for their solid faith in the market mechanism. They overlooked the fact that the market can only realise an optimal use and efficiency of priced inputs in the production process, which implies that the unpriced input of natural and environmental resources were not given full attention. They did not really have 'a close relationship' with the environment. Many people were confronted with an economic recession: their jobs were on the line and they were forced to concentrate on 'bread and butter' issues. In other words, the general social and political climate was not very favourable for the environment.

Environmental catastrophes

Perhaps the most radical event in the field of the environment in the '80s was the Chernobyl nuclear plant disaster of 1986. The radiation fallout caused a lot of human suffering and environmental damage in the immediate vicinity as well as in more remote areas. Radioactive clouds spread to other parts of the world and had an impact on the environment there as well. People were informed by radio and television about the location of the radioactive clouds. It was a morbid experience that before could only have been seen in a science fiction movie. As a result of this accident, the debate over nuclear power was (re)opened and the effects of the acclaimed technology were up for discussion. This was not an isolated incident. There were many others as well, including the sinking of oiltankers; the Amoco Cadiz, stranded on the coast of France and disrupting marine life for hundreds of miles, became an infamous case for fishermen as well as environmentalists. The effects were visible and tangible: a thick smear of oil on the beaches and dying birds and fish. Soil pollution scandals occupied local and regional politics. The number of sites that had been polluted added up to many tens of thousands. In many cases, it became almost impossible to use the site. In a number of cases, the soil had to be removed and cleaned as soon as possible. The exhaustion of natural resources continued: there was a general concern about the depletion of fossil fuels and raw materials; many animal and floral species became extinct; deserts claimed more and more land; the tearing down of tropical rainforests continued; the ozone layer got thinner and thinner and the greenhouse effect became apparent; the pollution of air and water increased in many areas of the globe. Scandinavian lakes and many forests in various parts of Europe were affected by acid rain.

Environment under pressure

The general social and political climate of the '80s was—as just illustrated—anything but a healthy one for the environment. The economic recession of the early 1980s focused people's attention on issues of unemployment, 'modern' poverty, decreasing purchasing power, bankruptcies in industry and trade, government cutbacks in expenditures, etc. Government policy was mainly focused on restructuring the economy, the fight against unemployment, stimulating free enterprise and on government regulating social life to a lesser degree. It was said that in the '70s, the welfare state had become a system that was no longer affordable and people on welfare were made into slaves of the state. It was in this social context that social and economic politics were primarily focused on increasing the profitability of corporations and improving the investment climate. Concern for the environment was not left completely out of the equation, but was (no longer) a primary political goal. Expenditures for the environment were endangered because of government cutbacks. Government was being too active, causing expenditures to rise sky-high. Budgetary restrictions were being enforced. These restrictions, however, did not lead to an absolute decrease in expenditures for tackling environmental problems, and in some cases, not even to a relative decrease. But the number of environmental problems exceeded the rate of public means to confront these problems.

Bread and butter

The public had other things to worry about. People's jobs were on the line; a number of people were made redundant. 'Bread and butter' problems rather than environmental issues became their primary concerns. It is striking, however, that the environmental awareness of the public in that period didn't really diminish. Surveys of Dutch citizens showed that environmental awareness in the period 1975-1985 remained reasonably high. Environmental behaviour, however, did not keep pace with environmental awareness. Although citizens generally held an environmentally correct position, they did not translate this into environmentally correct action, in part because the infrastructure did not encourage environmentally sound behaviour.

Change of strategy by environmentalists

Environmentalists were very active in this period, but changed their strategy. Many environmentalist organizations chose the path of deliberation. Representatives of these organizations joined committees, panels, advisory councils, etc. They tried to influence decision-making in matters concerning the environment. This made it possible to actually work side-by-side on the further details and implementation of environmental policies. This caused a great deal of controversy within environmentalist organizations. Many organizations didn't want to be absorbed into mainstream deliberations and meetings. They feared being smothered by the establishment and were concerned that too many concessions had to be made at the expense of the environment. The environment would not be helped by compromise. A number of environmentalist groups chose to move in a different direction. Various members of the environmentalist movement (Greenpeace, among others) continued to find strength in rallying and campaigning. They expected the best results from protest campaigns, sit-ins, petitions, and spectacular operations. Other parties chose the alternative path and practised an environmentally correct lifestyle in relatively small groups, keeping to themselves. They tried to live modestly and showed us that a lifestyle of addictive consumption is not inevitable. It was a comfortable (and healthy) life with non-pesticide vegetables, environmentally correct products, recycling, less packaging, less meat, etc. Generally speaking, the role of the environmentalists was less prominent in this period than in the '60s and '70s.

Stability of environmental sector

Despite the relatively unfavourable social and political climate, the 1980s were not without value to the environment and environmental policy. Environmental issues were permanently placed on the political and policy agenda of many Western countries. In many countries, a department or bureau was created to develop environmental policy. These institutions were an important counterbalance for the ruling conservative ideology of the '80s. The degree of institutionalization of environmental policy was so high that it could no longer be ignored. The number of people that had become directly or indirectly professionally dependent on the 'environmental sector' had grown. This group was a stable factor in the promotion of environmental interests. They made sure that money and other resources were reserved for the environment. Environmental policy kept its leverage, despite the fact that the political climate was all but favourable.

Expanding environmental policy

In spite of the economic crisis, environmental policy was expanding. Initially this was done by introducing a system of environmental legislation to curb various forms of air, water, and soil pollution. In many countries, a relatively comprehensive system of legislation and rules in the field of the environment was enforced. Particularly in the United States, but also in a large number of Western European countries, the enforcement of legislation was viewed as a necessary action. At first people felt confident that the system of legislation and rules would have a regulating effect. They believed that if a law was enforced, the problems would disappear. This optimism proved to be short-lived. During the '80s the ever increasing number of rules in the field of the environment came under heavy criticism. The efficacy of laws and other formal rules was also criticized. More legislation did not necessarily mean a cleaner environment; it meant that craftier methods for dodging environmental laws were developed. As a consequence of the explosive increase in environmental legislation and rules, many countries reduced the number of environmental regulations. Deregulation became a popular concept: it was necessary to reduce the number of rules. They needed to be comprehensible and, above all, rules needed to be feasible and enforceable.

New policy instruments

Environmental problems were generally seen as a problem of public health in the 1950s and 1960s. This implied that new policies were generally based on the legal format of existing public health regulations, in which 'command and control' was the normal practice. Politicians and policy-makers, however, became increasingly aware of the fact that ecological problems were becoming dominant in this field. In this new situation, the rationale for physical regulation became more or less marginal. Furthermore, these types of environmental policies did not lead to the desired situation, as many polluters developed a strong strategy to bargain with authorities about the level of the norms and to weaken them, arguing that international competition made it impossible to introduce strict norms in their sector. Additionally, many environmental economists argued that the implementation of economic instruments would undoubtedly lead to better environmental results and substantially lower costs. Emphasis was put on the behaviour-regulatory effect of market-conform policy instruments in the form of taxes and subsidies. In the course of the '80s, a number of financial incentives were enforced to encourage polluters to act in a more environmentally responsible way. These incentives came in many forms, for example, subsidies to stimulate environmentally correct behaviour, the manufacturing of environmentally correct products and production, but also taxes to reduce environmentally burdening activities. The business community was very much against the taxes that were being imposed, but in many countries government acted with vigour and left business no choice. Although many white papers and governmental documents gave a high profile to economic in- struments, one cannot overlook the fact that environmental policies in Western countries are mainly based on the implementation of some form of emission or deposit standards and that legal instruments are still dominant. Consequently, industries and businesses invested in environmental technology and other en- vironmental measures. The view that pollution prevention pays matured.

Pollution prevention pays

External (governmental) pressure, as well as a growing sense of responsibility on the part of the business community towards the environment, helped create new systems such as environmental accounting and corporate environmental care. The belief that 'pollution prevention pays' became more than just a slo- gan; it became one of the cornerstones of modern business policy. This meant that business was more actively—internally as well as externally—involved with the environment. This active concern of business for the environment took at least two forms. On the one hand, there was an explosive growth of commercial activities in industry as well as service organizations that were of- fering environmentally friendly products and services. All sorts of 'eco-pro- ducts' appeared on the market, and businesses were profiling themselves with slogans that stressed their environmentally friendly way of producing the pro- ducts. Consultancies mushroomed and made money on the increasing interest of business in the environment. On the other hand, a kind of 'ecologization' of existing branches took place. A process was introduced to make production operate in a more environmentally friendly way. Existing processes were put to the test in terms of their environmental friendliness, and if it turned out that they did not meet the required standards, measures were taken to attain the en- vironmental goals that had been set. This image of pro-active business ignores the many environmental scandals of the '80s. Some businesses did not take en- vironmental issues seriously, risking high fines and damage to their image and prestige.

Economy versus ecology

None of the issues presented earlier can be separated from the escalating de- bate over the relationship between economy and ecology that took place in the 1980s. Apparently, there were two conflicting opinions. On the one hand, there were those who believed that economy and ecology were at odds and therefore an expanding economy would automatically mean more pollution, more de- pletion of natural resources, and more damage to ecosystems. On the other hand, there were those who thought that economy and ecology are dependent on each other and that confronting environmental problems requires sufficient resources which can only be obtained by economic growth. Market-oriented economists shared that second opinion. They believed that economic growth was essential for ecological recovery. Environmentalists and environmental economists contested this view by pointing out that economic growth—even if it were to be selective—would eventually lead to the exhaustion of resources and deterioration of ecological systems. The growth of the economy requires the expansion of space-consuming activities such as the construction of industrial sites and the expansion of the infrastructure.

Institutionalization of environmental education

Furthermore, it is worth mentioning that the '80s had formed the basis of the institutionalization of environmental education at different levels. In this period, many universities organized courses concerning nature and the environment. In a number of cases, these efforts led to the establishment of a separate branch of science. What happened in universities often also took place in professional education, where courses and training programmes were created. The position of these courses and programmes was, however, quite marginal; environmental studies weren't really fully appreciated by other academic disciplines. The practical character of these studies gave it an aura of an applied science; a label which, according to the academic establishment, disqualified the studies. Concern for nature and environmental issues in primary and secondary education was meagre at first. In the course of the '80s, it changed somewhat, but establishing these studies in educational programmes just needed more time.

The Global 2000 Report to the President

In 1980 *The Global 2000 Report to the President* was published in the United States. The Global 2000 Study was shaped by a very brief directive in President Carter's May 23, 1977, Environmental Message: '. . . I am directing the Council on Environmental Quality and the Department of State, working in cooperation with . . . other appropriate agencies, to make a one-year study of probable changes in the world's population, natural resources, and environment through the end of the century. This study will serve as the foundation of our longer-term planning.' The central message of the Global 2000 Report is, if public policies around the world continue unchanged through the end of the century, a number of serious world problems will become worse: the world in the year 2000 will be more crowded, more polluted, less stable ecologically, and more vulnerable to disruption than the world we live in now. The world's population will have grown from 4 billion in 1975 to 6.35 billion in 2000. The gross national product per capita in most less developed countries will remain low. The food situation of the less developed countries will scarcely improve or will actually decline below present inadequate levels. The world's finite fuel resources are theoretically sufficient for centuries, but they are not evenly distributed. Regional water shortages will become more severe. Significant losses of world forests will continue. Serious deterioration of agricultural soils will occur worldwide. Atmospheric concentrations of carbon dioxide and ozone-depleting chemicals are expected to increase at rates that could alter the world's climate and upper atmosphere significantly by 2050. Extinctions of plant and animal species will increase dramatically. Yet there is reason for hope. Policies are beginning to change. The needed changes, however, go far beyond the capability and responsibility of a single nation. An era of unprecedented cooperation and commitment is essential. If decisions are delayed until the problems become worse, options for effective action will be severely reduced.

Global 2000 gives an outline of an alarming situation and calls for immediate action. This publication did even more than the report of the Club of Rome to raise the awareness of the American public to the fact that the earth is in a critical condition. This study convinced, above all, the US government and congress of the seriousness of the environmental issue. Moreover, the report emphasized the important role the American government could have in the attack on world environmental problems. This entailed a moral and political appeal to the American government to take up this 'world task' of saving the planet. The US was seen as the saviour of our planet, which really charmed the American public: the US is not just the protector of world peace, but guardian of the global environment as well.

World Conservation Strategy

Private environmentalist organizations had been demanding attention for the critical situation of the environment for over 10 years. Through protest marches and rallies they had put pressure on local, regional, and national governments to do something about environmental problems. They did this by means of various action groups all over the world that devoted their energy to the conservation of nature and stopping the decay of ecosystems, the depletion of natural resources, and further deterioration of the environment. Greenpeace drew attention to environmental scandals by spectacular actions and tried to prevent further degradation of the environment. Publications by various co-ordinating environmentalist organizations also helped raise public consciousness. The 'International Union for Conservation of Nature and Natural Resources' published an important document entitled, *World Conservation Strategy; Living Resource Conservation for Sustainable Development* (1980).

This publication emphasized the earth's limited supply of natural resources. The decay of nature and the environment is seen as a great threat for the future. The strategy suggested has all the necessary ingredients for finding a solution to the problem. In the first chapters, a connection is explicitly made between natural resources, the number of people that use these resources, and the polluting techniques of production. Consequently, an accurate inventory and analysis of the problems for each field is made and policies are developed for implementation at the national and international level. The general thrust of the argument is that generations to come must be able to live a comfortable life. The document also addresses the differences between the wealthy West and the developing countries, and argues that, in many instances, people in developing countries are forced to deplete the natural resources. In the Strategy, the concept 'sustainable development' plays a major role, put forward as a concept that invites us to define and analyze it.

The latter raises many different points. Firstly, it is a misconception that the notion of 'sustainable development' was first propagated by the Brundtland Committee. The concept had already been used in the Strategy, almost ten years earlier. Why did it get so little attention in 1980, while in 1987 everyone fought to get a copy of the Brundtland report? Was it the status of the messenger? The World Conservation Strategy had been written by the IUCN and can therefore be seen as an attempt by environmentalists to influence policy. The Brundtland Committee was formed under the auspices of the United Nations, which gave the deliberations more credibility. What role did the spirit of the age play? Was society in 1980 not up to discussing an issue that was only put on the international agenda in the second half of the 1980s? Whatever the reason, the concept 'sustainable development' only became popular years later.

Our Common Future

In the last decades, prosperity in the world has increased, but this increase in wealth is not equitably distributed. The developing countries became—in a global sense—poorer and poorer and the developing countries became more prosperous. At first this developmental issue was viewed as separate from the environmental issue. Thanks to the United Nation's World Commission on Environment and Development, the issues of environment and development have been related to each other. Because of the famous publication *Our Common Future* (1987), the relation between environment and development have become the object of our awareness and action.

Our Common Future is unmistakably one of the most influential studies of the 1980s. This report was drawn up by the Norwegian prime minister Brundtland. The document became particularly well-known because it used the concept 'sustainable development'; a concept which—as we have noted before—had been introduced in the World Conservation Strategy in 1980. The definition of sustainable development was: 'to ensure that it [=development] meets the needs of the present without compromising the ability of future generations to meet their own needs.' This concept proved to be cathartic in environmental policy. In a short time the concept was embraced and used in every possible instance. The meaning of the concept appealed to everyone and was of great value, also because the United Nations established a systematic relationship between environmental issues and developmental issues for the very first time. Since that time, people have become more aware of the fact that the environment is not an isolated issue, but is in essence linked to problems of food supply, poverty, safety, etc.

Our common future does not offer a detailed blueprint for action, but instead a pathway by which the peoples of the world may enlarge their spheres of cooperation. It gives notice that the time has come for a marriage of economy and ecology, so that governments and their people can take responsibility not just for environmental damage, but for the policies that cause that damage. Sustainable development is not a fixed state of harmony, but rather a process of change in which the exploitation of resources, the direction of investments, the orientation of technological development, and institutional change are made consistent with future as well as present needs. The commission does not pretend that the process is easy or straightforward. Painful choices have to be made. Thus, in the final analysis, sustainable development must rely on political will.

Shiva: Staying Alive; Women, Ecology and Development

One aspect of the environmental problem that had been ignored became prominent at the end of the eighties, namely, the feminist aspect. Environmental problems were and are often analysed by men and were and are seen as 'male problems'. Thanks to Shiva's *Staying Alive; Women, Ecology and Development* (1988), the interpretation of environmental problems has changed. As a physicist, philosopher, and feminist, she underscores the importance of a 'female look' at environmental problems. She examines the position of women in relation to nature (the forests, the food chain, and water supplies) and links the violation of nature to the violation and marginalization of women, especially in the Third World. Both violations arise from assumptions in economic development. This system is inherently exploitative. There is only one path to survival and liberation for nature, women and men, and that is the ecological path of harmony, sustainability, and diversity.

She elaborates the feminine principle that becomes an oppositional category of non-violent ways of conceiving the world and of acting in it to sustain all life by maintaining the interconnectedness and diversity of nature. It allows an ecological transition from violence to non-violence, from destruction to creativity, from anti-life to life-giving processes, from uniformity to diversity, and from fragmentation and reductionism to holism and complexity. Modern reductionist science turns out to be a patriarchal discipline, which has excluded women as experts and has simultaneously excluded ecological and holistic ways of knowing which understand and respect nature's processes and interconnectedness as science.

Concern for Tomorrow; a National Environmental Survey 1985-2010

Many countries adopted the views of *Our Common Future* as a starting-point in drawing up their national environmental policy. 'Sustainable development' became the main goal of environmental policy. In the Netherlands, a small European country in the delta of several great European rivers, a pioneering national environmental policy was formulated in the National Environmental Policy Plan. This policy plan was based on a pilot study done by an important Dutch environmental research institute: the National Institute of Public Health and Environmental Protection (RIVM). This institute had done a study commissioned by the Dutch government with the title *Concern for Tomorrow; a National Environmental Survey 1985-2010* (1989). The findings of this report were to become the basis of the nature and content of the National Environmental Policy Plan. What kind of a study was it, and what were its original aspects?

The 'Rijksinstituut voor Volksgezondheid en Milieuhygiëne (RIVM)' (National Institute of Public Health and Environmental Protection) is a Dutch research institute financed by the government. Basing their work on the idea of 'sustainable development', the institute did a national environmental survey for the period 1985-2010. This survey paved the way for the National Environmental Policy Plan. An analysis of the nature of ecological cycles shows that five levels can be distinguished within which characteristic material and energy flows occur. These are local, regional, fluvial, continental, and global levels. The environmental problems are labeled within the context of these five levels. The present magnitude of the environmental problems are presented along with the results of the long-term survey. The general idea of the report is that a far-reaching adjustment of our management of the environment must not to be delayed, even if we are still faced with numerous uncertainties.

The study served as a model for Western countries: environmental policy based on extensive research on the condition and development of the environment. This same effort is now being made in many countries by scientific research institutions that have some connection with the national government (though this does not mean that they do not conduct independent scientific research). These institutions have become environmental braintrusts and have vast quantities of specialized knowledge of many aspects of environmental issues at their disposal. The material they produce is of great importance to the basic justification of policy documents in environmental areas such as those presented by different national states.

Concluding remarks

Looking back on the 1980s, we can say that—despite the relatively unfavourable social and political climate—studies were published that were of great importance in tackling environmental issues. The most prominent document from this period, *Our Common Future*, was a conceptual and programmatic foundation for the environmental policy of the nineties. It was generally seen in many parts of society that the disruption of nature and the environment would have a very high pricetag, and that it would be better and cheaper to introduce policy based more on prevention than on restoring nature and the environment after the damage has taken place.

References

Committee for Long-Term Environmental Policy, 1994. *The environment; Towards a Sustainable Future*, Dordrecht.

Nelissen, N.J.M. et al., 1987. *De Nederlanders en hun Milieu; een Onderzoek naar het Milieubesef en Milieugedrag van Nederlanders*, Zeist.

Nelissen, N.J.M. et al., 1988. *Het Milieu; Vertrouw, maar Weet wel wie je Vertrouwt*, Zeist.

Nelissen, N.J.M, 1989. Environmental Consciousness in the Netherlands, in: *Netherlands Journal for Housing and the Built Environment*, 1989, p. 199-215.

Sandbach, F., 1980. *Environment, Ideology and Policy*, Oxford.

CHAPTER 20

The Global 2000 Report to the President

Major Findings and Conclusions

If present trends continue, the world in 2000 will be more crowded, more polluted, less stable ecologically, and more vulnerable to disruption than the world we live in now. Serious stresses involving population, resources, and environment are clearly visible ahead. Despite greater material output, the world's people will be poorer in many ways than they are today.

For hundreds of millions of the desperately poor, the outlook for food and other necessities of life will be no better. For many it will be worse. Barring revolutionary advances in technology, life for most people on earth will be more precarious in 2000 than it is now—unless the nations of the world act decisively to alter current trends.

This, in essence, is the picture emerging from the US Government's projections of probable changes in world population, resources, and environment by the end of the century, as presented in the Global 2000 Study. They do not predict what will occur. Rather, they depict conditions that are likely to develop if there are no changes in public policies, institutions, or rates of technological advance, and if there are no wars or other major disruptions. A keener awareness of the nature of the current trends, however, may induce changes that will alter these trends and the projected outcome.

Principal Findings

Rapid growth in world population will hardly have altered by 2000. The world's population will grow from 4 billion in 1975 to 6.35 billion in 2000, an increase of more than 50 per cent. The rate of growth will slow only marginally, from 1.8 per cent a year to 1.7 per cent. In terms of sheer numbers, population will be growing faster in 2000 than it is today, with 100 million people added each year compared with 75 million in 1975. Ninety per cent of this growth will occur in the poorest countries.

While the economies of the less developed countries (LDCs) are expected to grow at faster rates than those of the industrialized nations, the gross national product per capita in most LDCs remains low. The average gross national product per capita is projected to rise substantially in some LDCs (especially in

Latin America), but in the great populous nations of South Asia it remains below $200 a year (in 1975 dollars). The large existing gap between the rich and poor nations widens.

World food production is projected to increase 90 per cent over the 30 years from 1970 to 2000. This translates into a global per capita increase of less than 15 per cent over the same period. The bulk of that increase goes to countries that already have relatively high per capita food consumption. Meanwhile per capita consumption in South Asia, the Middle East, and the LDCs of Africa will scarcely improve or will actually decline below present inadequate levels. At the same time, real prices for food are expected to double.

Arable land will increase only 4 per cent by 2000, so that most of the increased output of food will have to come from higher yields. Most of the elements that now contribute to higher yields—fertilizer, pesticides, power for irrigation, and fuel for machinery—depend heavily on oil and gas.

During the 1990s world oil production will approach geological estimates of maximum production capacity, even with rapidly increasing petroleum prices. The Study projects that the richer industrialized nations will be able to command enough oil and other commercial energy supplies to meet rising demands through 1990. With the expected price increases, many less developed countries will have increasing difficulties meeting energy needs. For the one-quarter of humankind that depends primarily on wood for fuel, the outlook is bleak. Needs for fuelwood will exceed available supplies by about 25 per cent before the turn of the century.

While the world's finite fuel resources—coal, oil, gas, oil shale, tar, sands, and uranium—are theoretically sufficient for centuries, they are not evenly distributed; they pose difficult economic and environmental problems; and they vary greatly in their amenability to exploitation and use.

Nonfuel mineral resources generally appear sufficient to meet projected demands through 2000, but further discoveries and investments will be needed to maintain reserves. In addition, production costs will increase with energy prices and may make some nonfuel mineral resources uneconomic. The quarter of the world's population that inhabits industrial countries will continue to absorb three-fourths of the world's mineral production.

Regional water shortages will become more severe. In the 1970-2000 period population growth alone will cause requirements for water to double in nearly half the world. Still greater increases would be needed to improve standards of living. In many LDCs, water supplies will become increasingly erratic by 2000 as a result of extensive deforestation. Development of new water supplies will become more costly virtually everywhere.

Significant losses of world forests will continue over the next 20 years as demand for forest products and fuelwood increases. Growing stocks of commercial-size timber are projected to decline 50 per cent per capita. The world's forests are now disappearing at the rate of 18-20 million hectares a year (an area half the size of California), with most of the loss occurring in the humid tropical forests of Africa, Asia, and South America. The projections indicate that by 2000 some 40 per cent of the remaining forest cover in LDCs will be gone.

Serious deterioration of agricultural soils will occur worldwide, due to erosion, loss of organic matter, desertification, salinization, alkalinization, and waterlogging. Already, an area of cropland and grassland approximately the size of Maine is becoming barren wasteland each year, and the spread of desert-like conditions is likely to accelerate.

Atmospheric concentrations of carbon dioxide and ozone-depleting chemicals are expected to increase at rates that could alter the world's climate and upper atmosphere significantly by 2050. Acid rain from increased combustion of fossil fuels (especially coal) threatens damage to lakes, soils, and crops. Radioactive and other hazardous materials present health and safety problems in increasing numbers of countries.

Extinctions of plant and animal species will increase dramatically. Hundreds of thousands of species—perhaps as many as 20 per cent of all species on earth—will be irretrievably lost as their habitats vanish, especially in tropical forests.

The future depicted by the US Government projections, briefly outlined above, may actually understate the impending problems. The methods available for carrying out the Study led to certain gaps and inconsistencies that tend to impart an optimistic bias. For example, most of the individual projections for the various sectors studied—food, minerals, energy, and so on—assume that sufficient capital, energy, water, and land will be available in each of these sectors to meet their needs, regardless of the competing needs of the other sectors. More consistent, better-integrated projections would produce a still more emphatic picture of intensifying stresses, as the world enters the twenty-first century.

Conclusions

At present and projected growth rates, the world's population would reach 10 billion by 2030 and would approach 30 billion by the end of the twenty-first century. These levels correspond closely to estimates by the US National Academy of Sciences of the maximum carrying capacity of the entire earth. Already the populations in sub-Saharan Africa and in the Himalayan hills of Asia have exceeded the carrying capacity of the immediate area, triggering an erosion of the land's capacity to support life. The resulting poverty and ill health have further complicated efforts to reduce fertility. Unless this circle of inter-linked problems is broken soon, population growth in such areas will unfortunately be slowed for reasons other than declining birth rates. Hunger and disease will claim more babies and young children, and more of those surviving will be mentally and physically handicapped by childhood malnutrition.

Indeed, the problems of preserving the carrying capacity of the earth and sustaining the possibility of a decent life for the human beings that inhabit it are enormous and close upon us. Yet there is reason for hope. It must be emphasized that the Global 2000 Study's projections are based on the assumption that national policies regarding population stabilization, resource conservation, and environmental protection will remain essentially unchanged through the end of the century. But in fact, policies are beginning to change. In some areas, forests are being replanted after cutting. Some nations are taking steps to reduce soil losses and desertification. Interest in energy conservation is growing, and large sums are being invested in exploring alternatives to petroleum dependence. The need for family planning is slowly becoming better understood. Water supplies are being improved and waste treatment systems built. High-yield seeds are widely available and seed banks are being expanded. Some wildlands with their genetic resources are being protected. Natural predators and selective pesticides are being substituted for persistent and destructive pesticides.

Encouraging as these developments are, they are far from adequate to meet the global challenges projected in this Study. Vigorous, determined new initiatives are needed if worsening poverty and human suffering, environmental degradation, and international tension and conflicts are to be prevented. There are no quick fixes. The only solutions to the problems of population, resources, and environment are complex and long-term. These problems are inextricably linked to some of the most perplexing and persistent problems in the world—poverty, injustice, and social conflict. New and imaginative ideas—and a willingness to act on them—are essential.

The needed changes go far beyond the capability and responsibility of this or any other single nation. An era of unprecedented cooperation and commitment is essential. Yet there are opportunities—and a strong rationale—for the United States to provide leadership among nations. A high priority for this Nation must be a thorough assessment of its foreign and domestic policies relating to population, resources, and environment. The United States, possessing the world's largest economy, can expect its policies to have a significant influence on global trends. An equally important priority for the United States is to cooperate generously and justly with other nations—particularly in the areas of trade, investment, and assistance—in seeking solutions to the many problems that extend beyond our national boundaries. There are many unfulfilled opportunities to cooperate with other nations in efforts to relieve poverty and hunger, stabilize population, and enhance economic and environmental productivity. Further cooperation among nations is also needed to strengthen international mechanisms for protecting and utilizing the "global commons"— the oceans and atmosphere.

To meet the challenges described in this Study, the United States must improve its ability to identify emerging problems and assess alternative responses. In using and evaluating the Government's present capability for long-term global analysis, the Study found serious inconsistencies in the methods and assumptions employed by the various agencies in making their projections. The Study itself made a start toward resolving these inadequacies. It represents the Government's first attempt to produce an interrelated set of population, resource, and environmental projections, and it has brought forth the most consistent set of global projections yet achieved by US agencies. Nevertheless, the projections still contain serious gaps and contradictions that must be corrected if the Government's analytic capability is to be improved. It must be acknowledged that at present the Federal agencies are not always capable of providing projections of the quality needed for long-term policy decisions.

While limited resources may be a contributing factor in some instances, the primary problem is lack of coordination. The US Government needs a mechanism for continuous review of the assumptions and methods the Federal agencies use in their projection models and for assurance that the agencies' models are sound, consistent, and well documented. The improved analyses that could result would provide not only a clearer sense of emerging problems and opportunities, but also a better means for evaluating alternative responses, and a better basis for decisions of worldwide significance that the President, the Congress, and the Federal Government as a whole must make.

With its limitations and rough approximations, the Global 2000 Study may be seen as no more than a reconnaissance of the future; nonetheless its conclusions are reinforced by similar findings of other recent global studies that were examined in the course of the Global 2000 Study (see Appendix). All these studies are in general agreement on the nature of the problems and on the threats they pose to the future welfare of humankind. The available evidence leaves no doubt that the world—including this Nation—faces enormous, urgent, and complex problems in the decades immediately ahead. Prompt and vigorous changes in public policy around the world are needed to avoid or minimize these problems before they become unmanageable. Long lead times are required for effective action. If decisions are delayed until the problems become worse, options for effective action will be severely reduced.

The Global 2000 Report to the President, Government Printing Office, Washington, 1980, p. 1-5.

CHAPTER 21

World Conservation Strategy

Living Resource Conservation for Sustainable Development

IUCN

Executive Summary

The World Conservation Strategy is intended to stimulate a more focussed approach to the management of living resources and to provide policy guidance on how this can be carried out by three main groups:
— government policy makers and their advisers;
— conservationists and others directly concerned with living resources;
— development practitioners, including development agencies, industry and commerce, and trade unions.

1 The aim of the World Conservation Strategy is to achieve the three main objectives of living resource conservation:
a *to maintain essential ecological processes and life-support systems*
(such as soil regeneration and protection, the recycling of nutrients, and the cleansing of waters), on which human survival and development depend;
b *to preserve genetic diversity*
(the range of genetic material found in the world's organisms), on which depend the functioning of many of the above processes and life-support systems, the breeding programmes necessary for the protection and improvement of cultivated plants, domesticated animals and microorganisms, as well as much scientific and medical advance, technical innovation, and the security of the many industries that use living resources;
c *to ensure the sustainable utilization of species and ecosystems*
(notably fish and other wildlife, forests and grazing lands), which support millions of rural communities as well as major industries.

2 These objectives must be achieved as a matter of urgency because:
a *the planet's capacity to support people is being irreversibly reduced in both developing and developed countries:*
— thousands of million of tonnes of soil are lost every year as a result of deforestation and poor land management;

– at least 3,000 km^2 of prime farmland disappear every year under buildings
and roads in developed countries alone;

b *hundreds of millions of rural people in developing countries, including 500 million
malnourished and 800 million destitute, are compelled to destroy the resources necessary to free them from starvation and poverty:*

– in widening swaths around their villages the rural poor strip the land of tress
and shrubs for fuel so that now many communities do not have enough wood
to cook food or to keep warm;

– the rural poor are also obliged to burn every year 400 million tonnes of dung
and crop residues badly needed to regenerate soils;

c *the energy, financial and other costs of providing goods and services are growing:*

– throughout the world, but especially in developing countries, siltation cuts
the lifetimes of reservoirs supplying water and hydroelectricity, often by as
much as half;

– floods devastate settlements and crops (in India the annual cost of floods
ranges from \$140 million to \$750 million);

d *the resource base of major industries is shrinking:*

– tropical forests are contracting so rapidly that by the end of this century the
remaining area of unlogged productive forest will have been halved;

– the coastal support system of many fisheries are being destroyed or polluted
(in the USA the annual cost of the resulting losses is estimated at \$86 million).

3 The main obstacles to achieving conservation are:

a *the belief that living resource conservation is a limited sector, rather than a process
that cuts across and must be considered by all sectors;*

b *the consequent failure to integrate conservation with development;*

c *a development process that is often inflexible and needlessly destructive,*

due to inadequacies in environmental planning, a lack of rational use allocation
and undue emphasis on narrow short term interests rather than broader longer
term ones;

d *the lack of a capacity to conserve,*

due to inadequate legislation and lack of enforcement; poor organization (notably government agencies with insufficient mandates and a lack of coordination); lack of trained personnel; and a lack of basic information on priorities,
on the productive and regenerative capacities of living resources, and on the
trade-offs between one management option and another;

e *the lack of support for conservation,*

due to a lack of awareness (other than at the most superficial level) of the
benefits of conservation and of the responsibility to conserve among those who
use or have an impact on living resources, including in many cases governments;

f *the failure to deliver conservation-based development where it is most needed,* notably the rural areas of developing countries.

4 The World Conservation Strategy therefore:

a *defines living resource conservation and explains its objectives,* its contribution to human survival and development and the main impediments to its achievement (sections 1-4);

b *determines the priority requirements for achieving each of the objectives* (sections 5-7);

c *proposes national and subnational strategies* to meet the priority requirements, describing a framework and principles for those strategies (section 8);

d *recommends anticipatory environmental policies, a cross-sectoral conservation policy and a broader system of national accounting* in order to integrate conservation with development at the policy making level (section 9);

e *proposes an integrated method of evaluating land and water resources, supplemented by environmental assessments,* as a means of improving environmental planning; and *outlines a procedure for the rational allocation of land and water uses* (section 10);

f *recommends reviews of legislation* concerning living resources; *suggests general principles for organization within government;* and in particular *proposes ways of improving the organizational capacities for soil conservation and for the conservation of marine living resources* (section 11);

g *suggests ways of increasing the number of trained personnel;* and *proposes more management-orientated research and research-orientated management,* so that the most urgently needed basic information is generated more quickly (section 12);

h *recommends greater public participation* in planning and decision making concerning living resource use; and *proposes environmental education programmes and campaigns* to build support for conservation (section 13);

i *suggests ways of helping rural communities to conserve* their living resources, as the essential basis of the development they need (section 14).

5 In addition, the Strategy recommends international action to promote, support and (where necessary) coordinate national action, emphasizing in particular the need for:

a *stronger more comprehensive international conservation law,* and *increased development assistance for living resource conservation* (section 15);

b *international programmes* to promote the action necessary to conserve *tropical forests and drylands* (section 16), to protect areas essential for the preservation of *genetic resources* (section 17), and to conserve the global "commons"—*the open ocean, the atmosphere, and Antarctica* (section 18);

c *regional strategies* to advance the conservation of *shared living resources* particularly with respect to *international river basins and seas* (section 19);

d The World Conservation Strategy ends by summarizing *the main requirements for sustainable development,* indicating conservation priorities for the third Development Decade (section 20).

International Union for Conservation of Nature and Natural Resources, *World Conservation Strategy; Living Resource Conservation for Sustainable Development,* IUCN, Gland, 1980.

CHAPTER 22

Our Common Future

WORLD COMMISSION ON ENVIRONMENT AND DEVELOPMENT

From One Earth to One World; An Overview

In the middle of the 20th century, we saw our planet from space for the first time. Historians may eventually find that this vision had a greater impact on thought than did the Copernican revolution of the 16th century, which upset the human self-image by revealing that the Earth is not the centre of the universe. From space, we see a small and fragile ball dominated not by human activity and edifice but by a pattern of clouds, oceans, greenery, and soils. Humanity's inability to fit its doings into that pattern is changing planetary systems, fundamentally. Many such changes are accompanied by life-threatening hazards. This new reality, from which there is no escape, must be recognized— and managed.

Fortunately, this new reality coincides with more positive developments new to this century. We can move information and goods faster around the globe than ever before; we can produce more food and more goods with less investment of resources; our technology and science gives us at least the potential to look deeper into and better understand natural systems. From space, we can see and study the Earth as an organism whose health depends on the health of all its parts. We have the power to reconcile human affairs with natural laws and to thrive in the process. In this our cultural and spiritual heritages can reinforce our economic interests and survival imperatives.

This Commission believes that people can build a future that is more prosperous, more just, and more secure. Our report, *Our Common Future*, is not a prediction of ever increasing environmental decay, poverty, and hardship in an ever more polluted world among ever decreasing resources. We see instead the possibility for a new era of economic growth, one that must be based on policies that sustain and expand the environmental resource base. And we believe such growth to be absolutely essential to relieve the great poverty that is deepening in much of the developing world.

But the Commission's hope for the future is conditional on decisive political action now to begin managing environmental resources to ensure both sustainable human progress and human survival. We are not forecasting a future; we are serving a notice—an urgent notice based on the latest and best scientific evidence—that the time has come to take the decisions needed to secure the resources to sustain this and coming generations. We do not offer a detailed blueprint for action, but instead a pathway by which the peoples of the world may enlarge their spheres of co-operation.

1   The Global Challenge

Successes and Failures

Those looking for success and signs of hope can find many: Infant mortality is falling; human life expectancy is increasing; the proportion of the world's adults who can read and write is climbing; the proportion of children starting school is rising; and global food production increases faster than the population grows.

But the same processes that have produced these gains have given rise to trends that the planet and its people cannot long bear. These have traditionally been divided into failures of 'development' and failures in the management of our human environment. On the development side, in terms of absolute numbers there are more hungry people in the world than ever before, and their numbers are increasing. So are the numbers who cannot read or write, the numbers without safe water or safe and sound homes, and the numbers short of woodfuel with which to cook and warm themselves. The gap between rich and poor nations is widening—not shrinking—and there is little prospect, given present trends and institutional arrangements that this process will be reversed.

There are also environmental trends that threaten to radically alter the planet, that threaten the lives of many species upon it, including the human species. Each year another 6 million hectares of productive dryland turns into worthless desert. Over three decades, this would amount to an area roughly as large as Saudi Arabia. More than 11 million hectares of forests are destroyed yearly, and this, over three decades, would equal an area about the size of India. Much of this forest is converted to low-grade farmland unable to support the farmers who settle it. In Europe, acid precipitation kills forests and lakes and damages the artistic and architectural heritage of nations; it may have acidified vast tracts of soil beyond reasonable hope of repair. The burning of fossil fuels puts into the atmosphere carbon dioxide, which is causing gradual global warming. This 'greenhouse effect' may by early next century have increased average global temperatures enough to shift agricultural production areas, raise sea levels to flood coastal cities, and disrupt national economies. Other industrial gases threaten to deplete the planet's protective ozone shield to such an extent that the number of human and animal cancers would rise sharply and the oceans' food chain would be disrupted. Industry and agriculture put toxic substances into the human food chain and into underground water tables beyond reach of cleansing.

The World Commission on Environment and Development first met in October 1984, and published its report 900 days later, in April 1987. Over those few days:
– The drought-triggered, environment-development crisis in Africa peaked, putting 35 million people at risk, killing perhaps a million.
– A leak from a pesticides factory in Bhopal, India, killed more than 2,000 people and blinded and injured over 200,000 more.
– Liquid gas tanks exploded in Mexico City, killing 1,000 and leaving thousands more homeless.
– The Chernobyl nuclear reactor explosion sent nuclear fallout across Europe, increasing the risks of future human cancers.
– Agricultural chemicals, solvents, and mercury flowed into the Rhine River during a warehouse fire in Switzerland, killing millions of fish and threatening drinking water in the Federal Republic of Germany and the Netherlands.
– An estimated 60 million people died of diarrhoeal diseases related to unsafe drinking water and malnutrition; most of the victims were children.

There has been a growing realization in national governments and multilateral institutions that it is impossible to separate economic development issues from environment issues; many forms of development erode the environmental resources upon which they must be based, and environmental degradation can undermine economic development. Poverty is a major cause and effect of global environmental problems. It is therefore futile to attempt to deal with environmental problems without a broader perspective that encompasses the factors underlying world poverty and international inequality.

These concerns were behind the establishment in 1983 of the World Commission on Environment and Development by the UN General Assembly. The Commission is an independent body, linked to, but outside the control of governments and the UN system. The Commission's mandate gave it three objectives: to re-examine the critical environment and development issues and to formulate realistic proposals for dealing with them; to propose new forms of international co-operation on these issues that will influence policies and events in the direction of needed changes; and to raise the levels of understanding and commitment to action of individuals, voluntary organizations, businesses, institutes, and governments.

Through our deliberations and the testimony of people at the public hearings we held on five continents, all the commissioners came to focus on one central theme: many present development trends leave increasing numbers of people poor and vulnerable while at the same time degrading the environment. How can such development serve next century's world of twice as many people relying on the same environment? This realization broadened our view of development. We came to see it not in its restricted context of economic growth in developing countries. We came to see that a new development path was required, one that sustained human progress not just in a few places for a few years, but for the entire planet into the distant future. Thus 'sustainable development' becomes a goal not just for the 'developing' nations, but for industrial ones as well.

The Interlocking Crises

Until recently, the planet was a large world in which human activities and their effects were neatly compartmentalized within nations, within sectors (energy, agriculture, trade), and within broad areas of concern (environmental, economic, social). These compartments have begun to dissolve. This applies in particular to the various global 'crises' that have seized public concern, particularly over the past decade. These are not separate crises: an environmental crisis, a development crisis, an energy crisis. They are all one.

The planet is passing through a period of dramatic growth and fundamental change. Our human world of 5 billion must make room in a finite environment for another human world. The population could stabilize at between 8 billion and 14 billion sometime next century, according to UN projections. More than 90 per cent of the increase will occur in the poorest countries, and 90 per cent of that growth in already bursting cities.

Economic activity has multiplied to create a $13 trillion world economy, and this could grow five- or tenfold in the coming half-century. Industrial production has grown more than fiftyfold over the past century, four-fifths of this growth since 1950. Such figures reflect and presage profound impacts upon the biosphere, as the world invests in houses, transport, farms, and industries. Much of the economic growth pulls raw material from forests, soils, seas and waterways.

A mainspring of economic growth is new technology, and while this technology offers the potential for slowing the dangerously rapid consumption of finite resources, it also entails high risks, including new forms of pollution and the introduction to the planet of new variations of life forms that could change evolutionary pathways. Meanwhile, the industries most heavily reliant on environmental resources and most heavily polluting are growing most rapidly in the developing world, where there is both more urgency for growth and less capacity to minimize damaging side effects.

These related changes have locked the global economy and global ecology together in new ways. We have in the past been concerned about the impacts of economic growth upon the environment. We are now forced to concern our-selves with the impacts of ecological stress—degradation of soils, water regimes, atmosphere, and forests—upon our economic prospects. We have in the more recent past been forced to face up to a sharp increase in economic in-terdependence among nations. We are now forced to accustom ourselves to an accelerating ecological interdependence among nations. Ecology and economy are becoming ever more interwoven—locally, regionally, nationally, and glo-bally—into a seamless net of causes and effects.

Impoverishing the local resource base can impoverish wider areas: Defore-station by highland farmers causes flooding on lowland farms; factory pollution robs local fishermen of their catch. Such grim local cycles now operate nation-ally and regionally. Dryland degradation sends environmental refugees in their millions across national borders. Deforestation in Latin America and Asia is causing more floods, and more destructive floods, in downhill, downstream nations. Acid precipitation and nuclear fallout have spread across the borders of Europe. Similar phenomena are emerging on a global scale, such as global warming and loss of ozone. Internationally traded hazardous chemicals enter-ing foods are themselves internationally traded. In the next century, the envi-ronmental pressure causing population movements may increase sharply, while barriers to that movement may be even firmer than they are now.

Over the past few decades, life-threatening environmental concerns have surfaced in the developing world. Countrysides are coming under pressure from increasing numbers of farmers and the landless. Cities are filling with people, cars, and factories. Yet at the same time these developing countries must operate in a world in which the resources gap between most developing and industrial nations is widening, in which the industrial world dominates in the rule-making of some key international bodies, and in which the industrial world has already used much of the planet's ecological capital. This inequality is the planet's main 'environmental' problem; it is also its main 'development' problem.

International economic relationships pose a particular problem for environ-mental management in many developing countries. Agriculture, forestry, en-ergy production, and mining generate at least half the gross national product of many developing countries and account for even larger shares of livelihoods and employment. Exports of natural resources remain a large factor in their economies, especially for the least developed. Most of these countries face enormous economic pressures, both international and domestic, to overexploit their environmental resource base.

The recent crisis in Africa best and most tragically illustrates the ways in which economics and ecology can interact destructively and trip into disaster. Triggered by drought, its real causes lie deeper. They are to be found in part in national policies that gave too little attention, too late, to the needs of smallholder agriculture and to the threats posed by rapidly rising populations. Their roots extend also to a global economic system that takes more out of a poor continent than it puts in. Debts that they cannot pay force African nations relying on commodity sales to overuse their fragile soils, thus turning good land to desert. Trade barriers in the wealthy nations—and in many developing ones—make it hard for Africans to sell their goods for reasonable returns, putting yet more pressure on ecological systems. Aid from donor nations has not only been inadequate in scale, but too often has reflected the priorities of the nations giving the aid, rather than the needs of the recipients. The production base of other developing world areas suffers similarly both from local failures and from the workings of international economic systems. As a consequence of the 'debt crisis' of Latin America, that region's natural resources are now being used not for development but to meet financial obligations to creditors abroad. This approach to the debt problem is short-sighted from several standpoints: economic political, and environmental. It requires relatively poor countries simultaneously to accept growing poverty while exporting growing amounts of scarce resources.

The Commission has sought ways in which global development can be put on a sustainable path into the 21st century. Some 5,000 days will elapse between the publication of our report and the first day of the 21st century. What environmental crises lie in store over those 5,000 days?

During the 1970s, twice as many people suffered each year from 'natural' disasters as during the 1960s. The disasters most directly associated with environmental development mismanagement—droughts and floods—affected the most people and increased most sharply in terms of numbers affected. Some 18.5 million people were affected by drought annually in the 1960s, 24.4 million in the 1970s. There were 5.2 million flood victims yearly in the 1960s, 15.4 million in the 1970s. Numbers of victims of cyclones and earthquakes also shot up as growing numbers of poor people built unsafe houses on dangerous ground.

The results are not in for the 1980s. But we have seen 35 million afflicted by drought in Africa alone and tens of millions affected by the better managed and thus less-publicized Indian drought. Floods have poured off the deforested Andes and Himalayas with increasing force. The 1980s seem destined to sweep this dire trend on into a crisis-filled 1990s.

A majority of developing countries now have lower per capita incomes than when the decade began. Rising poverty and unemployment have increased pressure on environmental resources as more people have been forced to rely more directly upon them. Many governments have cut back efforts to protect the environment and to bring ecological considerations into development planning.

The deepening and widening environmental crisis presents a threat to national security—and even survival—that may be greater than well-armed, ill-disposed neighbours and unfriendly alliances. Already in parts of Latin America, Asia, the Middle East, and Africa, environmental decline is becoming a source of political unrest and international tension. The recent destruction of much of Africa's dryland agricultural production was more severe than if an invading army had pursued a scorched-earth policy. Yet most of the affected governments still spend far more to protect their people from invading armies than from the invading desert.

Globally, military expenditures total about $1 trillion a year and continue to grow. In many countries, military spending consumes such a high proportion of gross national product that it itself does great damage to these societies' development efforts. Governments tend to base their approaches to 'security' on traditional definitions. This is most obvious in the attempts to achieve security through the development of potentially planet-destroying nuclear weapons systems. Studies suggest that the cold and dark nuclear winter following even a limited nuclear war could destroy plant and animal ecosystems and leave any human survivors occupying a devastated planet very different from the one they inherited.

The arms race—in all parts of the world—pre-empts resources that might be used more productively to diminish the security threats created by environmental conflict and the resentments that are fuelled by widespread poverty.

Many present efforts to guard and maintain human progress, to meet human needs, and to realize human ambitions are simply unsustainable—in both the rich and poor nations. They draw too heavily, too quickly, on already over-drawn environmental resource accounts to be affordable far into the future without bankrupting those accounts. They may show profits on the balance sheets of our generation, but our children will inherit the losses. We borrow environmental capital from future generations with no intention or prospect of repaying. They may damn us for our spendthrift ways, but they can never collect on our debt to them. We act as we do because we can get away with it: future generations do not vote; they have no political or financial power; they cannot challenge our decisions.

But the results of the present profligacy are rapidly closing the options for future generations. Most of today's decision makers will be dead before the planet feels the heavier effects of acid precipitation, global warming, ozone depletion, or widespread desertification and species loss. Most of the young voters of today will still be alive. In the Commission's hearings it was the young, those who have the most to lose, who were the harshest critics of the planet's present management.

Sustainable Development

Humanity has the ability to make development sustainable—to ensure that it meets the needs of the present without compromising the ability of future generations to meet their own needs. The concept of sustainable development does imply limits—not absolute limits but limitations imposed by the present state of technology and social organization on environmental resources and by the ability of the biosphere to absorb the effects of human activities. But technology and social organization can be both managed and improved to make way for a new era of economic growth. The Commission believes that widespread poverty is no longer inevitable. Poverty is not only an evil in itself, but sustainable development requires meeting the basic needs of all and extending to all the opportunity to fulfil their aspirations for a better life. A world in which poverty is endemic will always be prone to ecological and other catastrophes.

Meeting essential needs requires not only a new era of economic growth for nations in which the majority are poor, but an assurance that those poor get their fair share of the resources required to sustain that growth. Such equity would be aided by political systems that secure effective citizen participation in decision making and by greater democracy in international decision making.

Sustainable global development requires that those who are more affluent adopt life-styles within the planet's ecological means—in their use of energy, for example. Further, rapidly growing populations can increase the pressure on resources and slow any rise in living standards; thus sustainable development can only be pursued if population size and growth are in harmony with the changing productive potential of the ecosystem.

Yet in the end, sustainable development is not a fixed state of harmony, but rather a process of change in which the exploitation of resources, the direction of investments, the orientation of technological development, and institutional change are made consistent with future as well as present needs. We do not pretend that the process is easy or straightforward. Painful choices have to be made. Thus, in the final analysis, sustainable development must rest on political will.

The Institutional Gaps

The objective of sustainable development and the integrated nature of the global environmental development challenges pose problems for institutions, national and international, that were established on the basis of narrow preoccupations and compartmentalized concerns. Governments' general response to the speed and scale of global changes has been a reluctance to recognize sufficiently the need to change themselves. The challenges are both interdependent and integrated, requiring comprehensive approaches and popular participation.

Yet most of the institutions facing those challenges tend to be independent, fragmented, working to relatively narrow mandates with closed decision processes. Those responsible for managing natural resources and protecting the environment are institutionally separated from those responsible for managing the economy. The real world of interlocked economic and ecological systems will not change; the policies and institutions concerned must.

There is a growing need for effective international co-operation to manage ecological and economic interdependence. Yet at the same time, confidence in international organizations is diminishing and support for them dwindling.

The other great institutional flaw in coping with environment/development challenges is governments' failure to make the bodies whose policy actions degrade the environment responsible for ensuring that their policies prevent that degradation. Environmental concern arose from damage caused by the rapid economic growth following to the Second World War. Governments, pressured by their citizens, saw a need to clean up the mess, and they established environmental ministries and agencies to do this. Many had great success—within the limits of their mandates—in improving air and water quality and enhancing other resources. But much of their work has of necessity been after-the-fact repair of damage: reforestation, reclaiming desert lands, rebuilding urban environments, restoring natural habitats, and rehabilitating wild lands.

The existence of such agencies gave many governments and their citizens the false impression that these bodies were by themselves able to protect and enhance the environmental resource base. Yet many industrialized and most developing countries carry huge economic burdens from inherited problems such as air and water pollution, depletion of ground-water, and the proliferation of toxic chemicals and hazardous wastes. These have been joined by more recent problems—erosion, desertification, acidification, new chemicals, and new forms of waste—that are directly related to agricultural, industrial, energy, forestry, and transportation policies and practices.

The mandates of the central economic and sectoral ministries are also often too narrow, too concerned with quantities of production or growth. The mandates of ministries of industry include production targets, while the accompanying pollution is left to ministries of environment. Electricity boards produce power, while the acid pollution they also produce is left to other bodies to clean up. The present challenge is to give the central economic and sectoral ministries the responsibility for the quality of those parts of the human environment affected by their decisions, and to give the environmental agencies more power to cope with the effects of unsustainable development.

The same need for change holds for international agencies concerned with development lending, trade regulation, agricultural development, and so on. These have been slow to take the environmental effects of their work into account, although some are trying to do so.

The ability to anticipate and prevent environmental damage requires that the ecological dimensions of policy be considered at the same time as the economic, trade, energy, agricultural, and other dimensions. They should be considered on the same agendas and in the same national and international institutions.

This reorientation is one of the chief institutional challenges of the 1990s and beyond. Meeting it will require major institutional development and reform. Many countries that are too poor or small or that have limited managerial capacity will find it difficult to do this unaided. They will need financial and technical assistance and training. But the changes required involve all countries, large and small, rich and poor.

World Commission on Environment and Development, *Our Common Future*, Oxford University Press, Oxford, 1987, p. 1-11.

CHAPTER 23

Staying Alive

Women, Ecology and Development

V. SHIVA

Development as a new project of western patriarchy

'Development' was to have been a post-colonial project, a choice for accepting a model of progress in which the entire world remade itself on the model of the colonising modern west, without having to undergo the subjugation and exploitation that colonialism entailed. The assumption was that western style progress was possible for all. Development, as the improved well-being of all, was thus equated with the westernisation of economic categories—of needs, of productivity, of growth. Concepts and categories about economic development and natural resource utilisation that had emerged in the specific context of industrialisation and capitalist growth in a centre of colonial power, were raised to the level of universal assumptions and applicability in the entirely different context of basic needs satisfaction for the people of the newly independent Third World countries. Yet, as Rosa Luxemberg has pointed out, early industrial development in western Europe necessitated the permanent occupation of the colonies by the colonial powers and the destruction of the local 'natural economy'.[1] According to her, colonialism is a constant necessary condition for capitalist growth: without colonies, capital accumulation would grind to a halt. 'Development' as capital accumulation and the commercialisation of the economy for the generation of 'surplus' and profits thus involved the reproduction not merely of a particular form of creation of wealth, but also of the associated creation of poverty and dispossession. A replication of economic development based on commercialisation of resource use for commodity production in the newly independent countries created the internal colonies.[2] Development was thus reduced to a continuation of the process of colonisation; it became an extension of the project of wealth creation in modern western patriarchy's economic vision, which was based on the exploitation or exclusion of women (of the west and non-west), on the exploitation and degradation of nature, and on the exploitation and erosion of other cultures. 'Development' could not but entail destruction for women, nature and subjugated cultures, which is why, throughout the Third World, women, peasants and tribals are struggling for liberation from 'development' just as they earlier struggled for liberation from colonialism.

The UN Decade for Women was based on the assumption that the improvement of women's economic position would automatically flow from an expansion and diffusion of the development process. Yet, by the end of the Decade, it was becoming clear that development itself was the problem. Insufficient and inadequate 'participation' in 'development' was not the cause for women's increasing under-development; it was rather, their enforced but asymmetric participation in it, by which they bore the costs but were excluded from the benefits, that was responsible. Development exclusivity and dispossession aggravated and deepened the colonial processes of ecological degradation and the loss of political control over nature's sustenance base. Economic growth was a new colonialism, draining resources away from those who needed them most. The discontinuity lay in the fact that it was now new national elites, not colonial powers, that masterminded the exploitation on grounds of 'national interest' and growing GNPs, and it was accomplished with more powerful technologies of appropriation and destruction.

Ester Boserup[3] has documented how women's impoverishment increased during colonial rule; those rulers who had spent a few centuries in subjugating and crippling their own women into de-skilled, de-intellectualised appendages, disfavoured the women of the colonies on matters of access to land, technology and employment. The economic and political processes of colonial under-development bore the clear mark of modern western patriarchy, and while large numbers of women and men were impoverished by these processes, women tended to lose more. The privatisation of land for revenue generation displaced women more critically, eroding their traditional land use rights. The expansion of cash crops undermined food production, and women were often left with meagre resources to feed and care for children, the aged and the infirm, when men migrated or were conscripted into forced labour by the colonisers. As a collective document by women activists, organisers and researchers stated at the end of the UN Decade for Women, 'The almost uniform conclusion of the Decade's research is that with a few exceptions, women's relative access to economic resources, incomes and employment has worsened, their burden of work has increased, and their relative and even absolute health, nutritional and educational status has declined.'[4]

The displacement of women from productive activity by the expansion of development was rooted largely in the manner in which development projects appropriated or destroyed the natural resource base for the production of sustenance and survival. It destroyed women's productivity both by removing land, water and forests from their management and control, as well as through the ecological destruction of soil, water and vegetation systems so that nature's productivity and renewability were impaired. While gender subordination and patriarchy are the oldest of oppressions, they have taken on new and more violent forms through the project of development. Patriarchal categories which understand destruction as 'production' and regeneration of life as 'passivity' have generated a crisis of survival. Passivity, as an assumed category of the 'nature' of nature and of women, denies the activity of nature and life. Fragmentation and uniformity as assumed categories of progress and development destroy the living forces which arise from relationships within the 'web of life' and the diversity in the elements and patterns of these relationships.

The economic biases and values against nature, women and indigenous peoples are captured in this typical analysis of the 'unproductiveness' of traditional natural societies:

> Production is achieved through human and animal, rather than mechanical, power. Most agriculture is unproductive; human or animal manure may be used but chemical fertilisers and pesticides are unknown . . . For the masses, these conditions mean poverty.[5]

The assumptions are evident: nature is unproductive; organic agriculture based on nature's cycles of renewability spells poverty; women and tribal and peasant societies embedded in nature are similarly unproductive, not because it has been demonstrated that in cooperation they produce *less* goods and services for needs, but because it is assumed that 'production' takes place only when mediated by technologies for commodity production, even when such technologies destroy life. A stable and clean river is not a productive resource in this view: it needs to be 'developed' with dams in order to become so. Women, sharing the river as a commons to satisfy the water needs of their families and society are not involved in productive labour: when substituted by the engineering man, water management and water use become productive activities. Natural forests remain unproductive till they are developed into monoculture plantations of commercial species. Development thus, is equivalent to maldevelopment, a development bereft of the feminine, the conservation, the ecological principle. The neglect of nature's work in renewing herself, and women's work in producing sustenance in the form of basic, vital needs is an essential part of the paradigm of maldevelopment, which sees all work that does not produce profits and capital as non or unproductive work. As Maria Mies[6] has pointed out, this concept of surplus has a patriarchal bias because, from the point of view of nature and women, it is not based on material surplus produced over and above the requirements of the community: it is stolen and appropriated through violent modes from nature (who needs a share of her produce to reproduce herself) and from women (who need a share of nature's produce to produce sustenance and ensure survival).

From the perspective of Third World women, productivity is a measure of producing life and sustenance; that this kind of productivity has been rendered invisible does not reduce its centrality to survival—it merely reflects the domination of modern patriarchal economic categories which see only profits not life.

Maldevelopment as the death of the feminine principle

In this analysis, maldevelopment becomes a new source of male-female inequality. 'Modernisation' has been associated with the introduction of new forms of dominance. Alice Schlegel[7] has shown that under conditions of subsistence, the interdependence and complementarity of the separate male and female domains of work is the characteristic mode, based on diversity, not inequality. Maldevelopment militates against this equality in diversity, and superimposes the ideologically constructed category of western technological man as a uniform measure of the worth of classes, cultures and genders. Dominant modes of perception based on reductionism, duality and linearity are unable to cope with equality in diversity, with forms and activities that are significant and valid, even though different. The reductionist mind superimposes the roles and forms of power of western male-oriented concepts on women, all non-western peoples and even on nature, rendering all three 'deficient', and in need of 'development'. Diversity, and unity and harmony in diversity, become epistemologically unattainable in the context of maldevelopment, which then becomes synonymous with women's underdevelopment (increasing sexist domination), and nature's depletion (deepening ecological crises). Commodities have grown, but nature has shrunk. The poverty crisis of the South arises from the growing scarcity of water, food, fodder and fuel, associated with increasing maldevelopment and ecological destruction. This poverty crisis touches women most severely, first because they are the poorest among the poor, and then because, with nature, they are the primary sustainers of society.

Maldevelopment is the violation of the integrity of organic, interconnected and interdependent systems, that sets in motion a process of exploitation, inequality, injustice and violence. It is blind to the fact that a recognition of nature's harmony and action to maintain it are preconditions for distributive justice. This is why Mahatma Gandhi said, 'There is enough in the world for everyone's need, but not for some people's greed.'

Maldevelopment is maldevelopment in thought and action. In practice, this fragmented, reductionist, dualist perspective violates the integrity and harmony of man in nature, and the harmony between men and women. It ruptures the co-operative unity of masculine and feminine, and places man, shorn of the feminine principle, above nature and women, and separated from both. The violence to nature as symptomatised by the ecological crisis, and the violence to women, as symptomatised by their subjugation and exploitation arise from this subjugation of the feminine principle. I want to argue that what is currently called development is essentially maldevelopment, based on the introduction or accentuation of the domination of man over nature and women. In it, both are viewed as the 'other', the passive non-self. Activity, productivity, creativity which were associated with the feminine principle are expropriated as qualities of nature and women, and transformed into the exclusive qualities of man. Nature and women are turned into passive objects, to be used and exploited for the uncontrolled and uncontrollable desires of alienated man. From being the creators and sustainers of life, nature and women are reduced to being 'resources' in the fragmented, anti-life model of maldevelopment.

Two kinds of growth, two kinds of productivity

Maldevelopment is usually called 'economic growth', measured by the Gross National Product. Porritt, a leading ecologist has this to say of GNP:

> *Gross* National Product—for once a word is being used correctly. Even conventional economists admit that the hey-day of GNP is over, for the simple reason that as a measure of progress, it's more or less useless. GNP measures the lot, all the goods and services produced in the money economy. Many of these goods and services are not beneficial to people, but rather a measure of just how much is going wrong; increased spending on crime, on pollution, on the many human casualties of our society, increased spending because of waste or planned obsolescence, increased spending because of growing bureaucracies: it's all counted.[8]

The problem with GNP is that it measures some costs as benefits (eg. pollution control) and fails to measure other costs completely. Among these hidden costs are the new burdens created by ecological devastation, costs that are invariably heavier for women, both in the North and South. It is hardly surprising, therefore, that as GNP rises, it does not necessarily mean that either wealth or welfare increase proportionately. I would argue that GNP is becoming, increasingly, a measure of how real wealth—the wealth of nature and that produced by women for sustaining life—is rapidly decreasing. When commodity production as the prime economic activity is introduced as development, it destroys the potential of nature and women to produce life and goods and services for basic needs. More commodities and more cash mean less life—in nature (through ecological destruction) and in society (through denial of basic needs). Women are devalued first, because their work cooperates with nature's processes, and second, because work which satisfies needs and ensures sustenance is devalued in general. Precisely because more growth in maldevelopment has meant less sustenance of life and life-support systems, it is now imperative to recover the feminine principle as the basis for development which conserves and is ecological. Feminism as ecology, and ecology as the revival of Prakriti, the source of all life, become the decentred powers of political and economic transformation and restructuring.

This involves, first, a recognition that categories of 'productivity' and growth which have been taken to be positive, progressive and universal are, in reality, restricted patriarchal categories. When viewed from the point of view of nature's productivity and growth, and women's production of sustenance, they are found to be ecologically destructive and a source of gender inequality. It is no accident that the modern, efficient and productive technologies created within the context of growth in market economic terms are associated with heavy ecological costs, borne largely by women. The resource and energy intensive production processes they give rise to demand ever increasing resource withdrawals from the ecosystem. These withdrawals disrupt essential ecological processes and convert renewable resources into non-renewable ones. A forest for example, provides inexhaustible supplies of diverse biomass over time if its capital stock is maintained and it is harvested on a sustained yield basis. The heavy and uncontrolled demand for industrial and commercial wood, however, requires the continuous overfelling of trees which exceeds the regenerative capacity of the forest ecosystem, and eventually converts the forests into non-renewable resources. Women's work in the collection of water, fodder and fuel is thus rendered more energy and time-consuming. (In Garhwal, for example, I have seen women who originally collected fodder and fuel in a few hours, now travelling long distances by truck to collect grass and leaves in a task that might take up to two days.) Sometimes the damage to nature's intrinsic regenerative capacity is impaired not by over-exploitation of a particular resource but, indirectly, by damage caused to other related natural resources through ecological processes. Thus the excessive overfelling of trees in the catchment areas of streams and rivers destroys not only forest resources, but also renewable supplies of water, through hydrological destabilisation. Resource intensive industries disrupt essential ecological processes not only by their excessive demands for raw material, but by their pollution of air and water and soil. Often such destruction is caused by the resource demands of non-vital industrial products. Inspite of severe ecological crises, this paradigm continues to operate because for the North and for the elites of the South, resources continue to be available, even now. The lack of recognition of nature's processes for survival as factors in the process of economic development shrouds the political issues arising from resource transfer and resource destruction, and creates an ideological weapon for increased control over natural resources in the conventionally employed notion of productivity. All other costs of the economic process consequently become invisible. The forces which contribute to the increased 'productivity' of a modern farmer or factory worker for instance, come from the increased use of natural resources. Lovins has described this as the amount of 'slave' labour presently at work in the world.[9] According to him each person on earth, on an average, possesses the equivalent of about 50 slaves, each working a 40 hour week. Man's global energy conversion from all sources (wood, fossil fuel, hydroelectric power, nuclear) is currently approximately 8×10^{12} watts. This is more than 20 times the energy content of the food necessary to feed the present world population at the FAO standard diet of 3,600 cal/day. The 'productivity' of the western male compared to women or Third World peasants is not intrinsically superior; it is based on inequalities in the distribution of this 'slave' labour. The average inhabitant of the USA for example has 250 times more 'slaves' than the average Nigerian. 'If Americans were short of 249 of those 250 "slaves", one wonders how efficient they would prove themselves to be?'

It is these resource and energy intensive processes of production which divert resources away from survival, and hence from women. What patriarchy sees as productive work, is, in ecological terms highly destructive production. The second law of thermodynamics predicts that resource intensive and resource wasteful economic development must become a threat to the survival of the human species in the long run. Political struggles based on ecology in industrially advanced countries are rooted in this conflict between *long term survival options* and *short term over-production and over-consumption*. Political struggles of women, peasants and tribals based on ecology in countries like India are far more acute and urgent since they are rooted in the *immediate threat to the options for survival* for the vast majority of the people, *posed by resource intensive and resource wasteful economic growth* for the benefit of a minority.

In the market economy, the organising principle for natural resource use is the maximisation of profits and capital accumulation. Nature and human needs are managed through market mechanisms. Demands for natural resources are restricted to those demands registering on the market; the ideology of development is in large part based on a vision of bringing all natural resources into the market economy for commodity production. When these resources are already being used by nature to maintain her production of renewable resources and by women for sustenance and livelihood, their diversion to the market economy generates a scarcity condition for ecological stability and creates new forms of poverty for women.

Two kinds of poverty
In a book entitled *Poverty: the Wealth of the People*[10] an African writer draws a distinction between poverty as subsistence, and misery as deprivation. It is useful to separate a cultural conception of subsistence living as poverty from the material experience of poverty that is a result of dispossession and deprivation. Culturally perceived poverty need not be real material poverty: subsistence economies which satisfy basic needs through self-provisioning are not poor in

the sense of being deprived. Yet the ideology of development declares them so because they do not participate overwhelmingly in the market economy, and do not consume commodities produced for and distributed through the market *even though they might be satisfying those needs through selfprovisioning mechanisms.* People are perceived as poor if they eat millets (grown by women) rather than commercially produced and distributed processed foods sold by global agribusiness. They are seen as poor if they live in self-built housing made from natural material like bamboo and mud rather than in cement houses. They are seen as poor if they wear handmade garments of natural fibre rather than synthetics. Subsistence, as culturally perceived poverty, does not necessarily imply a low physical quality of life. On the contrary, millets are nutritionally far superior to processed foods, houses built with local materials are far superior, being better adapted to the local climate and ecology, natural fibres are preferable to man-made fibres in most cases, and certainly more affordable. This cultural perception of prudent subsistence living as poverty has provided the legitimisation for the development process as a poverty removal project. As a culturally biased project it destroys wholesome and sustainable lifestyles and creates real material poverty, or misery, by the denial of survival needs themselves, through the diversion of resources to resource intensive commodity production. Cash crop production and food processing take land and water resources away from sustenance needs, and exclude increasingly large numbers of people from their entitlements to food. 'The inexorable processes of agriculture-industrialisation and internationalisation are probably responsible for more hungry people than either cruel or unusual whims of nature. There are several reasons why the high-technology-export-crop model increases hunger. Scarce land, credit, water and technology are pre-empted for the export market. Most hungry people are not affected by the market at all. . . . The profits flow to corporations that have no interest in feeding hungry people without money.'[11]

The Ethiopian famine is in part an example of the creation of real poverty by development aimed at removing culturally perceived poverty. The displacement of nomadic Afars from their traditional pastureland in Awash Valley by commercial agriculture (financed by foreign companies) led to their struggle for survival in the fragile uplands which degraded the ecosystem and led to the starvation of cattle and the nomads.[12] The market economy conflicted with the survival economy in the Valley, thus creating a conflict between the survival economy and nature's economy in the uplands. At no point has the global marketing of agricultural commodities been assessed against the background of the new conditions of scarcity and poverty that it has induced. This new poverty moreover, is no longer cultural and relative: it is absolute, threatening the very survival of millions on this planet.

The economic system based on the patriarchal concept of productivity was created for the very specific historical and political phenomenon of colonialism. In it, the input for which efficiency of use had to be maximised in the production centres of Europe, was industrial labour. For colonial interest therefore, it was rational to improve the labour resource *even at the cost of wasteful use of nature's wealth*. This rationalisation has, however, been illegitimately universalised to all contexts and interest groups and, on the plea of increasing productivity, labour reducing technologies have been introduced in situations where labour is abundant and cheap, and resource demanding technologies have been introduced where resources are scarce and already fully utilised for the production of sustenance. Traditional economies with a stable ecology have shared with industrially advanced affluent economies the ability to use natural resources to satisfy basic vital needs. The former differ from the latter in two essential ways: first, the same needs are satisfied in industrial societies through longer technological chains requiring higher energy and resource inputs and excluding large numbers without purchasing power; and second, affluence generates new and artificial needs requiring the increased production of industrial goods and services. Traditional economies are not advanced in the matter of non-vital needs satisfaction, but as far as the satisfaction of basic and vital needs is concerned, they are often what Marshall Sahlins has called 'the original affluent society'. The needs of the Amazonian tribes are more than satisfied by the rich rainforest; their poverty begins with its destruction. The story is the same for the Gonds of Bastar in India or the Penans of Sarawak in Malaysia.

Thus are economies based on indigenous technologies viewed as 'backward' and 'unproductive'. Poverty, as the denial of basic needs, is not necessarily associated with the existence of traditional technologies, and its removal is not necessarily an outcome of the growth of modern ones. On the contrary, the destruction of ecologically sound traditional technologies, often created and used by women, along with the destruction of their material base is generally believed to be responsible for the 'feminisation' of poverty in societies which have had to bear the costs of resource destruction.

The contemporary poverty of the Afar nomad is not rooted in the inadequacies of traditional nomadic life, but in the *diversion of the productive pasture-land of the Awash Valley*. The erosion of the resource base for survival is increasingly being caused by the demand for resources by the market economy, dominated by global forces. The creation of inequality through economic activity which is ecologically disruptive arises in two ways: first, inequalities in the distribution of privileges make for unequal access to natural resources—these include privileges of both a political and economic nature. Second, resource intensive production processes have access to subsidised raw material on which a substantial number of people, especially from the less privileged economic groups, depend for their survival. The consumption of such industrial raw material is determined purely by market forces, and not by considerations of the social or ecological requirements placed on them. The costs of resource destruction are externalised and unequally divided among various economic groups in society, but are borne largely by women and those who satisfy their basic material needs directly from nature, simply because they have no purchasing power to register their demands on the goods and services provided by the modem production system. Gustavo Esteva has called development a permanent war waged by its promoters and suffered by its victims.'[13]

The paradox and crisis of development arises from the mistaken identification of culturally perceived poverty with real material poverty, and the mistaken identification of the growth of commodity production as better satisfaction of basic needs. In actual fact, there is less water, less fertile soil, less genetic wealth as a result of the development process. Since these natural resources are the basis of nature's economy and women's survival economy, their scarcity is impoverishing women and marginalised peoples in an unprecedented manner. Their new impoverishment lies in the fact that resources which supported their survival were absorbed into the market economy while they themselves were excluded and displaced by it.

The old assumption that with the development process the availability of goods and services will automatically be increased and poverty will be removed, is now under serious challenge from women's ecology movements in the Third World, even while it continues to guide development thinking in centres of patriarchal power. Survival is based on the assumption of the sanctity of life; maldevelopment is based on the assumption of the sacredness of 'development'. Gustavo Esteva asserts that the sacredness of development has to be refuted because it threatens survival itself. 'My people are tired of development', he says, 'they just want to live'.[14]

The recovery of the feminine principle allows a transcendance and transformation of these patriarchal foundations of maldevelopment. It allows a redefinition of growth and productivity as categories linked to the production, not the destruction, of life. It is thus simultaneously an ecological and a feminist political project which legitimises the way of knowing and being that create wealth by enhancing life and diversity, and which legitimises the knowledge and practise of a culture of death as the basis for capital accumulation.

Shiva, V., *Staying Alive; Women, Ecology and Development*, Zed Books Ltd., New Delhi, 1988, p. 1-13.

Notes and References

1   Rosa Luxemberg, *The Accumulation of Capital*, London: Routledge and Kegan Paul, 1951.
2   An elaboration of how 'development' transfers resources from the poor to the well-endowed is contained in J. Bandyopadhyay and V. Shiva, 'Political Economy of Technological

Polarisations' in *Economic and Political Weekly*, Vol. XVIII, 1982, pp. 1827-32; and J. Bandyo-padhyay and V. Shiva, 'Political Economy of Ecology Movements', in *Economic and Political Weekly*, forthcoming.

3 Ester Boserup, *Women's Role in Economic Development*, London Allen and Unwin, 1970.

4 DAWN, *Development Crisis and Alternative Visions: Third World Women's Perspectives*, Bergen: Christian Michelsen Institute, 1985, p. 21.

5 M. George Foster, *Traditional Societies and Technological Change*, Delhi: Allied Publishers, 1973.

6 Maria Mies, *Patriarchy and Accumulation on a World Scale*, London: Zed Books, 1986.

7 Alice Schlegel (ed.), *Sexual Stratification: A Cross-Cultural Study*, New York: Columbia University Press, 1977.

8 Jonathan Porritt, *Seeing Green*, Oxford Blackwell, 1984.

9 A. Lovins, cited in S.R. Eyre, *The Real Wealth of Nations*, London: Edward Arnold, 1978.

10 R. Bahro, *From Red to Green*, London: Verso, 1984, p. 211.

11 R.J. Barnet, *The Lean Years*, London: Abacus, 1981, p. 171.

12 U.P. Koehn, 'African Approaches to Environmental Stress: A Focus on Ethiopia and Nigeria', in R.N. Barrett (ed.), *International Dimensions of the Environmental Crisis*, Colorado: Westview, 1982, pp. 253-89.

13 Gustavo Esteva, 'Regenerating People's Space', in S.N. Mendlowitz and R.B.J. Walker, *Towards a Just World Peace: Perspectives From Social Movements*, London: Butterworths and Committee for a Just World Peace, 1987.

14 G. Esteva, Remarks made at a Conference of the Society for International Development, Rome, 1985.

CHAPTER 24

Concern for Tomorrow

A National Environmental Survey 1985-2010

F. LANGEWEG (ED.)

1 Formulation of the problem for the long-term survey

In 1989, the government will publish a national environmental policy plan (NMP). This plan focuses on the strategic decisions to be taken over the next 8 to 10 years. Many environmental issues such as, for example, the greenhouse effect and the acidification of the soil are of a long-term nature which can far exceed the planning period of the NMP. It is therefore important to place the strategic decisions to be taken in the NMP also within the framework of the developments anticipated in the still longer term. The RIVM has been requested, in collaboration with other institutes, to draw up a background document aimed at the long-term exploration of the development of environmental problems. The period chosen is from 1985 up to about the year 2010. The Nuisance Act of 1875, aimed at local environmental problems, was for almost a century the only law available for environmental policy. During the past few decades in particular, the increase in scale of environmental problems has become manifest, from the built environment and the landscape to the river basins, the continent and the plane, as is evident from the damage to the ozone layer in the stratosphere and the anticipated climatic change. These problems are caused to an increasing degree by the accumulation of poorly degradable substances in the natural cycles and their reservoirs.

An analysis of the nature of the cycles shows that five levels can be distinguished within which characteristic material and energy flows occur. These are the local, regional, fluvial, continental and global levels.

The local level comprises the built environment within which man spends a large part of his time. Here the characteristic environmental problems are disturbance by stench and noise, occupational pollution, and air pollution in the town centres especially.

On a regional scale, the slow material flow in soil and ground water is the determinative factor. Problems arising from the accumulation of nutrients and persistent substances such as phosphate, nitrate, heavy metals and pesticides are of particular importance in the very regions where they are being introduced into the environment (overfertilization and dispersion), because of the very low transport rates. As a result of the extraction of water, the hydrological cycle is also affected on this scale. The disposal of waste, by dumping it legally or otherwise, leads to a threat to soil and ground water.

On a fluvial scale, transport via surface water is the carrier of two kinds of problems for the river basins and the contiguous epicontinental seas. The first stems from the systematic loading of the water with persistent substances. These also include occasional discharges of very toxic substances. The other problem concerns the supply of nutrients to the surface water, whereby aquatic ecosystems are damaged. Both types of problems can lead to the accumulation of substances in the waterbottoms.

On a continental scale, the circulation of air in the boundary layer (at an altitude of 0-3 km) is decisive for the dispersion of substance from within a few days to several weeks at most. It concerns aerosols, photo-oxidants, acidifying substances and radionuclides. The actual problems need not arise in the atmosphere itself but can also, as is the case with acidification, occur in the soil or water.

On the global scale, the circulation in the higher layers of the atmosphere leads worldwide to the mixing of additions to the atmosphere within one or two years. A host of substances with a lifetime of several years or longer is currently accumulating in the upper atmosphere. Greenhouse gases and ozone destroying substances are active on a global scale (climatic change, depletion of the ozone layer).

The time factor plays a part in the development of environmental problems. When substances build up in reservoirs it can take a long time before an environmental problem becomes apparent and often even longer before that problem has been solved once the emissions have ceased (lag.). In many instances there is also a limited buffer capacity of reservoirs, which is finite. Only when the buffer capacity has been exceeded the environmental problem will manifest itself. Acidifying emissions from the burning of fossil fuels have been increasing for decades. Only in recent years have the effects become noticeable. Once the emissions have stopped it will still be a long time before the original situation has recovered, even if this is possible. Quicker recognition of this type of process is essential (early warning).

In addition to the current central environmental problems (acidification, overfertilization, dispersion, disturbance and disposal) as well as the recently observed environmental problems (climatic change, damage to the ozone layer and the indoor environment), other, hitherto undetected new environmental problems may yet arise in the next few decades. For example, the risk associated with the use of genetically modified micro-organisms and the incursion of harmful plant or animal species as well as viruses can be mentioned here. The consequences of utilizing the oceans for human purposes, and the consequences of the introduction of new energy systems and new materials can also cause problems for the environment.

2   The present magnitude of the environmental problems considered above

Global environmental problems

The greenhouse effect is brought about by the emissions of CO_2 (energy, deforestation), methane (agriculture), nitrous oxide (energy), CFCs (industrial products) and ozone. These gases lead in the troposphere (at an altitude of 2-10 km) to a change in the radiation of heat from the atmosphere into space. As a result, the atmosphere will warm, the sea level will rise and climates will change worldwide. At present CO_2 is responsible for half of the greenhouse effect, methane for a quarter and other gases for the remainder. When the current trend in the emission of these gases continues, the temperature on Earth will have increased on average by 8 per cent by the year 2100. The sea level could then have risen by 70 centimetres at most over and above the autonomous trend of 15 to 20 centimetres per century. The Dutch contribution to the emissions is less than 1 per cent in absolute terms but considerably larger in relative terms (for example, per inhabitant). In the Netherlands, the particular effect which will arise concern coastal defense, the water management, food supply and natural ecosystems. It is still difficult to gain a comprehensive view of all the effects but they will be very far-reaching. How the Earth will respond to climatic change within a century, which in the past occurred on a much larger timescale, is still difficult to predict.

Chlorofluorocarbons (CFCs) and halons damage the ozone layer in the stratosphere. The ozone layer absorbs UV-B radiation which is harmful to man and nature. Increasing UV-B radiation can result in serious damage to public health from skin cancer and a weakening of the immune system, so that infectious diseases may develop. Satellite observations show that the decrease in ozone becomes greater every year. In 1887, a decrease of 95 per cent was observed locally and temporarily over the Antarctic. The loss of ozone is smaller with increasing distance from the poles. There has been a large growth in the production of CFCs over the past few decades, They are used as a refrigerant in refrigerators and airconditioning units, a propellant in aerosol sprays and as a blowing agent in the production of synthetic materials. Halons are used in fire extinguishers.

Continental environmental problems

The deposition of acidifying substances in The Netherlands has fallen by about 20 per cent since 1980, but the levels are still much too high. As a result, the vitality of the forests declines, heathland grasses up, and the fens acidify. The acidification is caused by SO_2, NO_x and NH_3. The SO_2 emissions by industry and the public utilities have fallen by some 40 per cent since 1980. However, the emissions of NO_x from the same sources and traffic have hardly decreased at all. The NH_3 emissions, mainly derived from agriculture, have even risen slightly.

The ozone concentrations in the living environment show a rising trend. A distinction should here be made between the peak concentrations occurring during episodes under certain conditions and the average background level. Peak concentrations are due to NO_x derived from the sources already mentioned and volatile organic compounds derived from industrial processes (solvents), refineries and traffic. The formation of peaks in the ozone concentrations occurs in the boundary layer near the Earth's surface. The background level arises in the free troposphere above this layer and reaches the biosphere through mixing into the boundary layer. Here, carbon monoxide (combustion processes) and methane (agriculture) play an important role. This problem even has a global dimension. The peak concentrations pose a threat to public health. The background concentration of ozone can damage the natural vegetation and agriculture products during the growing season.

Aerosols (dust) is the collective term for a diversity of solid particles. The fine fraction of this is generally of anthropogenic origin. In a warning sense, attention is being paid to metal aerosols (derived from high-temperature processes such as waste incinerators, energy generation and metal production), and acid aerosols in connection with the acidification issue. The magnitude of the aerosol problem cannot yet be satisfactorily quantified. The deposition of heavy metal via aerosols is in many instances an important burdening factor for the fluvial and regional systems. Acid aerosols and the "free" acid present in the atmosphere during episodes of air pollution seem to have a not inconsiderable effect on public health.

Nuclear accidents can give rise to the dispersal of radionuclides on a continental scale. There are at present 115 nuclear power stations in the countries around us. The accident with the nuclear reactor at Chernobyl has shown that even far-away nuclear power stations represent a risk to our country. The accident at Chernobyl currently contributes 2.5 per cent to the average radiation dose to the Dutch population, amounting to 2.4 mSv per year. The total risk to our country from all nuclear power stations in Europe is not yet known. Therefore, no long-term survey was made of this.

Fluvial environmental problems

The pollution of the Dutch inland water bodies by nutrients is still a factor of 10-25 too high compared with the natural situation. The concentrations of nutrients in a zone several tens of kilometres wide along the Dutch and German North Sea coast are now three to four times higher than they were 50 or more years ago. These excessive nutrient loads lead to a higher density of vegetable organisms (algal bloom). This can destabilize the aquatic ecosystems, resulting in a decline in the number of species.

There has in general been an improvement in the quality of the salt and fresh surface waters in recent years with respect to persistent substances such as metal and pesticides, not counting occasional discharges. These substances accumulate in the water bottom, causing standards set for a multi-functional soil to be far exceeded. Persistent substances can also build up in the food chain and produce effects especially in the fish and mammals at the top of this chain. The disappearance of salmon and sturgeon from the Rhine and the drop in the number of seals may be partly due to this. Continuation of the current pollution levels from domestic and industrial waste water offers no prospect for a recovery of the aquatic ecosystems.

Regional environmental problems

The groundwater, soil and small surface waters in The Netherlands are heavily polluted by the nutrients nitrogen, phosphorus and potassium, chiefly derived from agriculture. Since 1920, the use of nitrogen fertilizer in agriculture has increased nearly twenty-fold. The use of phosphorus fertilizer has remained about the same. The amount of animal manure put on the land has more than quadrupled since 1900. The deposition of nitrogen from the air has also increased. As a result, the nitrate concentration in groundwater has risen in the sandy regions of our country. This constitutes a threat to the drinking water supply and can lead to the destabilization of ecosystems. Phosphate accumulates in the soil. When the soil reservoir has filled up, breakthrough can occur so that the small surface waters are in jeopardy. This is already taking place, or will happen in the short term, in areas in North Brabant, South Limburg, Gelderland and Overijssel. The nutrient load is currently too high in a large number of small surface waters.

Persistent substances such as metals, pesticides and other anorganic compounds likewise constitute a major threat to the quality of soil and ground water. The soil is increasingly loaded with metals from the use of fertilizers and deposition from the air. The cadmium and copper concentrations in the soil are rising. The consumption of pesticides in agriculture has grown to about 25 million kilograms of active matter per year. As a result, the concentrations of pesticides adsorbing to the soil particles, such as lindane and organotin compounds, are increasing. Mobile pesticides, such as atrazine and dichloropropane, lead into groundwaters. They pose a threat to the drinking water supply. Continuation of the current pollution levels of nutrient and persistent substances jeopardizes the drinking water supply and the production of healthy food in agriculture, and can lead to the destabilization of ecosystems because a few species will eventually predominate.

Drying out has increasingly become a problem. Signs of moderate and severe drying out can be observed in large parts of the higher-lying regions of our country and in the dunes. Drying out arises when the water table is lowered by tapping underground water supplies by measures to control water levels when planning land use, and infrastructural measures. Drying out also contributes to the overfertilization and acidification problem and to the impoverishment of ecosystems. A negative side effect of drying out is that water extraneous to the region is let in by way of compensation, also involving adverse ecological consequences. The future extent of this drying out is still hard to predict so that a long-term survey of this was not made.

The dumping of waste poses a threat to the quality of soils and ground water. Large quantities of waste have been dumped either legally or illegally. This has already resulted in 1600 cases of soil contamination to be cleaned up. There are indications that this is only the tip of the iceberg. A rough analysis points to a significantly larger number of locations to be cleaned up, for the most part industrial sites.

Local environmental problems

Noise pollution is caused by road and rail traffic, aviation, industry and other economic activities. The current nuisance from noise is extensive. Road traffic is being experienced as a nuisance in 38 per cent of all dwellings, aircraft in 19 per cent, industry in 14 per cent and rail traffic in 6 per cent of the houses. About 300,000 hectares have been designated as areas of silence in The Netherlands.

Stench is a nuisance in about 19 per cent of all dwellings. In general, there are many substances responsible for this. Industry (9 per cent), agriculture (3 per cent) and traffic (7 per cent) are the principle sources.

The standards for air pollution by carbon monoxide, nitrous oxide, lead, formaldehyde, benz(a)pyrene and black smoke (soot) are exceeded on about 1000 kilometres of pavement in town centres because of traffic. With the introduction of unleaded petrol, the lead problem is now rapidly diminishing.

In the home itself (the indoor environment) numerous substances are present which could pose a threat to the public health. The principal agents are nitrogen dioxide arising from geysers and cooking appliances, noise caused by neighbours, radon entering via the crawl space, suspended particles from smoking, and allergens and fungi due to moisture in the home. Standards for one or more substances are exceeded in almost all dwellings. Primary determinants here are the characteristics of the house, the way the house is used and pollutants entering the home from outside.

3   Premises for the long-term survey

The long-term nature of environmental problems necessitates timely recognition of undesired social developments. Long-term surveys provide environmental policy-makers with an indication of the extent to which, by adjusting social processes, set goals can be realized. Previous studies dealt with scenarios for the future, for example, the 'Report of the Club of Rome' (1971), the OECD study 'Facing the Future' (1979) and 'Our Common Future' (Brundtland report, 1987). The last two reports found a sustainable economic development to be possible in principle. Its realization, however, will draw heavily upon the willingness of the present generation for initiating, partly through environmental policy, such a development in the longer term.

Combatting environmental pollution after it has arisen, by 'end of pipe' measures, is extremely costly and is more or less part of current practice. Such emission reducing and effect-directed measures have in general also been chosen in this long-term survey for solving environmental problems. A sustainable alternative will have to be aimed at more far-reaching prevention, re-use development of the associated technology. This will result in lower pollution abatement costs and a less rapid depletion of natural resources. Economic growth in this sustainable alternative could even be stronger than when the currently accepted practice is continued. Such structural changes are discussed only to a limited degree in this long-term survey because of their innovative nature.

Dutch environmental policy on its own can only promote a sustainable development on a local and regional scale. At the fluvial, continental and global levels, the international dimension becomes increasingly larger in the policy to be pursued.

In the presentation of the scenarios for the future, the middle variant was used whenever a number of alternatives was available. Furthermore, a distinction was made between scenarios in which either measures already established, or additional measures, currently considered to be technically feasible, are taken.

With high economic growth, an increase in global energy consumption of 2-2.5 per cent annually is anticipated. This increase will occur chiefly in developing countries, where growth figures of 5 per cent to 6 per cent are expected. In the middle scenario, energy consumption in The Netherlands is expected to grow by 1 per cent annually in the period 1985 to 2010.

As a result of the growing demand for electricity, the generating capacity of power plant will have to increase from 15,470 MW in 1985 to 22,410 MW in 2010. Under the present plans this growth will be met by new coal-fired power stations but no nuclear plants. In Europe, the planned expansion of nuclear power stations is expected to be 80 per cent by 2000.

By 2050, the world population is envisaged to have increased to 10 billion, of which 8,5 billion will inhabit the Third World. The population of The Netherlands will then be approaching 15 million.

To feed the growing population in developing countries, the area under cultivation in the world will have to be expanded further and productivity increased. Both measures give rise to environmental problems such as deforestation, the formation of deserts and a strong growth in the use of pesticides. The application of pesticides worldwide is increasing by 5 per cent to 6 per cent per year.

The Netherlands is one of the largest exporters of food and agricultural products in the world and because of this is also a major importer of livestock feed, vegetable oils, cacao and timber. In addition to the more than 20,000 square kilometres of land used at home, three times as large an area is used for Dutch agriculture in other countries, especially the Third World. Imports from the Third World are not infrequently accompanied by soil depletion, followed by soil erosion. The Netherlands itself, owing to the considerable import surplus, is confronted with the well-known manure issue.

The high productivity of Dutch agriculture is partly founded on large doses of (chemical) fertilizer and the application of pesticides, these being the highest in the world. Intensive livestock population has nearly doubled between 1970 and 1985, chiefly from an increase in pigs and poultry. The number of cattle will fall by almost 30 per cent between now and 2010 while the number of pigs and poultry will rise slightly. The use of fertilizers and the production of animal manure will remain about the same up to 2010 unless far-reaching measures are taken.

The drinking- and industrial-water supply is faced with growing water consumption. The combined water consumption by households, industry and agriculture will increase by 40 to 50 per cent between now and 2010.

The upward trend of traffic and transport is expected to continue in the coming years. The number of car kilometres travelled will have increased by about 60 per cent in 2010. Meanwhile, The Netherlands has the highest spatial car density in the world (128 cars per square kilometre). As a result of the increase in traffic and transport and cars becoming more fuel-efficient, fuel consumption is expected to grow by 40 per cent compared with 1985.

4   Results of the long-term survey

Global environmental problems

To limit climatic change in the next century to a rise in temperature of minimally 1.5 to possibly 4.5 degrees centigrade, 0.6 tonnes of carbon may be emitted for each global citizen per year. For The Netherlands this means an emission reduction of about 80 per cent. It is not yet clear how this reduction should be achieved. Energy conservation and the development of other energy systems should be stimulated. When no far-reaching emission-controlling measures are taken, there is a chance of large-scale disruption of the global biosphere. This can ultimately have consequences for the survival of mankind on Earth.

When the Montreal protocol is implemented, leading to a potential cut in CFC emissions of 50 per cent of the 1986 level, the ozone layer could decrease globally by on average 5-10 per cent. Only the essation of the CFC emissions, as agreed upon with industry in The Netherlands, wil lead to a fall in the CFC concentrations in the atmosphere, and a process of recovery of the ozone layer slower by several decades. At present, The Netherlands still emits 50,000 tonnes of halogenated hydrocarbons each year, with CFCs accounting for 20 per cent of this. Other halogenated hydrocarbons too, such as dichloromethane, are suspected of contributing to the depletion of the ozone layer.

Continental environmental problems

The acid-neutralizing capacity of 30 per cent of the soil in Central Europe will be impaired for a long time despite maximum control of acid emissions.

To prevent the most serious damage in The Netherlands, the emissions of SO_2 will have to fall by 90 per cent and those of NO_x by 70 per cent compared with 1980, in both our country and the countries around us. The NO_x emissions in Eastern Europe will then have to decrease by 30 per cent. The emission of NH_3 in The Netherlands should decline by 80 per cent and in the neighbouring countries by 60 per cent compared with 1980. Reductions of about 80 per cent for SO_2, 55-60 per cent for NO_x and 65-75 per cent for NH_3 are feasible in The Netherlands with the available technical control measures.

Even when extra measures are taken, the ozone concentrations on Earth will exceed the standard for peak concentration on 5-12 days per year. The standard for the seasonally averaged ozone concentration in the lower atmosphere has already been approximately reached. To prevent damage from ozone to life on Earth, the cuts in emissions should be even more drastic than those for combatting acidification, amounting to 70-90 per cent for NO_x and volatile organic substances (VOC) in Europe, and similar reductions in CO_2 and CH_4 on a global scale. About 50 per cent of the VOC emissions can be controlled in The Netherlands with the available technical means.

Fluvial environmental problems

Despite implementation of the Rhine Action Plan, the standard for phosphorus in the tidal river area, the Ketel Lake and the Gooi Lake, and that for nitrogen and phosphorus in the Westerschelde and the Wadden Sea will be exceeded after the year 2000. To avoid eutrophication in these fresh and salt surface waters, the discharge of nitrogen and phosphorus compounds should be reduced by at least 75 per cent instead of the 50 per cent included in the Rhine Action Plan.

Because of the Rhine Action Plan, standards for several accumulating substances (cadmium, copper, lindane, benz(a)pyrene) in fresh and salt surface waters, will rarely be exceeded. Standards for the water bottoms will still be exceeded, despite a cut in emission of 70-80 per cent, in a few sedimentation areas (Haringvliet, Hollands Diep, the Rotterdam harbour area and Ketel Lake). It is uncertain whether the future clean sediment deposited on the 70 million cubic meters of heavily polluted sediment in the freshwater sedimentation areas will adequately protect biota from taking up the pollutants. The same is true for the sedimentation areas in the Wadden Sea. It will take decades before the sediment in the North Sea will have attained a new equilibrium which lies below the standards.

The capacity of waste-water purification plants will increase only moderately, from the present 22 million inhabitant equivalents to 25 million in 2010. Phosphate and nitrate will, however, have to be removed. In a sustainable development, purification should aim at total recovery of raw materials and energy. Non-recyclable and poorly degradable substances must be excluded from the waste water.

Regional environmental problems

The cycle of nitrogen in agriculture can in principle not be closed, but the measures are aimed at minimizing the difference between supply and removal. The present standard for nitrate in drinking water can be met if production of a fertilizer from manure is combined with integral nitrogen management in agriculture. This conclusion doesn't hold for the target value for quality of drinking water. As a result, 10-20 ground-water pumping stations which draw water from shallow depths are now at risk, but this number could fall again in the next century if extra measures are taken. Putting too high a burden on the small surface waters can then be prevented. Even when the leaching of nitrogen is prevented completely, it will take until far into the next century before the required quality of the pumped water for the drinking water supply is attained.

The cycle of phosphorus in agriculture can be more or less closed by the measures planned. As a result, further accumulation of phosphate in the soil can be arrested. Phosphate breakthrough may occur locally on a limited scale. Because of subsequent delivery from the water bottom and phosphate runoff, the quality of the small surface waters will not improve sufficiently for some time.

The emission of copper and cadmium into the soil can be reduced by 35 per cent and 50 per cent respectively, which is not enough. The standards for copper will be exceeded on the long-term on a scale in arable soils. About 10 per cent of the agricultural land will fluctuate around the standard for cadmium by 2010. For grassland, this proportion will increase from the present 25 per cent to 40 per cent in 2010. The standard for cadmium will be exceeded in the long term in several polder areas in the west of The Netherlands. Depositions from the air will here also be an important factor.

Poorly degradable pesticides pose a threat to the drinking water supply obtained from ground water in sandy areas. It can take more than 100 years before too high concentrations have disappeared from the pumped water after the use of these pesticides has ceased.

In a preventive sense, the disposal of waste products should be aimed at recovery of raw material and energy, and reduction of the volumes. To this end, non-recyclable and poorly degradable substances should be excluded from the waste. These objectives are as yet only partly achieved. By 2010, 65 per cent of the waste products can be reused or find a useful application, 20 per cent can be incinerated and 15 per cent will have to be dumped. At present, about 40 per cent of the waste is dumped. There still remains an extensive soil cleanup operation, partly due to carelessly discarded waste products.

Local environmental problems

For large segments of the population, noise pollution wil not lessen with the planned measures because of the growth of traffic, aviation and industry. Additional measures could lead to a reduction of 40 per cent in the number of persons affected by traffic noise and of 90 per cent in those disturbed by noise from civil aircraft, while the annoyance caused by industry and trains will roughly stay at the current level.

The planned measures will reduce the proportion of dwellings bothered by stench from the present 19 per cent to 16 per cent in 2010. This can be cut to 8 per cent by additional measures.

The total length of pavement in towns where standards for air pollution are exceeded could be reduced by additional measures from 1000 kilometres in

1985 to 260 kilometres in 2010. There will be a slight increase of around 10 per cent with the planned measures, due to the growth in traffic. The extra technically feasible emission-reducing measures lead to a cut in emissions of 50-75 per cent. Volume and structural measures are required to prevent limits being exceeded.

At present, one or more standards for agents such as NO_x, radon (excluding building materials), neighbours' noise and airborne particles are exceeded in a large number of houses. Additional measures could cut this back to 70 per cent by 2010. Continuation of this policy will reduce the percentage to between 8 and 18 in the 21st century. This includes reduction in the potential exposure per house, which could be as much as 90 per cent or more.

Public health

Because of insufficient knowledge, the influence of environmental factors on health receives too little attention and is furthermore underestimated. A complicating factor here is that it concerns effects caused by long-term exposure to many factors at the same time.

A significant reduction in health risks from the burdening of the environment with pollutants can be achieved by reducing air pollution indoors and outdoors, limiting exposure to radon, avoiding increased exposure to UV-B radiation, limiting exposure to cadmium and controlling noise pollution.

Regions

The higher-lying regions in The Netherlands are the most vulnerable to various kinds of environmental pollution. Social and ecological functions are influenced decrementally by acidification, over-fertilization, toxic substances and dry-out. Regions with delta characteristics are vulnerable to accumulation of pollutants deteriorating the environmental quality. If the burden is not lightened, many species will disappear in the vulnerable regions in particular. This leads, among other things, to an impoverishment of the gene reservoir. These regions still seem to have recuperative powers. There will be no real improvement in the vulnerable regions unless the continental and global problems are solved. Effects on a regional scale, on the other hand, can enhance the effects caused by problems on a larger scale considerably and possibly also irreparably.

The situation can be improved significantly in regions currently heavily burdened with because of their susceptibility, on the sandy soils in Utrecht, Gelderland, Overijssel, North Brabant and Limburg as well as the loess areas in Limburg, the dune areas, the peat and peat-cutting areas, and the isolated lateral moraines. The Peelhorst, the old river-terrace landscape, the loess region and the chalk landscape in Limburg, and the isolated lateral moraines will, despite additional national and international measures, still be burdened considerably.

Costs and Benefits

With the planned measures, the percentage of the environmental costs of the Gross National Product (GNP) stays constant, up to 2010, at 2 per cent. The most stringent technical measures, currently believed to be feasible, will increase the environmental costs to 3-3.5 per cent of the GNP by 2010. This percentage is a maximum estimate as technological developments will probably make less expensive structural measures possible. About 15 per cent of the costs cannot be avoided because they result from effect-directed measures related to the cleanup or mitigation of the effects of the already existing environmental pollution.

The benefits of environmental measures accrue from the avoidance of damage in the future. These benefits cannot usually be expressed directly in monetary units. It is in fact a question of how much money the present generation is prepared to spend in order to hand down a clean environment to future generations. The quality of the environment should be such that a sustainable economic development is also attainable in the distant future. Structural adaptations aimed at prevention and efficiency are a prerequisite, making it possible to avoid high clean-up expenses and high costs of additional measures. Furthermore, it ensures a more prudent management of depletable resources.

The more large-scale the environmental problems become, the more urgently an international and structural preventive approach is required. From a global point of view, climatic change and depletion of the ozone layer can pose great risks to the environment and public health, and result in very extensive damage. This means that the present generation will have to consider the benefits anticipated in the long term from the measures planned as more important than the costs to be incurred in the decades ahead.

Langeweg, F. (ed.), *Concern for Tomorrow; A National Environmental Survey 1985-2010*, Rijksinstituut voor Volksgezondheid en Milieuhygiëne, Bilthoven, 1989, p. 11-20.

Part v

THE CURRENT STATE
OF AFFAIRS

Introduction to Part V: The Current State of Affairs

Where do we stand today? Have all the alarming reports had an effect? Has environmental consciousness actually led to environmentally sound behaviour? Do governmental and non-governmental organisations take the environment into account when implementing their policies? How much evidence needs to be presented before concrete action is taken? And if these actions are taken, will they be in time? Will enough means be put to use to turn the tide? Will the polluting society make way for an 'ecological community'?

Goodbye to the polluting society?

The process of change is in full swing. Slowly but surely, we are saying goodbye to the polluting society. But this process is not an easy one. There is often resistance to change. Fundamental social change processes are influenced by at least three factors: desire, pain, and force.

Lack of desire

Change must be desired. But in the case of the environment, the desire for change is often lacking. The new ecological order is not always communicated to people as an attractive future perspective as it is often associated with limitations on behaviour, increased taxes, and less prosperity. Sustainable development needs to be accompanied by an appealing picture of social and economic development. Since this imagery isn't always available or it is seen as unappealing or unrealistic, change is sometimes not desired and people want to continue their present behaviour.

Lack of pain

In other cases, the pain factor is missing. People fear change. An example is the introduction of the European CO_2 tax. Nearly everyone agrees with the goals of this tax, but many of them disapprove of the way it is to be attained. Therefore, more general environmental goals are under pressure, as in the case of the greenhouse effect and eutrofication. The pain factor is missing: in many cases, environmentally unfriendly behaviour is not attacked by measures which produce pain.

Lack of force

Sometimes the force needed to start change and keep it going is missing. Licensing and environmental auditing do not correspond with the actual environmental situation in companies. In many cases, however, the government doesn't even have the capacity to check up on the licences that were issued, let alone to adequately investigate and prosecute these companies in cases of non-compliance. In other words, the force factor too often exists only on paper. In these circumstances, there is a lack of force to change.

Phase and rate of change differences

In the process of social change, phase and rate of change differences occur. These differences are obvious between first and third world countries. In developing countries, issues other than environmental problems are the motor behind social change. Moreover, the speed of modernization of the economic system differs enormously between developing and developed countries. Differences in change phase and rate of change can also be seen, for example, in the collect and return systems in the different countries. The collecting and recycling percentages of various materials differ substantially. Even within one country, one can see that there are differences in change phase and speed between different economic sectors. Some industrial branches, such as the chemical industry, enforce their own responsible care programme to prevent unwanted environmental effects within their own sector. Other sectors have not yet started or are just at the beginning of the process of environmental care programmes.

Challenge for governments

The implementation of environmental policy imposes new demands regarding the required timetable and the dramatic changes needed. Environmental policy must be quantified as carefully as possible if we are to monitor the process. This means specifying the (interim) targets and deadlines for reaching them.

Governmental instruments

Only explicit descriptions of targets enable us to draw clear conclusions on the impact of environmental policy. In fact, only explicit targets make it possible to arrive at a true verification of the environmental policy performance considered. It can be concluded that the quality of a performance depends on the clearness and transparancy of the targets formulated. Environmental performance indicators can be used as a tool to make the targets formulated clearer and more transparant. What is fairly unique is that, in the Netherlands, indicators have been developed to measure the degree to which environmental policy targets, as presented in the National Environmental Policy Plan, are reached. For this purpose, 'environmental indicators' are elaborated by the National Institute for Environmental Care. This means that environmental problems such as acidification, the greenhouse effect, and the depletion of the ozone layer can be monitored by the measurement of environmental indicators. This enables the government to make up the balance of the existing environmental policy and to intensify environmental policy if necessary.

Challenge for business

Increasingly, modern industry is working together with government at all levels. A tendency towards joint responsibility and joint action is recognizable. In the past, solving environmental problems was seen by industry as additional costs. But nowadays this is being taken up as a challenge, separating tomorrow's winners from tomorrow's losers. Companies are encouraged to enhance the environmental performance of their processes and their products and they are giving greater consideration to their environmental performance as critical success factors.

Business tools

All kinds of environmental tools were developed to help companies be environmentally responsible. Tools like Environmental Management Systems, Environmental Auditing, Life-Cycle Assessment, Ecodesign, Environmental Performance Indicators, Environmental Labelling, Environmental Cost Accounting and Integrated Chain Management are being integrated into business as usual. Not only the government but also other actors were stimulating this development. Customers, suppliers, consumers and environmental organisations, banks, and insurance companies were formulating new requirements with respect to the environmental behaviour of the company. All these factors stimulated business' involvement with environmental problems.

Concepts like 'dematerialisation' (reducing the use of energy and matter per unit of product) and 'eco-efficiency' appeared. Such ideas as using less fossil energy sources and more sustainable, solar energy-based fuels, using biogenic renewable raw materials (comprising organic substrate) and mineralogenic raw materials (comprising inorganic substrate) and extending the life of products, creating as many recycling levels and loops as possible, were considered and integrated into decision-making processes.

This means that the actual state of the environment and the actual governmental and business environmental policy are relatively hopeful. There are signs of hope. Nearly all the actors in society are undertaking action to solve environmental problems. Industry is producing green products. Consumers are buying green products. Agriculture is reshaping its methods. Educational systems are teaching environmentally sound behaviour. Governments are implementing environmental action programmes. All these signs of hope, however, must not blind us to the fact that the number of signs of despair at the moment are much greater. The greenhouse effect is still with us. Air and water pollution are still going on. The destruction of rain forests is a daily event. The extinction of species has reached a critical level. The use of energy is still growing. Urban environmental problems are growing.

Brown: State of the World

Since 1984, *State of the World; a Worldwatch Institute Report on Progress toward a Sustainable Society* has been issued with regularity. In 1995, it was published for the 12th time. It deals almost always with issues at a global level, but occasionally it focuses on individual geographic regions as it did in State of the World 1985 for Africa and State of the World 1991 for Eastern Europe and the Soviet Union. As the dust from the cold war settles, the battle to save the planet will replace the battle over ideology as the organising theme of the new world order. In the twenty years since the first Earth Day in 1970, the earth has lost tree cover on an area nearly as large as the United States east of the Mississippi River. Desertification claimed more land than is devoted to crops in China. Thousands of plant and animal species with which we shared the planet in 1970 no longer exist. The world has lost as much topsoil as covers India's cropland. And more people joined the world's population than inhabited the planet in 1900. How can we design a vibrant world economy that does not destroy the natural resources and environmental systems on which it depends? That is *the* question of the 1990s and the question that *State of the World 1991* (1991) set out to answer.

State of the World 1991 lucidly examined the options for restoring our planet's health. It dealt with everything from energy production to urban transportation, and from forest management to the reuse of common materials like glass and paper. *State of the World 1991* details how we can provide the energy and goods the world needs in a way that is sustainable—that does not consume the resource base of future generations. The report concluded that partially replacing income taxes with environmental taxes is the key to quickly transforming our unsustainable global economy into one that is sustainable. Such green taxes would add charges to the costs of burning fossil fuels, using non-recyclable materials, and discharging toxic waste while generating income for environmentally sound development.

In 1995, *State of the World* devoted an entire chapter to China, in recognition that the surging demands of its 1.2 billion people could alter the global supply/demand balances for grain, petroleum, minerals, forest products, and fibres while complicating efforts to control such global pollutants as carbon dioxide. Never before have the incomes of such a large number of people risen so rapidly as in China today. The economic repercussions of this huge nation's emergence as a modern consumer economy will be experienced around the world for decades to come. This 12th edition of State of the World appeared at a time when more and more people were becoming aware of the effects of our rapidly growing population. The seemingly sudden collapse of the oceanic fishing economy, the rising expectations of China's 1.2 billion consumers, deforestation of the Indian subcontinent, and the increase in refugees around the world all point to a need for understanding the interrelationship of population and environmental forces. Clearly, the signs of despair are more abundant than the signs of hope.

King and Schneider: The First Global Revolution

In their book *The First Global Revolution* (1991), two leading members of the Council of The Club of Rome described the problems of our planet and indicated how they can be resolved. They start with a description of 'the great transition'. They claim we are in the early stages of the formation of a new type of world society which will be as different from today's as was that of the world ushered in by the Industrial Revolution from the society of the long agrarian period that preceded it.

As the title suggests, we are at the beginning of the first global revolution. The seeds of this revolution have been germinating slowly over many years, during which conditions of complexity, uncertainty, and rapid change have appeared as never before and are beginning to overwhelm the capacity of the world governance system. Great changes are taking place in economic life, in the relationships between nations, in the awakenings of minorities and nationalism, in values and norms, in population and urban growth, in technology, in the environment, and in the development of countries. Of serious concern are such themes as population, the environment, food, and energy. All these acute problems are being mismanaged by nations and international institutions.

We are in need of broader joint responsible action. In the shifting situations

of the present, the need to develop methods to make decisions in conditions of
uncertainty is paramount. The authors point out three solutions.

1 The first of these is the reconversion from a military to a civil economy.
2 The second theme is global warming and the energy problems. Delay in dealing with these problems could be tragic and catastrophic.
3 Third, there is the development issue. There is a need for new strategies for
   world harmony.

The book is part of the tradition of the Club of Rome. King and Schneider emphasize the importance of international action and co-operation between the
public and private sector. Their resolutions look simple, but are very difficult
to implement. In this sense, the book can be placed in the category of activism
for a better world.

Meadows, Meadows and Randers: Beyond the Limits

Beyond the Limits (1991) is a replication study of the well-known document *The
limits to growth*. It was published 20 years after the presentation of the report to
the Club of Rome. The Club of Rome report showed that if growth trends continued unchanged, the limits to physical growth on the planet would be
reached within 100 years. Many refused to accept its conclusions, yet the global scientific evidence since then has confirmed them.

Three of the four original authors have marshalled that evidence to write *Beyond the limits*. They show that the world has already overshot some of its limits
and, if present trends continue, we face the prospect of a global collapse—perhaps within the lifetimes of children alive today. But that is not the only option.

The book presents in clear terms the choice we have between a rapid and uncontrolled decline in food production, industrial capacity, population, and life
expectancy, or a sustainable future. The authors are well known in the field of
systems dynamics and the use of computer models to project the future. They
describe a range of possible outcomes and show that a sustainable society is
technically and economically feasible.

On the one hand, it seems surprising that the conclusions of 'Beyond the
Limits' are nearly the same as those of the report 'The Limits to Growth'. On
closer inspection, however, it would have been a surprise if there were great
differences because by looking at global problems for a longer time period, it is
possible to detect general trends that don't change in a few years. A time difference of 20 years for such research is a small one.

The Fifth Action Programme on the Environment

The European Community (now European Union) became active in environmental policy during the 1970s. The EU policy programme is formulated in the so-called 'EU action programmes'. Since the beginning, five action programmes have been published. The most recent one is *The Fifth Action Programme on the Environment* (1992). This Community Programme of Policy and Action in relation to the Environment and Sustainable Development is based on the general goal of reaching sustainability. In the accompanying document to this Fifth action programme, entitled *The State of the Environment in the European Community*, an overview is given of the situation in relation to air, water, soil, waste, quality of life, high-risk activities, and biological diversity. The causes of environmental degradation are mentioned and the economic aspects of the environmental problems are discussed in detail. A number of practical requirements for sustainable development are mentioned.

The EU is aware that sustainable development is not something which will be achieved in a period as short as that covered by this Programme (five years). A new strategy has been developed to reach sustainability. It focuses on the agents and activities which deplete natural resources and otherwise damage the environment. It endeavours to initiate changes in current trends and practices which are detrimental to the environment. Its aim is to achieve these changes in society's patterns of behaviour through the optimum involvement of all sectors of society in a spirit of shared responsibility. This responsibility will be shared through a significant broadening of the range of instruments to be applied to the resolution of the problems. The target sectors are industry, energy, transport, agriculture, and tourism. The existing legislative instruments are being broadened with market-based instruments, horizontal supporting instruments, and financial support mechanisms.

Many European countries are trying to bring their national environmental policy in line with the EU policy. The idea of sustainable development has been taken over, but there are various national interpretations of this idea. Within Europe there is a big difference between the national environmental policy of, for example, Greece, Italy, Spain, and Portugal, on the one hand, and the policy of Germany and Holland, on the other hand. The EU needs to harmonize the European environmental policy in a situation in which a great many national environmental problems and policy differences exist.

Harrison: The Third Revolution

In *The Third Revolution; Environment, Population and a Sustainable World* (1992), Harrison pleads for a multi-disciplinary theory development and action programme. The growing environmental crisis is so deep that we need the broadest analyses and strategy to cope with it. Yet, most theories in the field stress single causes: population, overconsumption, technology, inequality, and denial of the importance of others. Single-minded analysis leads to single-minded measures. According to Harrison, three factors are having a direct impact on the environment: population (the number of people), consumption (the amount each person consumes), and technology, which dictates how much space and resources are used and how much waste is produced to meet consumption needs.

Environmental impact is the result of all three multiplied by each other: $I = P \times C \times T$. Many other factors affect the environment directly. They include factors such as poverty and inequality, as often stressed by left wing political groups, or the degree of democracy, free enterprise, property rights, as often stressed by right wing political groups, or women's rights, as stressed by women's rights groups.

People are not remaining passive in the face of environmental problems. They are adapting. They are responding by changing technologies, fertility levels, and consumption patterns. They are also changing how they manage the environment. According to Harrison, the most promising approach is to reduce the impact of consumption levels by working on technology and population. This means that his study can be placed within the context of macro-social studies in which environmental problems are seen in direct relationship with the social order. Although Harrison pays a lot of attention to the factors of population, consumption and technology, other factors are not neglected.

Concluding remarks

It is apparent that in this period the central concepts are those of change, transition, and revolution. This is a remarkable statement because the period in which the first classics were written was also characterized by rapid and violent events. Rapid, owing to the short time in which the transformations took place. Violent, not in the sense of violence between people or groups, but between man and the environment. The general idea of sustainable development is the leading principle in environmental policy. The concept has forced us to look at environmental problems in a broader context, which means in relation to the existing social order and in relation to the north-south problems. Time will show us whether a sustainable future is possible. The time needed for this process is not a few years, but most probably some decades. Even if things go well, in many cases this will mean 'two steps forward and one step back.' If things do not go so well, we will come to a standstill, or worse: a 'one step forward and two steps back.'

References

Adriaanse, A., 1993. *Environmental policy performance indicators*, The Hague.

Cramer, J., 1995. *Ecological products, processes and life cycle issues*, Kiruna.

Hafkamp, W., 1995. *Contouren van een schone economie*, Tilburg.

Ministerie VROM, 1989. *Milieubeleidsplan*, 's-Gravenhage.

Nelissen, N.J.M., 1990. *Afscheid van de vervuilende samenleving?* Zeist.

CHAPTER 25

State of the World 1991

L.R. BROWN ET AL.

The New World Order

As the nineties begin, the world is on the edge of a new age. The cold war that dominated international affairs for four decades and led to an unprecedented militarization of the world economy is over. With its end comes an end to the world order it spawned.

The East-West ideological conflict was so intense that it dictated the shape of the world order for more than a generation. It provided a clear organizing principle for the foreign policies of the two superpowers and, to a lesser degree, of other governments as well. But with old priorities and military alliances becoming irrelevant, we are now at one of those rare points in history—a time of great change, a time when change is as unpredictable as it is inevitable.[1]

No one can say with certainty what the new order will look like. But if we are to fashion a promising future for the next generation, then the enormous effort required to reverse the environmental degradation of the planet will dominate world affairs for decades to come. In effect, the battle to save the planet will replace the battle over ideology as the organizing theme of the new world order.

As the dust from the cold war settles, both the extent of the environmental damage to the planet and the inadequacy of efforts to cope with it are becoming all too apparent. During the 20 years since the first Earth Day, in 1970, the world lost nearly 200 million hectares* of tree cover, an area roughly the size of the United States east of the Mississippi River. Deserts expanded by some 120 million hectares, claiming more land than is currently planted to crops in China. Thousands of plant and animal species with which we shared the planet in 1970 no longer exist. Over two decades, some 1.6 billion people were added to the world's population—more than inhabited the planet in 1900. And the world's farmers lost an estimated 480 billion tons of topsoil, roughly equivalent to the amount on India's cropland.[2]

This planetary degradation proceeded despite the environmental protection efforts of national governments over the past 20 years. During this time nearly all countries created environmental agencies. National legislatures passed thousands of laws to protect the environment. Tens of thousands of grassroots

---

* Units of measure are metric unless common usage dictates otherwise.

environmental groups sprung up in response to locally destructive activities. Membership in national environmental organizations soared. But as Earth Day 1990 chairman Denis Hayes asks, "How could we have fought so hard, and won so many battles, only to find ourselves now on the verge of losing the war?"[3]

One reason for this failure is that although governments have professed concern with environmental deterioration, few have been willing to make the basic changes needed to reverse it. Stabilizing climate, for example, depends on restructuring national energy economies. Getting the brakes on population growth requires massive changes in human reproductive behavior. But public understanding of the consequences of continuously rising global temperatures or rapid population growth is not yet sufficient to support effective policy responses.

The goal of the cold war was to get others to change their values and behavior, but winning the battle to save the planet depends on changing our own values and behavior.

The battle to save the earth's environmental support systems will differ from the battle for ideological supremacy in some important ways. The cold war was largely an abstraction, a campaign waged by strategic planners. Except for bearing the economic costs, which were very real, most people in the United States and the Soviet Union did not directly take part. In the new struggle, however, people everywhere will need to be involved: individuals trying to recycle their garbage, couples trying to decide whether to have a second child, and energy ministers trying to fashion an environmentally sustainable energy system. The goal of the cold war was to get others to change their values and behavior, but winning the battle to save the planet depends on changing our own value and behavior.

The parallel with the recent stunningly rapid changes in Eastern Europe is instructive. At some point, it became clear to nearly everyone that centrally planned economies were not only not working, but that they are inherently unworkable. Empty shelves in shops and long lines outside them demonstrated all too convincingly that a centrally controlled socialist economy could not even satisfy basic needs, much less deliver the abundance it promised. Once enough people, including Mikhail Gorbachev realized that socialist planners could not resolve this contradiction within the existing system, reform became inevitable.

Likewise, the contradiction between the indicators that measure the health of the global economy and those that gauge the health of its environmental support systems is becoming more visible. This inherent conflict affects all economic systems today: the industrialized economies of the West, the reforming economies of the East, and the developing economies of the Third World. As with the contradictions in Eastern Europe, those between economic and environmental indicators can be resolved only by economic reform, in effect by reshaping the world economy so that it is environmentally sustainable. (See Chapter 10.)

Two Views of the World

Anyone who regularly reads the financial papers or business weeklies would conclude that the world is in reasonably good shape and that long-term economic trends are promising. Obviously there are still problems—the US budget deficit, Third World debt, and the unsettling effect of rising oil prices—but to an economist, things appear manageable. Even those predicting a severe global recession in 1991 are bullish about the longer term economic prospects for the nineties.

Yet on the environmental front, the situation could hardly be worse. Anyone who regularly reads scientific journals has to be concerned with the earth's changing physical condition. Every major indicator shows a deterioration in natural systems: forests are shrinking, deserts are expanding, croplands are losing topsoil, the stratospheric ozone layer continues to thin, greenhouse gases are accumulating, the number of plant and animal species is diminishing, air pollution has reached health-threatening levels in hundreds of cities, and damage from acid rain can be seen on every continent.

These contrasting views of the state of the world have their roots in economics and ecology—two disciplines with intellectual frameworks so different that their practitioners often have difficulty talking to each other. Economists interpret and analyze trends in terms of savings, investment, and growth. They are guided largely by economic theory and indicators, seeing the future more or less as an extrapolation of the recent past. From their vantage point, there is little reason to worry about natural constraints on human economic activity; rare is the economic text that mentions the carrying capacity principle that is so fundamental to ecology. Advancing technology, economists believe, can push back any limits. Their view prevails in the worlds of industry and finance, and in national governments and international development agencies.[4]

In contrast, ecologists study the relationship of living things with each other and their environments. They see growth in terms of S-shaped curves, a concept commonly illustrated in high school biology classes by introducing a few algae into a petri dish. Carefully cultured at optimum temperature and with unlimited supplies of food, the algae multiply slowly at first, and then more rapidly, until growth eventually slows and then stops, usually because of waste accumulation. Charting this process over time yields the familiar S-shaped curve to which all biological growth processes in a finite environment conform.

Ecologists think in terms of closed cycles—the hydrological cycle, the carbon cycle, and the nitrogen cycle, to name a few. For them, all growth processes are limited, confined within the natural parameters of the earth's ecosystem. They see more clearly than others the damage to natural systems and resources from expanding economic activity.

Although the intellectual foundations of this view originate in biology, other scientific fields such as meteorology, geology, and hydrology also contribute. The ecological perspective prevails in most national academies of science, in international scientific bodies, and in environmental organizations. Indeed, it is environmentalists who are actively voicing this view, urging the use of principles of ecology to restructure national economies and to shape the emerging world order.

These divergent views of the world are producing a certain global schizophrenia, a loss of contact with reality. The events of 1990 typify this unhealthy condition. The celebration of Earth Day 1990 symbolized the growing concern for the environmental health of the planet. Estimates indicate that at least 100 million people in 141 countries participated in events on Sunday, April 22. Soon after, at the Group of Seven economic summit in Houston, national leaders from Europe, reflecting the mounting concern with global warming, urged the United States to adopt a climate-sensitive energy policy.[5]

A few weeks later, Iraq invaded Kuwait, unsettling oil markets. Almost overnight, concerns about energy shifted from the long-term climatic consequences of burning oil and other fossil fuels to a short-term preoccupation with prices at the local gasoline pump. More traditional views of energy security surfaced, eclipsing, at least temporarily, the concern with fossil fuel use and rising global temperatures.

The ecological view holds that continuing the single-minded pursuit of growth will eventually lead to economic collapse.

This schizophrenic perspective is translating into intense political conflict in economic policymaking. To the extent that constraints on economic expansion are discussed on the business pages, it is usually in terms of inadequate demand growth rather than supply-side constraints imposed by the earth's natural systems and resources. In contrast, the ecological view, represented by the environmental public interest community, holds that continuing the single minded pursuit of growth will eventually lead to economic collapse. Ecologists see the need to restructure economic systems so that progress can be sustained.

Both visions are competing for the attention of policymakers and, as more environmentally minded candidates run for office, for the support of voters. The different views are strikingly evident in the indicators used to measure progress and assess future prospects. The basic evidence cited by economists shows a remarkable performance over the last decade. (See table 1-1.) The value of all goods produced and services rendered grew steadily during the eighties, expanding some 3 per cent a year and adding more than $4.5 trillion to the gross world product by 1990, an amount that exceeded the entire world product in 1950. In other words, growth in global economic output during the eighties was greater than that during the several thousand years from the beginning of civilization until 1950.[6]

International trade, another widely used measure of global economic progress, grew even more rapidly, expanding by nearly half during the eighties. This record was dominated by the expanding commerce in industrial products, while growth in the trade of agricultural commodities and minerals lagged. Although the exports of some countries, such as those in East Asia, increased much more than others, all but a relatively small number of nations contributed to the rising tide of commerce.[7]

On the employment front, the International Labour Organization reports that the economically active population increased from 1.96 billion to 2.36 billion during the decade. Although impressive gains in employment were made in some regions, the growth in new jobs in the Third World did not keep pace with the number of new entrants, making this one of the least satisfying of the leading economic indicators.[8]

Using stock prices as a gauge, the eighties was a remarkable decade. Investors on the New York Stock Exchange saw the value of their portfolios growing by leaps and bounds, a pattern only occasionally interrupted, as in October 1987. The Standard and Poor Index of 500 widely held stocks showed stock values nearly tripling during the decade. Pension funds, mutual funds, and individual investors all benefited. (See figure 1-1.) The value of stock traded on the Tokyo Exchange Climbed even more rapidly.[9]

The contrast between these basic global economic indicators and those measuring the earth's environmental health could not be greater. While these particular leading economic measurements are overwhelmingly positive, all the principal environmental indicators are consistently negative. As the need for cropland led to the clearing of forests, for example, and as the demand for firewood, lumber, and paper soared, deforestation gained momentum. By the end of the decade, the world's forests were shrinking by an estimated 17 million hectares each year. Some countries, such as Mauritania and Ethiopia, have lost nearly all their tree cover.[10]

TABLE 1-1 Selected Global Economic and Environmental Indicators

Indicator	Observation
The Economy	
Gross World Product	Global output of goods and services totalled roughly $20 trillion in 1990, up from $15.5 trillion in 1980 (1990 dollars).
International Trade	Worlds exports of all goods—agricultural commodities, industrial products, and minerals—expanded 4 percent a year during the eighties, reaching more than $3 trillion in 1990.
Employment	In a typical year, growth of the global economy creates millions of new jobs, but unfortunately job creation lags far behind the number of new entrants into the labor force.
Stock prices	A key indicator of investor confidence, prices on the Tokyo and New York stock exchanges climbed to all-time highs in late 1989 and early 1990, respectively.
The Environment	
Forests	Each year the earth's tree cover diminishes by some 17 million hectares, an area the size of Austria. Forest are cleared for farming, harvest of lumber and firewood exceed sustainable yields, and air pollution and acid rain take a growing toll on every continent.
Land	Annual losses of topsoil from cropland are estimated at 24 billion tons, roughly the amount on Australia's wheatland. Degradation of grazing land is widespread throughout the Third World, North America, and Australia.
Climate System	The amount of carbon dioxide, the principal greenhouse gas in the atmosphere, is now rising 0.4 percent per year from fossil fuel burning and deforestation. Record hot summers of the eighties may well be exceeded during the nineties.
Air Quality	Air pollution reached health-threatening levels in hundreds of cities and crop-damaging levels in scores of countries.
Plant and Animal Life	As the number of humans inhabiting the planet rises, the number of plant and animal species drop. Habitat destruction and pollution are reducing the earth's biological diversity. Rising temperatures and ozone layer depletion could add to losses.

Source *Worldwatch Institute, based on sources documented in endnote 6.*

FIG I-I   Index of Stock Prices, 500 Common Stocks, 1950-90

1941-43 = 10

Source: U.S. Dept. of Commerce

Closely paralleling this is the loss of topsoil from wind and water erosion, and the associated degradation of land. Deforestation and overgrazing, both widespread throughout the Third World, have also led to wholesale land degradation. Each year, some 6 million hectares of land are so severely degraded that they lose their productive capacity, becoming wasteland.[11]

During the eighties, the amount of carbon pumped into the atmosphere from the burning of fossil fuels climbed to a new high, reaching nearly 6 billion tons in 1990. In a decade in which stock prices climbed to record highs, so too did the mean temperature, making the eighties the warmest decade since recordkeeping began more than a century ago. The temperature rise was most pronounced in western North America and western Siberia. Preliminary climate data for 1990 indicate it will be the hottest year on record, with snow cover in the northern hemisphere the lightest since the satellite record began in 1970.[12]

Air and water pollution also worsened in most of the world during the last 10 years. By 1990, the air in hundreds of cities contained health-threatening levels of pollutants. In large areas of North America, Europe, and Asia, crops were being damaged as well. And despite widespread reduction in water pollution in the United States, the Environmental Protection Agency reported in 1988 that groundwater in 39 states contained pesticides. In Poland, at least half the river water was too polluted even for industrial use.[13]

These changes in the earth's physical condition are having a devastating effect on the biological diversity of the planet. Although no one knows how many plant and animal species were lost during the eighties, leading biologists estimate that one fifth of the species on earth may well disappear during this century's last two decades. What they cannot estimate is how long such a rate of extinction can continue without leading to the wholesale collapse of ecosystems.[14]

How can one set of widely used indicators be so consistently positive and another so consistently negative? One reason the economic measures are so encouraging is that national accounting systems—which produce figures on gross national product—miss entirely the environmental debts the world is incurring. The result is a disguised form of deficit financing. In sector after sector, we are consuming our natural capital at an alarming rate—the opposite of an environmentally sustainable economy, one that satisfies current needs without jeopardizing the prospects of future generations. As economist Herman Daly so aptly puts it, "there is something fundamentally wrong in treating the earth as if it were a business in liquidation".[15]

To extend this analogy, it is as though a vast industrial corporation quietly sold off a few of its factories each year, using an incomplete accounting system that did not reflect these sales. As a result, its cash flow would be strong and profits would rise. Stockholders would be pleased with the annual reports, not realizing that the profits were coming at the expense of the corporation's assets. But once all the factories were sold off, corporate officers would have to inform stockholders that their shares were worthless.

In effect, this is what we are doing with the earth. Relying on a similarly incomplete accounting system, we are depleting our productive assets, satisfying our needs today at the expense of our children.

New Measures of Progress

Fortunately, there is a growing recognition of the need for new ways of measuring progress. Ever since national accounting systems were adopted a half-century ago, per capita income has been the most widely used measure of economic progress. In the early stages of economic development, expanded output translated rather directly into rising living standards. Thus it became customary and not illogical to equate progress with economic growth.

Over time, however, average income has become less satisfactory as a measure of well-being: it does not reflect either environmental degradation or how additional wealth is distributed. Mounting dissatisfaction has led to the development of alternative yardsticks. Two interesting recent efforts are the Human Development Index (HDI) devised by the United Nations and the Index of Sustainable Economic Welfare (ISEW) developed by Herman Daly and theologian John Cobb. A third indicator, grain consumption per person, is a particularly sensitive measure of changes in wellbeing in low-income countries.[16]

Average income does not reflect either environmental degradation or
how additional wealth is distributed.

The Human Development Index, measured on a scale of 0 to 1, is an aggregate
of three indicators: longevity, knowledge, and the command over resources
needed for a decent life. For longevity, the UN team used life expectancy at
birth. For knowledge, they used literacy rates, since reading is the key to acquiring information and understanding. And for the command over resources,
they used gross domestic product (GDP) per person after adjusting it for purchasing power. Because these indicators are national averages, they do not deal
directly with distribution inequality, but by including longevity and literacy
they do reflect indirectly the distribution of resources. A high average life expectancy, for example, indicates broad access to health care and to adequate
supplies of food.[17]

    A comparison of countries ranked by both adjusted per capita gross domes-
tic product and HDI reveals some wide disparities: some with low average incomes have relatively high HDIs, and vice versa. In Sri Lanka, for instance, per
capita GDP is only $2,053, while the HDI is 0.79. But in Brazil, where GDP is
twice as high at $4,307 per person, the HDI is 0.78, slightly lower. This is be-
cause wealth is rather evenly distributed in Sri Lanka, along with access to food
and social services, whereas in Brazil it is largely concentrated among the wealthiest one fifth of the population. The United States, which leads the world in
adjusted income per capita at $17,615, is 19th in the HDI column, below such
countries as Australia, Canada, and Spain.[18]

Per capita grain consumption looks at the satisfaction of a basic human need and
is far less vulnerable to distortion by inequities of purchasing power.

While the HDI represents a distinct improvement over income figures as a
measure of changes in human wellbeing, it says nothing about environmental
degradation. As a result, the HDI can rise through gains in literacy, life expectancy, or purchasing power that are financed by the depletion of natural support systems, setting the stage for a longer term deterioration in living conditions.

    The Daly-Cobb Index of Sustainable Economic Welfare is the most com-
prehensive indicator of well-being available, taking into account not only average consumption but also distribution and environmental degradation. After
adjusting the consumption component of the index for distributional inequality, the authors factor in several environmental costs associated with

FIG 1-2 GNP and Index of Sustainable Economic Welfare (ISEW) Per Capita, United States, 1950-88

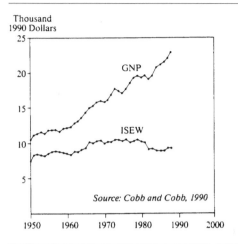

economic mismanagement, such as depletion of nonrenewable resources, loss of farmland from soil erosion and urbanization, loss of wetlands, and the cost of air and water pollution. They also incorporate what they call "long-term environmental damage", a figure that attempts to take into account such large-scale changes as the effects of global warming and of damage to the ozone layer.[19]

Applying this comprehensive measure to the United States shows a rise in welfare per person of some 42 per cent between 1950 and 1976. (See Figure 1-2.) But after that the ISEW began to decline, falling by just over 12 per cent by 1988, the last year for which it was calculated. Simply put, about 15 years ago the net benefits associated with economic growth in the United States fell below the growth of population, leading to a decline in individual welfare.[20]

The principal weakness of the ISEW, which has been calculated only for the United States, is its dependence on information that is available in only a handful of nations. For example, few developing countries have comprehensive data on the extent of air and water pollution, not to mention information on year-to-year changes. The same drawback applies to the HDI, since life expectancy data depend heavily on infant mortality information that is collected at best once a decade in most of the Third World.

A measure in many ways more relevant to well-being in low-income countries is per capita grain consumption. It looks at the satisfaction of a basic human need and is far less vulnerable to distortion by inequities of purchasing power. The distribution of wealth between the richest one fifth of a country and the poorest one fifth can be as great as 20 to 1, as indeed it is in Algeria,

Brazil, or Mexico, but the per capita consumption of grain by these same groups cannot vary by more than 4 to 1. Among more affluent countries, this figure peaks at about 800 kilograms a year, with the limit set by the quantity of grain-fed livestock products that can be consumed. At the lower end, people cannot survive if annual grain consumption drops much below 180 kilograms (about 1 pound a day) for an extended period. Thus, a gain in average grain consumption in a country typically means a gain in welfare.[21]

At the top end of this scale, the figure can be used to measure threats to health. Beyond a certain point—a point well below the level of consumption in the more affluent countries—rising grain consumption per person, most of it in the form of fat-rich livestock products, leads to increases in heart disease, certain types of cancer, and an overall reduction in life expectancy.

Grain production is also a more sensitive barometer of environmental degradation than income is, since it is affected more immediately by environmentally destructive activities outside agriculture, such as air pollution, the hotter summers that accompany global warming, and increased flooding as a result of deforestation.

In summary, the Index of Sustainable Economic Welfare is by far the most sophisticated indicator of progress now available, although its use is constrained by lack of data. In low-income countries where the relevant data to calculate the ISEW are not available, changes in grain consumption per person can tell more than income figures about improvements—or deterioration—in well-being.

Brown, L. R. et al., *State of the World 1991*, W.W. Norton & Company, New York, 1990, p. 4-11.

Notes and References

1   For a more detailed discussion of this time of change, see Charles William Maynes, "America Without the Cold War", *Foreign Policy*, Spring 1990, and Paul H. Nitze, "America: An Honest Broker", and Robert Tucker, "1989 and All That", both in *Foreign Affairs*, Fall 1990.

2   Jean-Paul Lanly, *Tropical Forest Resources* (Rome: UN Food and Agricultural Organization (FAO), 1982); H.E. Dregne, *Desertification of Arid Land* (New York; Harwood Academic Publishers, 1983); UN Environment Programme, *General Assessment of Progress in the Implementation of the Plan of Action to Combat Desertification 1978-1984* (Nairobi: 1984); species loss from E.O. Wilson, ed., *Biodiversity* (Washington, DC: National Academy Press, 1988); UN Department of International Economic and Social Affairs (DIESA), *World Population Prospects 1988* (New York; 1989); Lester R. Brown and Edward C. Wolf, *Soil Erosion: The Quiet Crisis in the World Economy*, Worldwatch Paper 60 (Washington, DC: Worldwatch Institute, September 1984).

3   Dennis Hayes, "Earth Day 1990: Threshold of the Green Decade", *Natural History*, April 1990.

4 For a more detailed discussion of the differences between economists and ecologists, see the writings of Hazel Henderson, one of the pioneers in this field, especially *The Politics of the Solar Age; Alternatives to Economics* (Indianapolis, Ind.: Knowledge Systems, Inc., rev. ed., 1988).

5 Earth Day 1990 participants and countries based on Christina L. Dresser, Earth Day 1990 Executive Director, San Francisco, Calif., private communication, October 1 1990.

6 Data in Table 1-1 based on the following: gross world economic output in 1990 from the 1988 gross world product from Central Intelligence Agency (CIA), *Handbook of Economic Statistics, 1989* (Washington DC: 1989), with Soviet and Eastern Europe gross national products extrapolated from Paul Marer, *Dollar GNP's of the USSR and Eastern Europe* (Baltimore: John Hopkins University Press, 1985), with adjustments to 1990 based on growth rates from International Monetary Fund (IMF), *World Economic Outlook* (Washington, DC: October 1990), and CIA, *Handbook of Economic Statistics*, and with the composite deflator from Office of Management and Budget, *Historical Tables, Budget of the United States Government, Fiscal Year 1990* (Washington, DC: US Government Printing Office, 1989); historical estimates based on Angus Maddison, *The World Economy in the 20th Century* (Paris: Organisation for Economic Co-operation and Development, 1989); international trade increase is Worldwatch Institute estimate based on IMF, *International Financial Statistics*, October 1990, and *Yearbook* (Washington, DC: 1990); US Department of Commerce, Bureau of Economic Analyses, "Standard and Poor Index of 500 Widely Held Stocks", Washington, DC, 1990; Tokyo Stock Exchange, *Monthly Statistics Report*, June 1990; deforestation figure from FAO, which is in the midst of preparing a new global forest assessment, according to "New Deforestation Rate Figures Announced", *Tropical Forest Programme* (IUCN Newsletter), August 1990; Brown and Wold, *Soil Erosion*; Dregne, *Desertification of Arid Land*; carbon dioxide estimate based on Gregg Marland et al., *Estimates of CO_2 Emissions from Fossil Fuel Burning and Cement Manufacturing, Based on the United Nations Energy Statistics and the US Bureau of Mines Cement Manufacturing Data* (Oak Ridge, Tenn.: Oak Ridge National Laboratory, 1989), on Gregg Marland, private communication and printout, Oak Ridge National Laboratory, Oak Ridge, Tenn., July 6, 1989, and on British Petroleum (BP), *BP Statistical Review of World Energy* (London: 1990).

7 IMF, *International Financial Statistics*.

8 International Labour Organization, *Economically Active Population Estimates, 1950-80, and Projections, 1985-2025, Vol. 5* (Geneva:1986).

9 US Department of Commerce, "Standard and Poor Index of 500 Widely Held Stocks;" Tokyo Stock Exchange, *Monthly Statistics Report*.

10 FAO, "New Deforestation Rate Figures Announced"; Erik P. Eckholm, *Losing Ground: Environmental Stress and World Food Prospects* (New York: W.W. Norton & Co., 1976); World Resources Institute, *World Resources, 1990-91* (New York: Oxford University Press, 1990); FAO, *Production Yearbook* (Rome: various years).

11 Dregne, *Desertification of Arid Land*.

12 Worldwatch Institute estimate based on Marland et al., *Estimates of CO_2 Emissions*, on Marland, private communication and printout, and on BP, *BP Statistical Review*; James E. Hansen, Goddard Institute for Space Studies, National Aeronautics and Space Administration (NASA), "The Green House Effect: Impacts on Current Global Temperature and Regional Heat Waves", Testimony before the Committee on Energy and Natural Resources, US Senate, Washington, DC, June 23, 1988; James E. Hansen et al., "Comparison of Solar and Other Influences on Long-Term Climate", Proceedings of Goddard Conference, NASA, 1990; P.D. Jones, climatic Research Unit, University of East Anglia, Norwich UK, "Testimony to the US

Senate on Global Temperatures", before the Commerce Committee, US Senate, Washington, DC, October 11, 1990.

13 UN Environment Programme and World Health Organization, *Assessment of Urban Air Quality* (Nairobi: Global Environment Monitoring System, 1988); W. Martin Williams et al., Office of Pesticide Programs, US Environmental Protection Agency (EPA), *Pesticides in Ground Water Data Base: 1988 Interim Report* (Washington, DC: 1988); Stanley J. Kabala, "Poland: Facing the Hidden Cost of Development", *Environment*, November 1985.

14 Ariel E. Lugo, "Estimating Reductions in the Diversity of Tropical Forest Species", in Wilson, *Biodiversity*.

15 Herman E. Daly, "Sustainable Development: From Concept and Theory Towards Operational Principles", *Population and Development Review* (Proceedings of Hoover Institution Conference on Population and Development), forthcoming special issue.

16 UN Development Programme (UNDP), *Human Development Report 1990* (New York: Oxford University Press, 1990); Herman E. Daly and John B. Cobb, Jr., *For the Common Good: Redirecting the Economy Toward Community, the Environment, and a Sustainable Future* (Boston: Beacon Press, 1989).

17 UNDP, *Human Development Report 1990*.

18 Ibid.; adjusted per capita gross domestic product is in 1987 US Dollars; note that income figures that are not adjusted for purchasing power can differ significantly—In Sri Lanka, for example, the unadjusted figure is $400.

19 Daly and Cobb, *For the Common Good*.

20 Ibid., Figure 1-2 is based on Clifford W. Cobb and John B. Cobb, Jr., Revised Index of Sustainable Economic Development, according to C.W. Cobb, Sacramento, Calif., private communication, September 28, 1990.

21 World Bank, *World Development Report 1990* (New York: Oxford University Press, 1990); US Department of Agriculture (USDA), Economic Research Service (ERS), *World Grain Database* (unpublished printouts) (Washington, DC: 1990). Based on the author's personal experience, to avoid starvation a person needs roughly 1 pound of grain per day, while 13 ounces is enough for survival with minimum physical activity.

CHAPTER 26

The First Global Revolution

A Report by the Council of The Club of Rome

A. KING & B. SCHNEIDER

Learning Our Way into a New Era

We shall make no attempt to summarise our conclusions; indeed the very nature of the problematique precludes such a possibility. Instead we shall make some observations and suggestions as to how to blaze a trail into the thicket of the future through learning, which is a leading feature of the resolutique. Before doing so, however, we shall restate a few guiding principles that are scattered throughout the book:
— the need for the involvement and participation of everyone in seeking a way through the intertwining complex of contemporary problems;
— recognition that the possibilities of positive change reside in the motivations and values that determine our behaviour;
— understanding that the behaviour of nations and societies reflects that of its individual citizens;
— acceptance of the postulate that dramatic solutions are unlikely to come from the leaders of governments, but that thousands of small, wise decisions, reflecting the new realization of millions of ordinary people, are necessary for securing the survival of society;
— effect given to the principle that privilege, whether individual or national, must always be complemented by a corresponding responsibility.

As stated in the introduction, the ideas and action proposals of this book are offered as a basis for learning our way into the future. It is not necessary—indeed it would be impossible to expect—for there to be complete agreement on all the thoughts we have expressed with regard to the world in revolution, or on the relative importance we have given to the various problems. The material presented here should rather be regarded as matters for widespread discussion and debate; it is intended to spark off a variety of examinations and reassessments on the part of those responsible for the management of society at all levels. Beyond this, it is hoped that the many whose contacts with governance are quite remote, but whose future is deeply involved in the changes to be foreseen, will begin to more clearly understand the significance of many of the topics presented such as the interdependence of the nations and the interaction of the problems. The time has come to show how every individual is more or less directly concerned with the problems of the world and the changes that are brewing, even if he or she can more easily perceive the symptoms than their causes. Even now, few remain untouched; one has only to mention the problems for some of co-existence with immigrants of different ethnic origins, the effect on children and adolescents of certain television programmes, the internationalization of automobiles or the international spread of the products for sale in supermarkets.

To learn our way through this period of transition and to identify sure points of reference, we have to modify our reasoning, our mental images, our behaviour and the realities on which we base our judgments so that we can understand this world mutation, with its array of global issues such as the environment, food security, development of the poor countries, the crises of governance and all the others we have attempted to describe.

The situation of complexity and uncertainty will condemn decision makers at all levels—and especially the politicians—to search for new approaches and to adopt untraditional attitudes. But it will not be possible to implement their decisions, no matter how brave and how pertinent, unless they succeed in obtaining wide public support. However, general resistance to change and fear of the unknown constitute an unfavourable environment to strong, but unfamiliar action. The dynamics of public opinion will not be able to operate usefully, unless the individuals who make it up have access to the nature of global phenomena and acquire, through their understanding of them, the conviction of what is at stake—the very survival of the human species. It is obvious, however, that the eloquence of the facts alone will be insufficient to convince individuals that these phenomena are of immediate concern to them. To most people they will seem distant, theoretical and too vast in comparison with the problems of everyday life, their family, professional, financial, health and day-to-day survival problems. The scope of these difficulties may well elicit a reaction of withdrawal, a refusal to understand or anxiety at the thought of the individual in his helplessness and isolation having to grapple with a set of facts that are mind-boggling in their variety and complexity.

Such doubts and alienation will have to be acknowledged and deliberately addressed so that they can be dispelled by shared fears and a familiarity with the facts gradually achieved through discussion with others. The situation must be seen in local and personal terms. This is one reason for the need for a revitalization of democracy on a more participative basis, stimulated by comprehension of the global concerns.

The need, then, is "to think globally and act locally". The Club of Rome has, from its beginning, realized the need for such an approach and there are a

multitude of ways in which it is or could be achieved. We offer a few examples
in the following.

Global-Local Interaction

On the initiative of Maurice Strong[1] and the Club of Rome, a meeting was held
in 1989 in Denver with some forty Colorado decision makers to discuss the following question: In what ways do the great world problems affect the economic
and social life of the state of Colorado and in what way can the political and
economic leaders of the state exercise an influence or have an impact on these
great problems? During the work and discussions of the meeting, interaction
became more and more evident in a number of areas and especially on environmental issues. If every inhabitant in Colorado makes energy-saving as well
as fighting against waste his or her daily duty the action will have repercussions
on the situation of Colorado, therefore of the United States, therefore of the
world. If the individual is alone, the result will be merely symbolic. If a number
of individuals join to act in the direction of better environmental protection
and if their influence in the community strengthens their fight, then the result
will be significant. The Denver meeting was followed by an open forum in
which the ideas and conclusions of the small, restricted meeting were shared
with a large audience of the general public. Similar meetings are being planned
in other countries, initially in Japan, and similar approaches are being taken by
other bodies and sometimes even by governments.

In a different area, that of development, we underscored[2] the role of local initiatives in the development process, often taken by nongovernmental organisations, groups of villagers and the like with action on farming, health and hygiene, education, etc. Corresponding activities are also spreading in the big-
city slums and both of these are contributing to modifying the concept and the
global vision of development policies, for which they have provided the reality
of experience in the field that is the reflection of the multiplicity of geographic,
cultural and human situations.

The Club of Rome, in disseminating its concerns and encouraging the emergence of global thinking in local action, has encouraged the creation of National Associations for the Club of Rome. These now exist in about thirty
countries in the five continents. The Associations are governed by a common
charter, some of the articles of which insist on the nature of the interaction of
the local and the global:

> Each Association shall approach the global problems in terms of the country's own
> cultural values and thus contribute to the general understanding of the human con-
> dition on the planet.

> It shall have the duty to disseminate locally to decision makers, academics, indus-

trial circles and the public at large, the reports, findings and attitudes of the Club. It shall contribute experience, creative ideas and proposals, towards the under-standing of the global problems, to the Club.

The National Associations for the Club of Rome have, therefore, the mission of bridging communication between the national realities and the proble-matique as seen nationally on the one hand and the global thinking of the Club on the other, and acting as relays for the circulation and dissemination of Club thinking in each country. Going from global to local and from local to global is a radical transformation in modes of thinking and reasoning which will become essential henceforth. It is a new intellectual exercise which we shall have to ex-tend and integrate.

Local Individual Interaction

The picture would be incomplete were we not to address the action possi-bilities of the individual human being who is at the centre of the entire edifice. In extreme cases such as the threat of war or natural disasters, individuals are immediately transformed into citizens, aware of their responsibilities and ready for cohesive action. Other, less spectacular but likewise significant examples bear witness to the fact that individuals are not inert and indifferent in the face of imminent dangers. When there is an environmental threat close by, or when a situation arises where people's interests are at stake and gross instances of ex-ploitation are revealed, we find that initiatives are taken in the most diverse fields by individuals and small groups prepared to fight for causes that affect them directly or indirectly and by which they feel motivated.

Suffice it to mention as examples transportation or telephone users' organi-sations, or in a different category, NGOs that care for disabled children, old people or battered wives; again there are NGOs in the fight against AIDS and a host of other diseases, NGOs in the struggle for human rights, ecological NGOs, peace groups and a multitude of developmental NGOs such as we have presented above. Neither must we forget the initiatives in many countries by the jobless to create employment for themselves or to set up their own busi-nesses as well as the NGOs that were founded to assist small businesses and to provide them with technical assistance.

Individual commitment to action of these sorts is therefore possible and al-ready widespread, which demonstrates that a link can be established between the human being and local or national actions, which, in some cases, flourish, extend and become international.

The Emergence of the Informal Sector

The success of grassroots NGO initiatives no longer needs to be demonstrated.

Very often, these movements are sparked off by individual men and women. Examples throughout the world are multitudinous. In the Indian state of Uttar Pradesch, the local people have rallied around a man called Sunderlal Bahuguna to stop the construction of a US $1.7 billion dam, which would have submerged their villages and seriously increased the danger of avalanches in the region. Several reports that questioned the technical feasibility of the project, added to an eleven-day fast by Bahuguna, led the government to back down on its plans. In Kenya a woman, Wangari Maathai—founder and President of the grass-roots Green Belt Movement and member of The Club of Rome—has led a successful battle to stop the construction of a sixty-two-floor office-building in a popular Nairobi public park. In Mexico City, where the problem of pollution has gone far beyond bearable limits, Marcos Chan Rodriguez mobilized his neighbourhood to form a grassroots group to reduce the operations of a cement factory that was pouring cement particles into the air. In the process, the group realized that to arouse the ruling party's interest, it had to appeal to the left-wing opposition, and therefore make the democratic system work.

The enormous proliferation of NGOs can be seen in every sector of national and international activity; some are strictly professional, others represent special interests; they may be single-issue groups or deal with general concerns; they may have a religious orientation or be based on a particular political ideology. The arising of this wide variety of NGOs is a healthy phenomenon which demonstrates that the human fabric is able to react to the rigidity and to an apparent impotence of national and international official structures in face of current problems. This new pattern of informal bodies has little coherence and often appears somewhat anarchistic because it is marked more by spontaneity and flexibility than by its concern with structure—something it is wary of. Most of the innumerable new NGOs are weak, financially and otherwise, but their lack of power is often compensated by vigour and enthusiasm. In a few cases such as that of the "greens" they may even try to penetrate the official structures by presenting candidates for parliament. In other cases, for example that of the Worldwatch Institute of Washington, surveys of world trends are scrutinized seriously by political personalities in many countries.

This new, so-called informal sector is beginning to be taken seriously by governments and international institutions, often with some reluctance and despite the apparent incompatibility of what is official with that which is not. However, some NGOs possess experience, insights and knowledge that governments lack, as well as representing significant elements of public concern which cannot be ignored. Thus some co-operation is emerging between the official and the informal and this is proving useful to the latter as different NGOs meet and discover their similarities and differences. We feel that similar co-operation is necessary also in the international fora. Intergovernmental discussions tend to be even more sterile and distanced from reality than those on the national level and hence a leavening of the debate by inclusion of a few carefully selected nonofficial experts in the committees could be inspiring. We have already suggested this in putting forward the idea of a UN Environment Security Council.

Despite more and more frequent meetings where several NGOs come together, their objectives and effectiveness remain scattered and generally unknown for many NGOs. Without meaning to advocate the structuring of the informal sector which might easily lose its soul in the process, a more effective system of mutual information would avoid much useless dispersion, encourage a fruitful exchange of experience and lead to the creation of alliances, thus increasing overall effectiveness.

This is one of the areas which the Club of Rome will pursue in its new initiatives. Although global efforts are essential to face some of the inescapably global issues, we must continue to operate at many levels—global, regional, national, provincial and local. At times we need not jump too quickly to the highest level, when local or regional efforts can be more successful. In fact, substantial impacts can be made, even on the largest-scale issues, through multiple actions on a small scale.

Innovation in Language, Analysis and Approach
In so many of the elements of this global revolution, we lack knowledge and indeed there is no guarantee that more research wil lead to greater certainty or that research will yield its results in time for them to influence decisions which are urgently required now.

We know a lot, but we understand very little.

We have therefore to learn to act in face of continuing uncertainty. Politics has always been the art of making decisions under conditions of uncertainty. The difference today is that the uncertainty is much deeper and is compounded with rapid change. This abiding uncertainty demands adaptation of our institutions and approaches to achieve greater flexibility and a greater capacity for reaction as we keep our sights on the moving targets of history.

A central challenge in this connection is how to reconcile the economic language and concepts that dominate today with environmental language and concepts. Two approaches are possible: environmental aspects can be added to conventional economic analysis, or economic approaches can be integrated within a broader ecological view. Great care and precise thinking are needed in this area, in which distinction must be made between different types of economics: macroeconomics, microeconomics and environmental or ecological economics. We must find ways of integrating environmental aspects more effectively with the established and powerful approaches of both macro- and microeconomics.

The role of the market and its relation to that of government is of vital importance in seeking to resolve and manage the environmental problems. No solutions based exclusively on the market exist in the real world. All Western countries, for example, have developed mixed economies in which governments provide a framework of regulations, incentives, support and guidelines to the private sector. It has been acknowledged that the market approach alone cannot handle problems of common property resources or issues of long-term common interest. Government must provide the boundary conditions in the public interest.

The problems we face are not only intellectual and analytical; real interests and the structure of power are always at stake. In the real world, contradictory interests are inevitably operating. In establishing a normative approach, arrangements on action have to be established between power groups and, indeed, between nations which will evidently continue to have different interests, values, norms and cultural traditions.

The Values on Which Action Is Based

We must be more explicit about the importance of values and ethics in the different areas of the problematique, for this will become a battleground for the future and a fundamental ingredient for the resolutique. If we accept concern for the prospects of future generations, we cannot escape consideration of how the problems and values of the present generation will influence these prospects. Our efforts to create a sustainable world society and economy demands that we diminish the profligate life-styles in the industrialized countries through a slow-down in consumption—which may, in any case be forced on us by environmental constraints. The ethical imperative also implies renewed efforts to eliminate poverty throughout the world.

The ethical approach has not thus far been a matter of major concern, to say the least, for decision makers in politics and business. At the very most we can find a cloudy ethical reaction among the public at large in their stands against corruption, against pollution and against a conception of the economy which appears to forget that it should be above all designed to serve men and women.

In the Western countries there exist, as we have said, frameworks of legislation to regulate the operations of the market forces, antitrust and antidumping laws, fair trading agreements credit controls and the like as well as codes of good practice often implicitly accepted by the business community. Such measures are necessary to ensure smooth and acceptable running of the capitalist society, to prevent fraud, to protect the work force and the public. While there is a degree of ethical motivation in the regulatory system, many of the measures are matters of convenience to provide propitious conditions for economic progress. Ecological disasters, sometimes causing death and destruction, raise these matters to a new level of importance and force industry to accept a degree of social responsibility in its own long-term self-interest, despite the fact that the costs involved threaten next year's bottom line. More and more it will be necessary to develop the recognizable ethical norms that society demands and with which industry can live, albeit uncomfortably. These should be matters of current concern for the Eastern European countries now accepting the system of the market forces with somewhat uncritical enthusiasm.

An ethical conception of international relations, which the world badly needs, cannot evolve unless it is also the inspiration on the national level and finally on the individual level as well. The development of such a conception will require much research and dialogue for a consistent, harmonious and dynamic menu for co-existence to be proposed and adopted as a common denominator by a variety of peoples into their history, their culture and their values.

King, A. & B. Schneider, *The First Global Revolution; A Report by the Council of The Club of Rome*, Pantheon Books, New York, 1991, p. 246-259.

Notes and References

1   Secretary General of the United Nations Conference on Environment and Development, member of the Club of Rome.

2   See Chapter 7, Development versus Underdevelopment.

CHAPTER 27

Beyond the Limits

Confronting Global Collapse; Envisioning a Sustainable Future

D. MEADOWS, D. MEADOWS & J. RANDERS

Transitions to a Sustainable System

The Sustainable Society

There are many ways to define sustainability. The simple definition is: A sustainable society is one that can persist over generations, one that is far-seeing enough, flexible enough, and wise enough not to undermine either its physical or its social systems of support.

The World Commission on Environment and Development put that definition into memorable words: A sustainable society is one that "meets the needs of the present without compromising the ability of future generations to meet their own needs".[1]

From a systems point of view a sustainable society is one that has in place informational, social and institutional mechanisms to keep in check the positive feedback loops that cause exponential population and capital growth. That means that birth rates roughly equal death rates, and investment rates roughly equal depreciation rates, unless and until technical changes and social decisions justify a considered and controlled change in the levels of population or capital. In order to be socially sustainable the combination of population, capital, and technology in the society would have to be configured so that the material living standard is adequate and secure for everyone. In order to be physically sustainable the society's material and energy throughputs would have to meet economist Herman Daly's three conditions:[2]

- Its rates of use of renewable resources do not exceed their rates of regeneration.
- Its rates of use of nonrenewable resources do not exceed the rate at which sustainable renewable substitutes are developed.
- Its rates of pollution emission do not exceed the assimilative capacity of the environment.

Whatever such a society would be like in detail, it could hardly be more different from the one in which most people now live. The collective human imagination is strongly imprinted by its recent experience either of poverty or of rapid material growth and of determined efforts to maintain that growth at all costs. Therefore many mental models are too full of growth-dominated notions to allow imagining a sustainable society. Before we can elaborate on what sustainability *could* be, we need to state what it *need not* be.

Sustainability does not mean no growth. A society fixated on perpetual growth tends to hear any criticism of growth as a total negation. But as Aurelio Peccei, founder of The Club of Rome, pointed out, that reaction just substitutes one oversimplification for another:

> All those who had helped to shatter the myth of growth . . . were ridiculed and figuratively hanged, drawn, and quartered by the loyal defenders of the sacred cow of growth. Some of those. . .accuse the [*Limits to Growth*] report . . . of advocating ZERO GROWTH. Clearly, such people have not understood anything, either about the Club of Rome, or about growth. The notion of zero growth is so primitive-as, for that matter, is that of infinite growth-and so imprecise, that it is conceptual nonsense to talk of it in a living, dynamic society.[3]

A sustainable society would be interested in qualitative development, not physical expansion. It would use material growth as a considered tool, not as a perpetual mandate. It would be neither for nor against growth, rather it would begin to discriminate kinds of growth and purposes for growth. Before this society would decide on any specific growth proposal, it would ask what the growth is for, and who would benefit, and what it would cost, and how long it would last, and whether it could be accommodated by the sources and sinks of the planet. A sustainable society would apply its values and its best knowledge of the earth's limits to choose only those kinds of growth that would actually serve social goals and enhance sustainability. And when any physical growth had accomplished its purposes, it would be brought to a stop.

A sustainable society would not freeze into permanence the current inequitable patterns of distribution. It would certainly not permit persistence of poverty. To do so would not be sustainable for two reasons. First, the poor would not and should not stand for it. Second, keeping any part of the population in poverty would not, except under dire coercive measures, allow the population to stabilize. For both moral and practical reasons any sustainable society must provide material sufficiency and security for all. To get to sustainability from here, the remaining material growth possible-whatever space there is for more resource use and pollution emissions, plus whatever space is treed up by higher efficiencies and lifestyle moderations on the part of the rich-would logically be allocated those who need it most.

A sustainable state would not be the society of despondency and stagnancy, high unemployment and bankruptcy that current market systems experience when their growth is interrupted. The difference between a sustainable society and a present-day economic recession is like the difference between stopping an automobile purposely with the brakes and stopping it by crashing into a brick wall. When the present economy overshoots, it turns around too fast and too unexpectedly for people or enterprises to retain, relocate, readjust. A transition to sustainability could take place slowly enough and with enough forewarning so that people and businesses could find their proper place in the new society.

There is no reason why a sustainable society need be technically or culturally primitive. Freed from both material anxiety and material greed, human society would have enormous possibilities for the expansion of human creativity in constructive directions. Without the high costs of growth for both human society and the environment, both technology and culture could bloom. John Stuart Mill, one of the first (and last) economists to take seriously the idea of an economy consistent with the limits of the earth, saw that what he called a "stationary state" could support an ever-evolving and improving society. More than a hundred years ago he wrote:

> I cannot . . . regard the stationary state of capital and wealth with the unaffected aversion so generally manifested towards it by political economists of the old school. I am inclined to believe that it would be, on the whole, a very considerable improvement on our present condition. I confess I am not charmed with the ideal of life held out by those who think that the normal state of human beings is that of struggling to get on; that the trampling, crushing, elbowing, and treading on each other's heels . . . are the most desirable lot of human kind . . . .. It is scarcely necessary to remark that a stationary condition of capital and population implies no stationary state of human improvement. There would be as much scope as ever for all kinds of mental culture and moral and social progress; as much room for improving the Art of Living, and much more likelihood of its being improved.[4]

A sustainable world would not and could not be a rigid one, with population or production or anything else held pathologically constant. One of the strangest assumptions of present-day mental models is the widespread idea that a world of moderation must be a world of strict, centralized, government control. We don't believe that kind of control is possible, desirable or necessary. A sustainable world would need to have rules, laws, standards, boundaries, and social agreements, of course, as does every human culture. Some of the rules for sustainability would be different from the rules people are used to now. Some of the necessary controls are already coming into being, as, for example, in the international ozone agreement.

But rules for sustainability, like every workable social rule, would not remove important freedoms; they would create them or protect them against those who would destroy them. A ban on bank-robbing inhibits the freedom of the thief in order to assure everyone's freedom to deposit and withdraw their money safely. A ban on overuse of a resource or a generation of pollution serves a similar purpose.

It doesn't take much imagining to come up with a minimum set of social structures-feedback loops that carry new information about costs, consequences, and sanctions-that would keep a society sustainable, allow evolution, fluctuation, creativity and change, and permit many more freedoms than would ever be possible in a world that continues to crowd against its limits.

Some people think that a sustainable society would have to stop using nonrenewable resources, since their use is by definition unsustainable. That idea is an overly rigid interpretation of what it means to be sustainable. Certainly a sustainable society would use gifts from the earth's crust more thoughtfully and efficiently than the present world does. It would price them properly and keep more of them available for future generations. But there is no reason not to use them, as long as their use meets the criterion of sustainability already defined, namely that renewable substitutes should be developed, so that no future society finds itself built around the use of a resource that is suddenly no longer available or affordable.

There is also no reason for a sustainable society to be uniform. Diversity is both a cause of and a result of sustainability in nature, and it would be in human society as well. Most people envision a sustainable world as decentralized, with boundary conditions keeping each locality from threatening the viability of another or of the earth as a whole. Cultural variety and local autonomy could be greater, not less, in such a world.

There is no reason for a sustainable society to be undemocratic, or boring, or unchallenging. Some games that amuse and consume people today, such as arms races and the accumulation of unlimited amounts of wealth, would no longer be played. But there still would be games, challenges, problems to solve, ways for people to prove themselves, to serve each other, to realize their abilities, and to live good lives, perhaps more satisfying than any that are possible today.

That was a long list of what a sustainable society is not. In the process of spelling it out, we have also, by contrast, indicated what we think a sustainable society could be. But the details of that society will not be worked out by one bunch of computer modelers; it will require the ideas, vision, and talents of billions of people.

From the structural analysis of the world system we have described in this book, we can contribute only a simple set of general guidelines for restructuring the world system toward sustainability. We list the guidelines below. Each one can be worked out in hundreds of specific ways at all levels from households to communities to nations to the world as a whole. Other people will see better than we can how to implement these changes in their own lives and cultures and political systems. Any step in any of these directions is a step toward sustainability.

– Improve the signals.

Learn more about and monitor both the welfare of the human population and the condition of local and planetary sources and sinks. Inform governments and the public as continuously and promptly about environmental conditions as about economic conditions. Include real environmental costs in economic prices; recast economic indicators like the GNP so that they do not confuse costs with benefits, or throughput with welfare, or the depreciation of natural capital with income.[5]

– Speed up response times.

Look actively for signals that indicate when the environment is stressed. Decide in advance what to do if problems appear (if possible, forecast them before they appear) and have in place the institutional and technical arrangements necessary to act effectively. Educate for flexibility and creativity, for critical thinking and the ability to redesign both physical and social systems. Computer modeling can help with this step, but more important would be general education in systems thinking.

– Minimize the use of nonrenewable resources.

Fossil fuels, fossil groundwaters, and minerals should be used only with the greatest possible efficiency, recycled when possible (fuels can't be recycled, but mineral and water can), and consumed only as part of a deliberate transition to renewable resources.

– Prevent the erosion of renewable resources.

The productivity of soils, surface waters, rechargeable groundwaters, and all living things, including forests, fish, game should be protected and, as far as possible, restored and enhanced. These resources should only be harvested at the rate they can regenerate themselves. That requires information about their regeneration rates, and strong social sanctions or economic inducements against their overuse.

– Use all resources with maximum efficiency.

The more human welfare can be obtained with the less throughput, the better the quality of life can be while remaining below the limits. Great efficiency gains are both technically possible and economically favorable. Higher efficiency will be essential, if current and future world populations are to be supported without inducing a collapse.

– Slow and eventually stop exponential growth of population and physical capital. There are real limits to the extent that the first five items on this list can be pursued. Therefore this last item is essential. It involves institutional and physiological change and social innovation. It requires defining levels of population and industrial output that are desirable and sustainable. It calls for goals defined around the idea of development rather than growth. It asks, simply but profoundly, for a vision of the purpose of human existence that does not require constant physical expansion.

We can expand on this last, most daunting, but most important step toward sustainability by pointing to the pressing problems that underlie much of the physiological and cultural commitment to growth: poverty, unemployment, and unmet nonmaterial needs. Growth as presently structured is in fact not solving these problems, or is solving them far too slowly and inefficiently. Until better solutions are in sight, however, society will never let go of its addiction to growth. These are the three areas where completely new thinking is most urgently needed.

– Poverty.
"Sharing" is a forbidden word in political discourse, probably because of the deep fear that real equity would mean not enough for anyone. "Sufficiency" and "solidarity" are concepts that can help structure new approaches to ending poverty. Everyone needs assurance that sufficiency is possible and that there is a high social commitment to ensure it. And everyone needs to understand that the world is tied together both ecologically and economically. We are all in this overshoot together. There is enough to go around, if we manage well. If we don't manage well, no one will escape the consequences.

– Unemployment.
Human beings need to work, to have the satisfaction of personal productivity, and to be accepted as responsible members of their society. That need should be not left unfulfilled, and it should not be filled by degrading or harmful work. At the same time, employment should not be a requirement for the ability to subsist. Considerable creativity is necessary here to create an economic system that uses and supports the contributions that all people are able and willing to make, that shares work and leisure equitably, and that does not abandon people who for reasons temporary of permanent cannot work.

– Unmet nonmaterial needs.
People don't need enormous cars; they need respect. They don't need closetsful of clothes; they need to feel attractive and they need excitement and variety and beauty. People don't need electronic entertainment; they need something

worthwhile to do with their lives. And so forth. People need identity, community, challenge, acknowledgement, love, joy. To try to fill these needs with material things is to set up an unquenchable appetite for false solutions to real and never-satisfied problems. The resulting psychological emptiness is one of the major forces behind the desire for material growth. A society that can admit and articulate its nonmaterial needs and find nonmaterial ways to satisfy them would require much lower material and energy throughputs and would provide much higher levels of human fulfilment.

How, in practice, can anyone attack these problems? How can the world evolve a social *system* that solves them? That is the real arena for creativity and choice. It is necessary for the present generation not only to bring itself below the earth's limits but to restructure its inner and outer worlds. That process will touch every arena of life. It will require every kind of human talent. It will need not only technical and entrepreneurial innovation, but also communal, social, political, artistic, and spiritual innovation. Lewis Mumford recognized fifty years ago not only the magnitude of the task, but also the fact that it is a particularly *human* task, one that will challenge and develop the *humanity*—in the most noble sense of that word—of everyone.

> An age of expansion is giving place to an age of equilibrium. The achievement of this equilibrium is the task of the next few centuries . . . The theme for the new period will be neither arms and the man: nor machines and the man: its theme will be the resurgence of life, the displacement of the mechanical by the organic, and the re-establishment of the person as the ultimate term of all human effort. Cultivation, humanization, co-operation, symbiosis: there are the watchwords of the new world-enveloping culture. Every department of life will record this change: it will affect the task of education and the procedures of science no less than the organization of industrial enterprises, the planning of cities, the development of regions, the interchange of world resources.[6]

The necessity to take the industrial world of growth to its next stage of evolution is not a disaster, it is an opportunity. How to seize the opportunity, how to bring into being a sustainable world that is not only functional but desirable is a question about leadership and ethics and vision and courage. Those are properties not of technologies, markets, government, corporations, or computer models but of the human heart and soul. To speak of them the authors need a chapter break here, to take off their computer modeling hats and put away their scientists' white coats and reappear as plain human beings.

Meadows, D., Meadows, D. & J. Randers, *Beyond the Limits; Confronting Global Collapse; Envisioning a Sustainable Future*, Earthscan Publications Ltd., London, 1991, p. 209-217.

Notes and references

1 WCED, *Our Common Future*, op. cit.

2 Herman Daly is one of the few people who have begun to think through what kinds of social institutions might work to maintain a desirable sustainable state. He comes up with a thought-provoking mixture of market and regulatory devices. See, for example, Herman Daly, 'Institutions for a Steady-State Economy' in *Steady State Economics* (Washington, DC: Island Press, 1991).

3 Aurelio Peccei, *The Human Quality* (Oxford: Pergamon Press, 1977). 85.

4 John Stuart Mill, *Principles of Political Economy*, first published 1848.

5 For an example of 'sustainable accounting,' sec Raul Solorzano et al., *Accounts Overdue: Natural Resource Depreciation in Costa Rica* (Washington, DC: World Resources Institute, December 1991).

6 Lewis Mumford, *The Condition of Man* (New York: Harcourt Brace Jovanovich, 1944), 598-99.

CHAPTER 28

Fifth Action Programme on the Environment

COMMISSION OF THE EUROPEAN COMMUNITY

Executive Summary

Introduction

1 Over the past two decades four Community action programmes on the environment have given rise to about 200 pieces of legislation covering pollution of the atmosphere, water and soil, waste management, safeguards in relation to chemicals and biotechnology, product standards, environmental impact assessments and protection of nature. The Community's 4th Action Programme on the Environment has not been completed—it runs up to the end of 1992—and its impact will not be known for some years to come. While a great deal has been achieved under these programmes and measures, a combination of factors calls for a more far-reaching policy and more effective strategy at this juncture:

(i) a new Report on the State of the Environment published in conjunction with this Programme[1] indicates a slow but relentless deterioration of the general state of the environment of the Community notwithstanding the measures taken over the past two decades, particularly as respects the issues referred to in para 16 below; the Report also shows up significant deficiencies in the quantity, quality and comparability of data which are crucial for environment-related policies and decisions. In this context it is of the utmost importance that the European Environment Agency become operational;

(ii) the present approach and existing measures are not geared to deal with the expected growth in international competition and the upward trends in Community activity and development which will impose even greater burdens on natural resources, the environment and, ultimately, the quality of life;

(iii) global concerns about the climate change/deforestation/energy crisis, the seriousness and persistence of problems of underdevelopment and the progress of political and economic change in Central and Eastern Europe add to the responsibility of the European Community in the international field.

2 The new Treaty on European Union, signed by all Member States on 7 February 1992 has introduced as a principle objective the promotion of sustainable growth respecting the environment (Article 2). It includes among the activities of the Union a policy in the sphere of the environment (Article 3k), specifies that this policy must aim at a high level of protection and that environmental protection requirements must be integrated into the definition and implementation of other Community policies (Article 130r (2)). The new Treaty also attaches special value to the principle of subsidiarity (Article 3b), and states that decisions should be taken as closely as possible to the citizens (Article A). Furthermore, the Community policy on the environment is required to contribute to promoting measures at international level to deal with regional or worldwide environmental problems (Article 130r (1)). In this latter context the Community will endeavour to find solutions in the field of development and environment at the United Nations Conference on Environment and Development (UNCED) in Rio de Janeiro in June 1992.

3 All human activity has an impact on the biophysical world and is, in turn, affected by it. The capacity to control this interrelationship conditions the continuity, over time, of different forms of activity and the potential for economic and social development. Within the Community, the long-term success of the more important initiatives such as the Internal Market and economic and monetary union will be dependent upon the sustainability of the policies pursued in the fields of industry, energy, transport, agriculture and regional development; but each of these policies, whether viewed separately or as it interfaces with others, is dependent on the carrying capacity of the environment.

4 The achievement of the desired balance between human activity and development and protection of the environment requires a sharing of responsibilities which is both equitable and clearly defined by reference to consumption of and behaviour towards the environment and natural resources. This implies integration of environment considerations in the formulation and implementation of economic and sectoral policies, in the decisions of public authorities, in the conduct and development of production processes and in individual behaviour and choice. It also implies effective dialogue and concerted action among partners who may have differing short-term priorities; such dialogue must be supported by objective and reliable information.

5 As used in the Programme, the word 'sustainable' is intended to reflect a policy and strategy for continued economic and social development without detriment to the environment and the natural resources on the quality of which continued human activity and further development depend. The Report of the World Commission on Environment and Development (Brundtland) defined sustainable development as '*development which meets the needs of the present without compromising the ability of future generations to meet their own needs.*' It entails preserving the overall balance and value of the natural capital stock, redefinition of short, medium and long-term cost/benefit evaluation criteria and instruments to reflect the real socio-economic effects and values of consumption and conservation, and the equitable distribution and use of resources between nations and regions over the world as a whole. In the latter context, the Brundtland Report pointed out that the developed countries, with only 26 per cent of the world population, are responsible for about 80 per cent of world consumption of energy, steel and other metals, and paper and about 40 per cent of the food.

6 Following are some of the practical requirements for achieving sustainable development:
— since the reservoir of raw materials is finite, the flow of substances through the various stages of processing, consumption and use should be so managed as to facilitate or encourage optimum reuse and recycling, thereby avoiding wastage and preventing depletion of the natural resource stock;
— production and consumption of energy should be rationalized; and
— consumption and behaviour patterns of society itself should be altered.

7 It is clear that sustainable development is not something which will be achieved over a period as short as that covered by this Programme. 'Towards Sustainability' should be seen, accordingly, as an important step only in a longer-term campaign to safeguard the environment and the quality of life of the Community and, ultimately, our planet.

The Community's role in the wider international arena
8 In the early stages, Community policy and action on the environment were mainly focussed on the solution of particularly acute problems within the *Community*. Later there was a clearer recognition that pollution did not stop at its frontiers and that it was necessary, therefore, to intensify co-operation with *third countries*. In recent years, the evolution has gone a step further and it is now generally accepted that issues of a *global nature*—climate change, ozone depletion, diminution of biodiversity, etc.—are seriously threatening the ecological balance of our planet as a whole.

9 These issues are to be addressed at the highest level at the United Nations Conference on Environment and Development (UNCED). Just as the 1972 UN Conference in Stockholm created a new awareness and concern about the environment at broad international level, so too can UNCED bring global political will and commitment to effective action into a new dimension. Apart from the expected adoption of framework conventions on climate change and biodiver-

sity and of principles on conservation and development of forests, UNCED should pave the way forward by adopting:

— an '*Earth Charter*' or Declaration of basic rights and obligations with respect to environment and development;

— an agenda for action, '*Agenda 21*' which will constitute an agreed work programme of the international community for the period beyond 1992 and into the 21st century.

10 In the declaration on the environment made in Dublin in June 1990 the European Council stressed the special responsibility of the Community and its constituent Member States in the wider international arena when it stated that '*the Community must use more effectively its position of moral, economic and political authority to advance international efforts to solve global problems and to promote sustainable development and respect for the global commons.*' In conformity with the said declaration, the Community and the Member States must increase their efforts to promote international action to protect the environment and to meet the specific needs and requirements of its partners in the developing world and in Central and Eastern Europe.

The credibility of the industrialised world, including the Community, from the viewpoint of developing countries will be commensurate with the extent to which it puts its own house in order. In adopting and implementing this Programme, the Community will be in a position to offer the leadership foreseen in the Dublin Declaration.

The new strategy for environment and development

11 The approach adopted in drawing up this new policy programme differs from that which applied in previous environmental action programmes:

—it focuses on the agents and activities which deplete natural resources and otherwise damage the environment, rather than wait for problems to emerge;

—it endeavours to initiate changes in current trends and practices which are detrimental to the environment, so as to provide optimal conditions for socio-economic well-being and growth for the present and future generations;

— it aims to achieve such changes in society's patterns of behaviour through the optimum involvement of all sectors of society in a spirit of shared responsibility, including public administration, public and private enterprise, and the general public (as both individual citizens and consumers);

—responsibility will be shared through a significant broadening of the range of instruments to be applied contemporaneously to the resolution of particular issues or problems.

12 For each of the main issues, *long-term objectives* are given as an indication of the sense of direction or thrust to be applied in the pursuit of sustainable development, certain *performance targets* are indicated for the period up to the year 2000 and a representative selection of *actions* is prescribed with a view to achieving the said targets. These objectives and targets do not constitute legal commitments but, rather, performance levels or achievements to be aimed at now in the interests of attaining a sustainable development path. Neither should all the actions indicated require legislation at Community or national level. (Note: Because of substantial disparities and short-comings in both the quantity and quality of data available, it has not been possible to have homogenous levels of precision in the objectives and targets included in the Programme.)

13 The Programme takes account of the diversity of situations in various regions of the Community and, in particular, of the need for the economic and social development of the less wealthy regions of the Community. It aims to protect and enhance the inherent advantages of these latter regions and to afford protection to their more valuable natural assets as a resource-base for economic development and social improvement and prosperity. In the case of the more developed regions of the Community, the aim is to restore or maintain the quality of their environment and natural resource base for their continued economic activity and quality of life.

14 The success of this approach will rely heavily on the flow and quality of information both in relation to the environment and as between the various actors including the general public. The role of the European Environment Agency is seen as crucial in relation to the evaluation and dissemination of information, distinction between real and perceived risks and provision of a scientific and rational basis for decisions and actions affecting the environment and natural resources.

15 In relation to the motivation of the general public, the main tasks will fall to levels other than the Community level. The Commission, for its part, will commit its information services to a campaign of environmental information and awareness-building.

The importance of education in the development of environmental awareness cannot be overstated and should be an integral element in school curricula from primary level onwards.

Environmental challenges and priorities

16 The Programme addresses a number of environmental *issues*: climate change, acidification and air pollution, depletion of natural resources and biodiversity, depletion and pollution of water resources, deterioration of the urban environment, deterioration of coastal zones, and waste. This list is not an exhaustive one but, pursuant to the principle of subsidiarity, it comprises matters of particular seriousness which have a Community-wide dimension, either because of Internal Market, crossboundary, shared resource or cohesion implications and because they have a crucial bearing on environmental quality and conditions in almost all regions of the Community.

17 These issues are addressed not so much as problems, but as *symptoms* of mismanagement and abuse. The real 'problems', which cause environmental loss and damage, are the current patterns of human consumption and behaviour. With this distinction in mind and with due respect to the principle of subsidiarity, priority will be given to the following fields of action with a view to achieving tangible improvements or changes during the period covered by the Programme:

— *Sustainable Management of Natural Resources*: soil, water, natural areas and coastal zones
— *Integrated Pollution Control and Prevention of Waste*
— *Reduction in the Consumption of Non-Renewable Energy*
— *Improved Mobility Management* including more efficient and environmentally rational location decisions and transport modes
— Coherent packages of measures to achieve improvements in *environmental quality in urban areas*
— *Improvement of Public Health and Safety*, with special emphasis on industrial risk assessment and management, nuclear safety and radiation protection.

Selected target sectors

18 Five target sectors have been selected for special attention under this Programme: Industry, Energy, Transport, Agriculture and Tourism. These are sectors where the Community as such has a unique role to play and where a Community approach is the most efficient level at which to tackle the problems these sectors cause or face. They are also chosen because of the particularly significant impacts that they have or could have on the environment as a whole and because, by their nature, they have crucial roles to play in the attempt to achieve sustainable development. The approach to the target sectors is designed not only for the protection of public health and the environment as such, but for the benefit and sustainability of the sectors themselves.

Industry
19 Whereas previous environmental measures tended to be proscriptive in

character with an emphasis on the 'thou shalt not' approach, the new strategy leans more towards a 'let's work together' approach. This reflects the growing realization in industry and in the business world that not only is industry a significant pan of the (environmental) problem but it must also be part of the solution. The new approach implies, in particular, a reinforcement of the dialogue with industry and the encouragement, in appropriate circumstances, of voluntary agreements and other forms of self-regulation.

Nevertheless, Community action is and will continue to be an important element in the avoidance of distortions in conditions of competition and preservation of the integrity of the Internal Market.

20 The three pillars on which the environment/industry relationship will be based will be:
— improved resource management with a view to both rational use of resources and improvement of competitive position;
— use of information for promotion of better consumer choice and for improvement of public confidence in industrial activity and controls and in the quality of products;
— Community standards for production processes and products.

In developing measures to ensure the sustainability of the industrial sector, special consideration will be given to the position of small and medium enterprises and to the matter of international competitiveness.

In mid-1992 the Commission will publish a comprehensive Communication on international competitiveness and protection of the environment.

Energy
21 Energy policy is a key factor in the achievement of sustainable development. While the Community's energy sector is making steady progress in dealing with local and regional environmental problems such as acidification, global issues are daily growing in importance. The challenge of the future will be to ensure that economic growth, efficient and secure energy supplies and a clean environment are compatible objectives.

22 The achievement of this balance requires a strategic perspective well beyond the period covered by this Programme. The key elements of the strategy up to 2000 will be improvement in energy efficiency and the development of strategic technology programmes moving towards a less carbon-intensive energy structure including, in particular, renewable energy options.

Transport
23 Transport is vital to the distribution of goods and services, to trade and to regional development.

Present trends in the Community's transport sector are all leading towards greater inefficiency, congestion, pollution, wastage of time and value, damage to health, danger to life and general economic loss. Transport demand and traffic are expected to increase even more rapidly with the completion of the Internal Market and the political and economic developments in Central and Eastern Europe.

24 A strategy for sustainable mobility will require a combination of measures which includes:

— improved land-use/economic development planning at local, regional, national and transnational levels;

— improved planning, management and use of transport infrastructures and facilities; incorporation of the real costs of both infrastructure and environment in investment policies and decisions and also in user costs:

— development of public transport and improvement of its competitive position;

— continued technical improvement of vehicles and fuels; encouraged use of less polluting fuels;

— promotion of a more environmentally rational use of the private car, including changes in driving rules and habits.

In conjunction with this Programme, the Commission has published a more comprehensive Communication dealing with transport and the environment and the need to aim for sustainable mobility.

Agriculture
25 The farmer is the guardian of the soil and of the countryside. Improvements in farming efficiency, increased mechanisation levels, improved transport and marketing arrangements, increased international trade in food products and feedstuffs have all contributed to the fulfillment of the original Treaty objectives of assuring the availability of food supplies at reasonable prices, the stabilization of markets and a fair standard of living for the agricultural community. At the same time, however, changes in farming practices in many regions of the Community have led to overexploitation and degradation of the natural resources on which agriculture itself ultimately depends: soil, water and air.

26 In addition to environmental degradation, serious problems have emerged in the case of commodity overproduction and storage, rural depopulation, the Community budget and international trade (both as regards agricultural products and wider trade agreements). It is not only environmentally desirable, therefore, but also makes sound agricultural, social and economic sense to seek to strike a more sustainable balance between agricultural activity, other forms of rural development and the natural resources of the environment.

27 The Programme builds on the Commission's proposals for reform of the CAP and for development of the Community's forests so as to work towards a balanced and dynamic development of the rural areas of the Community which will meet the sector's productive, social and environmental functions.

Tourism

28 Tourism is an important element in the social and economic life of the Community. It reflects the legitimate aspirations of the individual to enjoy new places and absorb different cultures as well as to benefit from activities or relaxation away from the normal home or work setting. It is also an important economic asset to many regions and cities of the Community and has a special contribution to make to the economic and social cohesion of the peripheral regions. Tourism represents a good example of the fundamental link which exists between economic development and environment, with all the attendant benefits, tensions and potential conflicts. If well planned and managed, tourism, regional development and environment protection can go hand in hand. Respect for nature and the environment, particularly in coastal zones and mountain areas, can make tourism both profitable and long-lasting.

29 The World Tourism Organisation predicts a significant increase in tourism activity to and within Europe during this decade. Most of this increase is likely to take place in the Mediterranean Region, and in particular types of locations such as historic towns and cities, mountain areas and coastal zones. UNEP's Blue Plan on the Mediterranean predicts a doubling, at least, of solid wastes and waste waters resulting from tourism by the year 2000, and a potential doubling in the land occupied by tourist lodgings.

30 The European Community supports tourism through its investments in necessary infrastructures; it can also serve as a 'facilitator' in relation to other interests. But, in a practical reflection of the principle of subsidiarity and the spirit of shared responsibility, it is mainly at levels other than that of the Community that the real work of reconciling tourism activity and development and the guardianship of natural and cultural assets must be brought into a sustainable balance, i.e. by Member States, regional and local authorities, the tourism industry itself and individual tourists.

The three main lines of action indicated in the Programme deal with

— diversification of tourism activities, including better management of the phenomenon of mass tourism, and encouragement of different types of tourism;

— quality of tourist services, including information and awareness-building, and visitor management and facilities;

— tourist behaviour, including media campaigns, codes of behaviour and choice of transport.

Broadening the range of instruments

31 Previous action programmes have relied almost exclusively on legislative measures. In order to bring about substantial changes in current trends and practices and to involve all sectors of society in a full sharing of responsibility, a broader mix of instruments is needed. The mix proposed can be categorised under four headings:

(i) *Legislative instruments,* designed to set fundamental levels of protection for public health and the environment, particularly in cases of high risk, to implement wider international commitments and to provide Community-wide rules and standards necessary to preserve the integrity of the Internal Market.

(ii) *Market-based instruments,* designed to sensitize both producers and consumers towards responsible use of natural resources, avoidance of pollution and waste by internalising of external environmental costs (through the application of economic and fiscal incentives and disincentives, civil liability, etc.) and geared towards 'getting the prices right' so that environmentally-friendly goods and services are not at a market disadvantage vis-à-vis polluting or wasteful competitors.

(iii) *Horizontal, supporting instruments* including improved base-line and statistical data, scientific research and technological development, (as respects both new less-polluting technologies and technologies and techniques for solving current environmental problems) improved sectoral and spatial planning, public/consumer information and education and professional and vocational education and training.

(iv) *Financial support mechanisms*: besides the budgetary lines which have direct environmental objectives, such as LIFE, the Structural Funds, notably ENVIREG, contribute significant amounts to the financing of actions for the improvement of the environment. Moreover, the new Cohesion Fund decided upon at the Maastricht Summit aims at cofinancing projects which are intended to improve the environment in Spain, Greece, Portugal and Ireland. Article 130r (2) of the new Treaty provides that environment policy must aim at a high level of protection based on the precautionary principle and preventive action, taking into account the diversity of situations in the various regions of the Community, and that environment policy must be integrated into the definition and implementation of other Community policies. In this context, it will be necessary to ensure that all Community funding operations, and in particular, these involving the Structural Funds, will be as sensitive as possible to environmental considerations and in conformity with environmental legislation. By way of qualification it must be recalled here that the new Treaty provides, in Article 130s (4), that without prejudice to certain measures of a Community nature, the Member States are responsible for financing and implementing environment policy.

The principle of subsidiarity

32 The principle of subsidiarity will play an important part in ensuring that the objectives, targets and actions are given full effect by appropriate national regional and local efforts and initiatives. In practice it should serve to take full account of the traditions and sensitivities of different regions of the Community and the cost-effectiveness of various actions and to improve the choice of actions and appropriate mixes of instruments at Community and/or other levels.

The objectives and targets put forward in the Programme and the ultimate goal of sustainable development can only be achieved by concerted action on the part of all the relevant actors working together in partnership. On the basis of the Treaty on the European Union (Article 3b), the Community will take action, in accordance with the principle of subsidiarity, only if and insofar as the objectives of the proposed action cannot be sufficiently achieved by the Member States and can therefore, by reason of the scale or effects of proposed action, be better achieved by the Community.

33 The Programme combines the principle of subsidiarity with the wider concept of shared responsibility; this concept involves not so much a choice of action at one level to the exclusion of others but, rather, a mixing of actors and instruments at the appropriate levels, without any calling into question of the division of competences between the Community, the Member States, regional and local authorities.

Table 18 of the document and the 'actors' column of the other tables indicate respectively the manner in which the various actors are intended to combine and the different actors considered most relevant for the implementation of specific measures.

Making the Programme work

34 Up to the present, environmental protection in the Community has mainly been based on a legislative approach ('top-down'). The new strategy advanced in this Programme implies the involvement of all economic and social partners ('bottom-up'). The complementarity and effectiveness of the two approaches together will depend, in great measure, on the level and quality of dialogue which will take place in pursuance of partnership.

35 Inevitably, it will take some considerable time for the current patterns of consumption and behaviour to turn in the direction of sustainability. In practical terms, the effectiveness of the strategy will depend, for the foreseeable future, on the inherent quality of the measures adopted and the practical arrangements for their enforcement. This will require better preparation of measures, more effective co-ordination with and integration into other policies, more systematic follow-up and stricter compliance-checking and enforcement.

36 For these reasons—but without prejudice to the Commission's right of initiative and its responsibility to ensure satisfactory implementation of Community rules—the following ad hoc dialogue groups will be convened by the Commission:

(i) *a General Consultative Forum* comprising representatives of enterprise, consumers, unions and professional organisations, non-governmental organisations and local and regional authorities;

(ii) *an Implementation Network* comprising representatives of relevant national authorities and of the Commission in the field of practical implementation with Community measures; it will be aimed primarily at exchange of information and experience and at the development of common approaches at practical level, under the supervision of the Commission;

(iii) *an Environmental Policy Review Group*, comprising representatives of the Commission and the Member States at Director-General level to develop mutual understanding and exchange of views on environment policy and measures.

37 These three dialogue groups will serve, in a special way, to promote greater sense of responsibility among the principal actors in the partenariat, and to ensure effective and transparent application of measures. They are not intended to duplicate the work of committees established by Community legislation for the purposes of follow-up in respect of specific measures, nor by the Commission in relation to specific fields of interest such as consumer protection, tourism development etc. nor by Member States for implementation and enforcement of policy at national level. Finally, they will not substitute the existing dialogue between industry and the Commission, which it is intended to strengthen in any event.

Review of Programme

38 While the Programme is essentially targeted towards the year 2000, it will be reviewed and 'rolled-over' at the end of 1995 in the light of improvements in relevant data, results of current research, and forthcoming reviews of other Community policies e.g. industry, energy, transport, agriculture, and the structural funds.

Conclusion

39 This Programme itself constitutes a turning point for the Community. Just as the challenge of the 1980s was completion of the Internal Market, the reconciliation of environment and development is one of the principal challenges facing the Community and the world at large in the 1990s. "Towards Sustainability" is not a programme for the Commission alone, nor one geared towards environmentalists alone. It provides a framework for a new approach to the environment and to economic and social activity and development and requires positive will at all levels of the political and corporate spectrums, and the involvement of all members of the public active as citizens and consumers in order to make it work.

40 The Programme does not purport to "get everything right". It will take a long time to change patterns of behaviour and consumption and to attain a sustainable development path. The Programme, accordingly, is intended primarily to *break the current trends*. The bottom line is that the present generation must pass the environment on to the next generation in a fit state to maintain public health and social and economic welfare at a high level. As an intermediate goal, the state of the environment, the level and quality of natural resources and the potential for further development at the end of this decade should reflect a marked improvement on the situation which obtains today. The road to sustainability may be long and difficult . . . but the first steps must be taken NOW!

Structure of the Document

41 The document is divided into three parts, the two main parts being related to internal and external actions. This distinction is made so as to reflect what can politically and legally be done within the Community itself in accordance with the powers and procedures incorporated in the Treaties, and what the Community and its constituent Member States can contribute or achieve in partnership with other developed and developing countries in relation to global or regional issues and problems.

42 Part I summarises the state of the environment in the Community and growing threats to its future health (Ch. 1) and sets out a new strategy designed to break the current trends and to set a new course for sustainable development (Ch. 2). The strategy entails active involvement of all the main actors in society (Ch. 3) using a broader range of instruments, including market-related instruments and improved information, education and training (Ch. 7) so as to achieve identifiable or quantifiable improvements in the environment or changes in consumption and behaviour (Ch. 5).

43 A special, concentrated, effort will be made in the case of five target sectors of Community-wide significance (Ch. 4) and in relation to the avoidance and management of risks and accidents (Ch. 6).

44 In an effort to be both concise and as clear as possible, the measures which together constitute the action programme are set out in a series of tables which are predominantly, though not entirely, homogenous.
These tables are structured so as to indicate:
– the long-term objectives in the various fields;
– the qualitative or quantitative targets to be attained by the year 2000;
– the specific actions required to be taken;
– the time-frame proposed for such actions;
– the actors or sectors of activity which will be called upon to play a part.
Pursuant to the principle of subsidiarity, the lead role is indicated by the use of an italic type-face e.g. *MS*.

45 Finally, Part I attempts to indicate how responsibility can in practice be shared (Ch. 8) and the measures proposed to ensure satisfactory implementation and enforcement (Ch. 9).

46 Part II summarises the environmental threats and issues in the wider international sphere (Ch. 10) and what will or can be done by the Community and its constituent Member States in the context of both general international and bilateral cooperation (Ch. 11 and 12, resp.) in relation to global and regional issues and to environment and development issues in developing countries and Central and Eastern Europe. Chapter 13 deals with the United Nations Conference on Environment and Development which will take place in June 1992. It also refers to the correlation between the internal and external dimensions of the Community's policy on the environment.

47 Part III is quite short and very general, dealing with the selection of priorities (Chapter 14), the question of costs (Chapter 15) and the intention to carry out a mid-term review of the Programme in 1995 (Chapter 16). While in a document which puts forward a policy and strategy aimed at breaking trends there is less a question of selecting priority actions than defining a *'critical path'*; nevertheless, the Programme does include a listing of horizontal measures and fields of action which require to be accorded priority. On the question of costs the document points to the difficulties of undertaking such exercise (partly because of the traditional practice of treating the environment as an infinite source of free raw materials and waste sinks, and partly because not enough has been done to determine the real costs of *'non-action'*) and puts forward a 5-point plan to devise appropriate costing mechanisms for the future.

Fifth Action Programme on the Environment, Commission of the European Community, Brussels, 1993, p. 11-18.

Note

1   This is not published here. See COM(92) 23 final—Vol. III.

CHAPTER 29

The Third Revolution

Environment, Population and a Sustainable World

P. HARRISON

That we would do, we should do when we would

It is often said that the 1990s will be the decade that decides the fate of the planet. That may be so. More probably, it is the second in a run of three or four decisive decades.

We wasted the first. The 1980s were our best chance to act. We passed it up. Debt made the developing countries into net donors to the rich countries. Social spending was cut back in Africa and Latin America. Progress in education and health halted and reversed. Fertility did not decline as expected. The cold war ended, but the peace dividend seemed likely to go towards cutting taxes, or helping former communist countries, rather than to helping the Third World in any significant way. In the 1980s we came upon Claudius at prayer—and we let him live to do more evil.

The contrast between Claudius and Hamlet is in many respects the core of Shakespeare's play. Where Hamlet is wavering, Claudius is decisive. He is not made coward by conscience, nor disabled by thinking too precisely on the event. For him there is no chink between though and deed. As soon as he senses danger he sends Hamlet to England. As soon as Hamlet returns, Claudius devises not one, but two plots to kill him.

We must learn to act swiftly like Claudius. We must break the fatal habit of acting too late.

To do so means working to reduce population growth and excess consumption, and to change damaging technologies for benign. It also means changing all those indirect factors that affect the three primary agents. We must strengthen our institutions, to shorten the delays in perceiving and acting on environmental problems (see Chapter 17). We must improve our capacity to monitor changes in the environment. We process the technology to do so: what we need is the commitment of funds and manpower. We must educate farmers, foresters and fishers to spot early warning signals of damage, so they will become concerned at an earlier stage.

We must spread democracy where it does not exist. Where it does, we must strengthen it by improving education, and creating the right of access to government and company information on environmental impacts. We must introduce free markets where they do not exist. And where they do, their blind spots should be removed: environmental and other social costs and benefits must be fully accounted.

We must increase security of tenure or ownership over land. And for forests, rangelands, rivers, oceans and atmosphere, we must strengthen community, national and international control.

We must improve our understanding of environmental processes. We are really only beginners in understanding the incredible complexity of ecosystems—including the biggest ecosystem of all, the planet Earth. This means devoting a great deal more resources to environmental sciences. And it means creating the framework for a much more interdisciplinary approach to research and to first degrees. The demands of the environment will break down the compartmentalization of knowledge. This process is already visible in research programmes on planetary change, which could well become the master-science overarching all others. As part of this we desperately need an overarching science of human interactions with the environment, combining demography, socioeconomic and technological studies with dynamic analysis of the physical environment.

Finding and spreading appropriate technologies is the final stage of the process. We must give far higher priority to research into environmentally benign technologies, especially in energy production and in agriculture.

Aspects of resilience

Life is more resilient than most people realize. It bubbles out of hot fissures in the ocean bed. It reddens polar snowfields. Life if hard to keep down. Desertified rangelands recover when rested. Logged forests regrow if they are left alone. Even human disturbance can create a mosaic of cleared areas, regenerating areas, wasteland, field, field edge. Some sites in Israel, after ten millennia of severe human disturbance, have a species diversity almost as high as moist tropical forest.[1]

Human culture is resilient too. Our dominance on earth is not due to rigidity, but our speed in adapting to changes in the environment. In every sphere we have seen adjustment processes at work. When population growth, consumption changes or technology create problems, people work to overcome those problems—though they do not always succeed.

There may be a higher-level adjustment process, working at the level of whole economic and value systems.

Hunter-gatherers have an almost religious reverence for nature. Animals, plants, even rocks and rivers, have souls. Plants and animals are taken as needed—but with due concern for sustainability, since overconsumption of wild food sources will deplete them and threaten group survival.

However, most early hunter-gatherers did eventually consume beyond the limits of their wild food supply. As population grew, agriculture spread, and allowed population to grow further. With agriculture comes a shift in attitudes. Wild nature is no longer the main source of food, but the cradle of weeds and pests and the den of predators. And it is potential land to be cleared for farming. Attitudes to wilderness change from reverence to revulsion, fear, desire to exterminate, tame and convert.

These negative views persisted up the mid eighteenth century in Britain. The Alps were considered 'hideous' in 1621 by traveller James Howell. In 1681 John Houghton denounced Hampstead Heath as a 'barren wilderness' in urgent need of cultivation. To Dr Johnson the Scottish Highlands were a 'wide extent of hopeless sterility'.[2]

Absence makes the heart grow fonder

It took complete separation from nature to bring back reverence for nature. Large scale urbanization, combining physical crowding with social isolation, drove the romantics to seed out the solace of unspoiled countryside—just as it had earlier driven the intelligentsia of imperial Rome to idealize the rustic simplicity of their ancestors.

To Wordsworth, London in 1791 was a 'huge fermenting mass of humankind', a 'perpetual whirl of trivial objects'. In lonely rooms, amid the din of town and cities, he consoled himself with memories of 'wild secluded scenes' and called himself a nature worshipper. 'Nature then', he wrote, 'to me was all in all'.[3]

As urbanization progressed, there was a reversal of earlier ideals of beauty. Gardens, once rigorously geometrical, became deliberately unkempt. In 1882 the painter Constable wrote of his aversion to landscaped parks: 'It is not beauty because it is not nature.'[4]

In more recent times further separations from nature have occurred. Chemicals replaced manure and compost on farmland. Processed foods supplanted raw. Synthetic fibres ousted natural ones. Fossil fuels took the place of wood and animal power. Plastic elbowed out cardboard. Cement and steel superseded timber and clay. In general, the artificial dislodged the natural. The inevitable reaction to this process was the rise of organic farming, of raw and unprocessed foods, of handicrafts.

High levels of industrialization and urbanization lead to a plateau in the human relations with the environment. There is a high density of waste, of pollution, of traffic congestion. At the same time democracy is relatively developed. Pressures to control pollution are pronounced, and heeded. Leisure and wide car ownership provide easy access to countryside. Pressures for conservation grow. An increasing proportion of land is set aside as national parks and nature reserves.

But like the lower-level adjustments, this macro-adjustment process is too slow to prevent large-scale damage to the environment: indeed it comes about only where damage has already occurred. In the most advanced developed countries it is beginning to reach an active phase, where controls are applied systematically and progressively on industrial and farming processes. But there is still a marked hesitancy in tackling pollution from individual domestic sources—since this could lose votes. In no country on earth has the adjustment come anywhere near the stage where environmental damage has ceased.

Most developing countries are in transition between phases two and three, agriculture and industrialization. At this stage the level of damage is truly massive. There is continued forest clearance for agriculture, reckless logging, industrialization without effective democratic controls on pollution.

Towards the third revolution

The agricultural and industrial revolutions were the response to pressures of population growth on the environment. Shortage of wild food resources led to the first, shortage of wood to the second. In the first case we adjusted technology and consumption (reducing the variety of our diet); but population growth increased. In the second we adjusted technology, shifting to fossil energy; but population growth accelerated further. Later, in response to the costs and opportunities of industrialization and urbanization, we finally began to adjust our populations.

We are now in the throes of the third revolution. This time the spur is not resource shortage, but the impact of waste and wasting. It will demand responses right across the board, in population, consumption and technology and everything that affects them. It will continue until we reach a sustainable balance with our natural environment.

There is one factor that could short-cut the adjustment process: it is the major shift in values that has begun. Such shifts are determined by historical forces. But once in motion on a large scale the new values motivate human action and themselves become historical forces. Witness the rise of Christianity or Islam, the democratic revolutions of 1776-1848, or the socialist revolutions of 1917-75.

The current value shift is possibly the most far-reaching in our attitude to nature since the rise of transcendental religions between 600 BC and AD 700. The new values have already gestated, in developed countries and among the intelligentsia of more urbanized Third World countries.

They are now spreading with the speed of a new religion. And the global nature of our environmental crisis—like the threat of hell fire in early Christianity—is the most persuasive of its evangelists.

Indeed the new values have a quasi-religious content. There is a tendency away from the concept of a God outside or above the universe, towards a divinity inherent in everything. Away from transcendentalism, toward immanence and pantheism. There is a parallel shift away from a simple mechanical and deterministic view of the physical world, towards a more holistic view. The world is not an engine, but an organism. Gaia is treated as a deity by some.

The concept that humans by right are masters of the earth, the pinnacle of creation, is already giving way to the idea that we are stewards, with a responsibility to look after the patrimony with which we have been entrusted. This idea in turn phases into the view that we are members of a community. If one member overreaches itself, the others rebel. If we damage nature, we damage ourselves. We cannot, try as we will, control nature over the long run: we can only co-operate with it.[5]

The new philosophy involves a new ethics of extended altruism. Altruism has prevailed within the primary group since hunter-gatherer times. The great universal religions extended altruism to cover whole societies. In the modern world it is applied theory—though not in practice—across national boundaries. The goal of equity, social justice and the abolition of absolute poverty within the present generation has wide influence among individuals, if not among governments.

The next ethical principle is altruism towards future generations. We must not prosper at the expense of poverty and constraint for our descendants. We ought to leave them a world that is not diminished in its richness. Development must be sustainable, or it is no more than a brief interlude between aeons of human misery.

The third principle is altruism towards other species. They have rights as well as humans. We should not thrive on earth by exterminating other organisms. We can hardly avoid confining them to smaller areas. But at least there we should work to preserve them and the ecosystems of which they are part. This concept is also inherent in the second. If we don't treat other species equitable, we impoverish the world that our children inherit.[6]

The new philosophy and the new ethics are in some respects not new at all. They are the resurrected wisdom of the hunter-gatherer, extended out from the local habitat to the entire planet. In the same way the new forestry and the new agriculture will emphasize the diversity that hunter-gatherers prized. Human attitudes to the environment have come full circle.

The readiness is all
Adjustment will occur. Sooner, with less damage, if we are wise. Later, with

more damage, if we are not. Hamlet had to kill Claudius: the only question was when, and whether Claudius would kill Hamlet first. If Hamlet had acted when he should, seven innocent lives would have been spared.

Much of the last chapter was aimed at policy makers. But environmental problems are the outcome of individual actions multiplied millions of times. That is one reason why they are so intractable, but at the same time it is a source of hope. Many changes in environmental behaviour do not have to pass through the whole political system to become effective. Of course political action helps: not just campaigning for wider changes, but for local ones. Anyone can change their own behaviour straight away, as an individual, as a member of a family, of a local community and a workplace. We don't have to wait till collective disaster forces us to change on a collective scale.

We can plant trees. Turn our garden into a wildlife retreat. Pay for others to conserve habitats and restore degraded ones—nature will need millions of midwives over the coming decades. If we are farmers or foresters, we can try to maintain production, or long-term income, in ways that preserve or enhance biological capital rather than reducing it.

Remember the waste monuments. Do we really wish to be commemorated by a garbage mausoleum one to four thousand times our bodyweight? Or by 240-tonne carbon balloons, three and a half thousand times our weight?[7]

We can turn lights off when we leave a room, starting now. Lower heating or air conditioning controls. Insulate the home. Walk, cycle or use stairs where we can. Wash hands in cold water. Next time we renew a light bulb or an appliance, make sure it is the most energy efficient we can afford. Choose a car for maximum fuel economy, not maximum power to impress.

Use consumer power to change what manufacturers provide. Eat organic, fresh and unprocessed foods if you can afford them. Eat less red meat. Recycle what you can. Refuse unnecessary packaging. Don't buy canned drinks. Buy recycled goods. Use things till they wear out. Repair anything that can be repaired. Don't trash anything that someone else could make use of. Have no more than two kids.

Try to break the habit of impressing other people through conspicuous consumption. Status should be determined by how little we damage the earth, or how much we enhance it—not by how much we contribute to its spoliation. Excess consumption should become a matter of scorn and shame, not pride.

The list is endless, and we can add to it each day. Every new addition is a new victory. For it is a war—one that will go on until we finally achieve a sustainable balance with our environment.

It is not only the Kalsakas and the Hatias that will suffer if we procrastinate. It will be our own children and grandchildren. The time is near when every child will ask its parent 'What did you do in the environment war, mum and dad? Were you one of those who helped destroy my future? Or were you one of those who helped save it?'

Remember, if your children are still at home, that they will still be alive in the year 2040 or 2050. Remember that your grandchildren will see the world of 2070 or 2080.

The future is really that close. And the future will judge us harshly if we do not change. For we risk being remembered as the generation which, like the Kwakiutl chiefs, sent for more possessions to heap on the potlatch fire while the earth burned. The generation that in the space of only three or four decades closed up the horizons, hemmed in our children's freedom, left them a duller, uglier, poorer world.

The generation that caught Claudius at prayer, and let him go on living.

The time is out of joint. Like Hamlet, we were born to set it right.

Should we succeed, the present age will be seen as the age of the Third Revolution. It will not, any more than the first two, be immediate. It began its infancy no more than two decades ago. In human terms, it is now an inexperienced and impulsive youth. Early adulthood is at least a decade away, maturity twenty years ahead, full fruition more distant still.

It will be known as the age when human numbers, consumption and technologies were shifted into sustainable balance with the environment. When we developed social arrangements to keep us in balance, despite inevitable and permanent change. When the needs of living humans, of future humans and of other species were reconciled.

In all of human history, no generation has ever borne such a responsibility on its shoulders.

Harrison, P., *The Third Revolution; Environment, Population and a Sustainable World,* Tauris, London, 1992, p. 297-305.

Notes and References

1 Lugo, Ariel, 'Estimating Reductions in the Diversity of Tropical Forest Species', and Mooney, Harold, 'Lessons from Mediterranean Climate Regions', Both Wilson, E.O., ed., *Biodiversity*, National Academy Press, Washington DC, 1988.

2 Thomas, Keith, *Man and the Natural World*, Allen Lane, London, 1983, pp. 254-68.

3 Wordsworth, *Prelude*, viii, 620-730; *Tintern Abbey*.

4 Thomas *op. cit.*, p. 266.

5 For a review of attitudes to nature in the world's religions see Regenstein, Lewis, *Replenish the Earth*, SCM Press, London, 1991. Regenstein considerably exaggerates the ecological concern of ancient religions. But he does show that every religion has some traditional material that can be emphasized to support the new attitudes to nature.

6 See Sone, Christopher, *Earth and Other Ethics*, Harper & Row, New York, 1987.

7 See Chapter 13 on waste. The carbon balloon is based on an average life expectancy of 76

Part VI

THE FUTURE

Introduction to Part VI: The Future

What does the future have in store for the environment and what does the environment have in store for the future? Many of the classical studies implicitly or explicitly address the future perspective. They were written, as it were, in the interest of the future. Concern about the future of the planet has prompted researchers to do studies to point out how serious the problem is, who is responsible, what can be done, etc. Today, as well, there are some researchers who are trying to create an image of the long-term future of the planet. At the end of the 20th century, it shouldn't really surprise us that some are starting to wonder what the 21st century will bring us in terms of the environment.

An obligation to think about the future

Writings on the future usually have an ambivalent character. On the one hand, they warn us of the many great dangers ahead that are almost unavoidable. On the other hand, they display an almost unlimited trust in the future, in the possibilities of turning the tide. The central assumption behind most of these studies is: 'we have an obligation to think about the future.' Thinking about the future is not an effort without engagement; nor is it entertainment for fortune tellers who are looking for easy profits. It is a very serious matter for those who want to take responsibility for future generations. Humanity has a moral obligation to think about the future. Not just to think about it, but also to act in a way that could benefit the planet in the long term.

Self destroying prophecy

Many studies of the future have one common trait: the expectations described are hardly ever met. At first, this may be seen as proof of the naivety of the authors of these books, but if we look closer, we can see that there is more to it. These studies are frequently written out of concern for the future and are often a plea to those responsible to act as quickly as possible to stop disasters from occuring. The authors are therefore hoping that their prophecies will prove untrue in that they predict terrible things if something isn't done soon. To turn the tide, they propose action. If these authors are heard, their predictions can be reversed. They are thus systematically destroying their own predictions. Academically, this may be disappointing; socially, it can only be applauded.

Agenda 21

Thinking about a future in which the environment and development go together in a lasting commitment was strongly promoted by the 1992 World Conference of the United Nations in Rio de Janeiro. The countries present, who committed themselves to the *Agenda 21* (1992), had the task of working out a plan of sustainable development: not only in the abstract and verbally, but by tangible means of local action. In this sense, one could say that the future in fact started yesterday. Every day that goes by without action is a lost one. But how do we design sustainable development? What does a sustainable future mean? What do we need to do to attain that goal?

Firstly, we need state regulations. They must curb existing unsustainable aspects of society in the direction of a sustainable society by means of focused programmes and actions. Furthermore, business also needs to get involved; they cannot stay out of the debate on the major world issues of environment and development. Business must take sides. They must actively work for a lasting future and be compelled to do this in order to stay in business. These actions must not be isolated actions by government and business, but joint ventures, co-productions. There is a joint responsibility and therefore the need for joint action.

No surprising decisions were made in Rio de Janeiro. In the phase preceding the conference, there was a lot of lobbying concerning the design texts. Room for new proposals had been diplomatically taken away. The standpoints were known and the parties (G77, G7, US, EC, Japan, etc.) didn't give an inch on their own agendas. The attitude of the United States in particular was heavily criticized because of delaying climatological negotiations and the Treaty on Biodiversity. The attempts of the US to intimidate some European countries into signing a supplementary declaration to actually stabilize CO_2 emissions offended many people. The use of the term 'environmental extremists' by president Bush didn't really go over well either. The non-governmental organizations named the US the most environmentally hostile nation on the planet. The results of the Rio conference—for the short term—are not very hopeful. A radical breakthrough in our approach to the environment and development is unlikely. Yet for the middle-long and long term, there are clear signs of hope. Never before were such complex and, for the future of our planet, such essential issues discussed. Never before were so many leading executives brought together to talk about these topics. Never before had there been such an opportunity to make an inventory of problems and possible solutions, to analyze and prepare them for political decision-making. It was in need of a finishing touch, but that didn't cause the issues to disappear from the political agenda. On the contrary, these issues will be given top priority on the political agenda in the coming decades. Rio has had an important cathartic function. We mustn't see it as either the end, or the beginning, but as an intermediate stage in the long journey ahead.

Ever since the first UN conference on the environment in Stockholm in 1972, a lot has happened in the field of the environment. In the last 20 years, a non-issue has become a hot issue. Even Rio will have its effect. The follow-up is just as important as the preparations. Many subsequent conferences were planned, and some have already been held, e.g. the Geneva conference, which had the goal of detailing the actual implementory measures of the treaties. It has been proposed by representatives of environmentalist organizations that more initiatives be developed with those countries that showed interest in taking further steps towards a sustainable society at the Rio Conference. Certain countries could have leading roles and in due course encourage others. In other words, a variant to the 'two-speed principle' is proposed. One mustn't forget that the Declaration of Human Rights is now approximately 200 years old and that we are still in the process of giving form and content to this declaration every single day. Agenda 21 is only a few years old and therefore cannot be expected to become reality overnight.

Gore: Earth in the Balance

One of the more influential documents concerning the future of the planet is *Earth in the Balance; Forging a New Common Purpose* (1992) by Al Gore. This book points out the major responsibilities of governments in protecting the future of the planet. American Vice-President Al Gore wrote *Earth in the Balance* with the subtitle 'ecology and the human spirit'. The book takes an unblinking look at the dimensions of the environmental crisis. It not only points out the root causes of this crisis, but offers a well-balanced strategy to avoid the destruction of the ecosphere. The book consists of three parts. The first part is a description of the balance of the earth at risk. Part Two concerns the search for balance and Part Three aims at striking the balance. A new common purpose is formulated: protecting the ecological system of which we are a part. We now face the prospect of a kind of global civil war between those who refuse to consider the consequences of civilization's relentless advance and those who refuse to be silent partners in the destruction. The book ends by formulating a global Marshall plan, aimed at stabilizing world population, developing and sharing appropriate technologies, creating a new global economy, introducing a new generation of treaties and agreements and a new global environmental consensus.

As is the case with many programmatic books, there is a shift between intentions and actual policy. When we analyse the US environmental policy in relation to Al Gore's *Earth in the Balance* there are at least some discrepancies. This is not only the case for the USA; one can see this phenomenon in many countries. The importance of Al Gore's study is that it functions as a leading principle for intensifying environmental policy.

Schmidheiny: Changing Course

Government needs total support from business in the field of environmental policy. Perhaps we should make it clearer that business is government's most important partner in guaranteeing sustainable development. Business is becoming more and more aware of this fact; they need a future too. In entrepreneurial circles, many have started to realize that the functioning of future enterprises is strongly dependent on a clean environment. The environment is no longer an external factor without meaning; it has become a crucial external factor. Everyone is dependent on resources. The necessity of a healthy living environment is understood. Environmental measures can prove profitable. Clients ask for clean products and environmentally friendly production processes. People in the neighbourhood demand a guarantee of safety from corporate executives. In short, business cannot be aloof to environmental issues, a view that is now almost commonplace in business life. One exemplary publication is Schmidheiny's *Changing Course; a Global Business Perspective on Development and the Environment* (1992).

The environmental challenge has grown from local pollution to global threats and opportunities. The business challenge has grown likewise—from relatively simple technical fixes and additional costs to a corporate-wide collection of threats and opportunities that are of central importance in separating tomorrow's winners from tomorrow's losers. Corporate leaders—so it is said—must take this into account when designing strategic plans for business and deciding the priorities of their own work. Business will play a vital role in the future health of this planet. As business leaders—Schmidheiny states—we are committed to sustainable development, to meeting the needs of the present without compromising the welfare of future generations. Corporations that achieve greater and greater efficiency while preventing pollution through good housekeeping, materials substitution, cleaner technology, and cleaner products and that strive for more efficient use and recovery of resources can be called 'eco-efficient'. The price of goods and services must increasingly parallel and reflect the environmental costs of their production, use, recycling, and disposal. This is fundamental and is best achieved by a synthesis of economic instruments designed to correct distortions and encourage innovation and continuous improvement, regulatory standards to direct performance, and voluntary initiatives by the private sector.

This is the first time that an important group of business leaders has looked at these issues from a global perspective and reached major agreements on the need for an integrated approach in confronting the challenges of economic development and the environment. This means breaking with business-as-usual mentalities and conventional wisdom that sidelines environmental and human concerns. Changing course is an invitation and a challenge to business leaders to choose the more promising and more rewarding option of participation.

Long-Term Environmental Policy

Attention to the long term is beginning to take form here and there. As usual, such attention is promoted by study committees, panels, advisory committees etc., that concern themselves with these issues. Yet the number of studies for the longer term is limited. There aren't that many committees around the world who are concerned with long-term environmental issues. This is understandable because it is hard enough to make tomorrow's weather forecast, let alone predict the future condition of the planet. Some organizations deserve mention, e.g., the International Institute for Applied System Analysis (IIASA), Laxenburg, Austria. In some Western European countries and in the United States, certain advisory committees are involved with long-term environmental policy. Among them is the Dutch Committee for Long-Term Environmental Policy.

The Dutch Committee for Long-Term Environmental Policy was an advisory board of the Dutch Minister of Housing, Spatial Planning and Environment from 1988-1993. Its task was to formulate well-documented advice on long-term developments that could influence the environment and indicate how environmental policy should respond to such developments. The committee published two important documents, one in Dutch, entitled: *Het Milieu; Denkbeelden voor de 21ste eeuw* (*The environment; Ideas for the 21st Century*, 1990) and one in English *The Environment; Towards a Sustainable Future* (1994). The latter, in essence, exemplifies the search for a new social order, an order in which economic development and environmental protection are considered inextricably interdependent. The need for fundamental societal transformations is underscored and the outlines of a sustainable future are described by giving a sketch of the basic institutions of a sustainable society. The study is based on several assumptions.

a The first is that in our present Western society the first signs of a sustainable future should be recognizable: hence it is worthwhile to look for *the signs of hope*.

b The second assumption is that in order to promote a sustainable future, it is necessary that some basic *transformations* take place: socialization, actor-ization and dialogue-ization concerning environmental problems.

c The third is that these transformations take place within the different levels of our daily life (local, regional, continental, global). Owing to the internationalization of our existence, we emphasize the importance of the process of *internationalization* of environmental policy.

d The fourth assumption is that a sustainable future implies radical changes in the *basic conditions and institutions* of society. But which basic institutions need radical changes?

The committee has selected and analyzed the following basic institutions of society: biosphere, population, international economic cooperation, education, (wo)men's emancipation, political order, science, technology, and spatial structure. At the end of the book, suggestions for achieving a sustainable society are given.

Concluding remarks

Mankind is entering a new stage of development in which a reformulation of the relation between society and the environment is taking place. There are signs of hope that can be seen as the first indications of a new era. The different levels of government, industry, agriculture, consumer organizations, non-governmental organizations, schools and universities, mass media, artists, youth organizations, etc., are mobilizing for a sustainable future. We can see shifts in the environmental consciousness of people, a growing awareness of the importance of environmental problems, an expanding willingness to act for the environment, and the growth of environmentally sound behaviour. We are also seeing more attention given to environmental problems in the mass media, the development of ecologically sound production techniques, the production of ecologically sound products, an agriculture based on the ecological expansion of environmental laws and regulations, university courses on environmental problems, and much more.

We are currently witnessing a process of social mobilization against the deterioration of the environment and the restructuring of society in the direction of an ecological society (Nelissen, 1990). But without a doubt, the signs of despair are more numerous than the signs of hope.

The signs of hope are also indications of the social reconstruction that is taking place. This social reconstruction implies a redefinition and remodelling of all that, up to now, was seen as 'normal' and 'usual'. It was 'normal' to think that growth could last forever, that natural resources could be used without limits and that industrial production existed in the way it did. 'Common' life was no longer common. Grocery shopping by car in the neighbourhood became 'uncommon'. The careless use of energy became 'anti-social'. The emission of environmentally polluting substances became 'deviant behaviour'. The separate collection of domestic waste became 'common'. The disposal of glass in the glass bin became a 'normal' part of the weekly routine. Thinking about the environmental implications of the use of certain products became 'normal', etc. The redefining and remodelling of 'normal' and 'common', and of 'abnormal' and 'deviant' is having an impact on all aspects of life: production, consumption, transport, leisure, housing, gardening etc.

This means that our thinking and action, along with our institutions and organizations, are involved. They are also in need of a redefinition and remodelling of their objectives and modes of conduct. The recognizable signs of hope do not mean that we have already reached the objective of a sustainable future. In that respect, we are only at the beginning; life still has to be permeated with the environmental dimension and considerable efforts have to be made to bring this about. Nevertheless, the social mobilization to stop the deterioration of the environment is on its way, at least in a number of countries. The process must not be seen as a small success that has little to do with the threats the earth is facing, but as a sign of a serious attack on these threats.

References

Adriaanse, A., 1993. *Environmental Policy Performance Indicators*, 's-Gravenhage.

Commissie Lange Termijn Milieubeleid, 1990. *Het Milieu; Denkbeelden voor de 21ste Eeuw*, Zeist.

Committee for Long-Term Environmental Policy, 1993. *The Environment; Towards a Sustainable Future*, Dordrecht.

Glasbergen, P. (ed.), 1994. *Milieubeleid; een Beleidswetenschappelijke Inleiding*, 's-Gravenhage.

Morin, E. and A.B. Kern, 1993. *Terre-Patrie*, Paris.

CHAPTER 30

Agenda 21

UNITED NATIONS CONFERENCE ON ENVIRONMENT AND DEVELOPMENT

Declaration of the United Nations Conference on Environment and Development

Rio de Janeiro, 3-14 June 1992

Reconfirming the Declaration of the United Nations Conference on the Human Environment, Stockholm, 16 June 1972[1], and striving to continue to build on this declaration,

With the goal of establishing a new and just global partnership by creating new forms of cooperation between states, important social organizations and people,

Striving for international agreements, which will honour the interest of every person and will protect the integrity of the global system of environment and development,

Acknowledging that the earth, our home, is, by nature, an all-comprising and coherent entity,

declares:

Principle 1
Human beings are at the centre of concerns for sustainable development. They are entitled to a healthy and productive life in harmony with nature.

Principle 2
States have, in accordance with the Charter of the United Nations and the principles of international law, the sovereign right to exploit their own resources pursuant to their own environmental and developmental policies, and the responsibility to ensure that activities within their jurisdiction or control do not cause damage to the environment of other States or of areas beyond the limits of national jurisdiction.

Principle 3
The right to development must be fulfilled so as to equitably meet developmental and environmental needs of present and future generations.

Principle 4

In order to achieve sustainable development, environmental protection shall constitute an integral part of the development process and cannot be considered in isolation from it.

Principle 5

All States and all people shall cooperate in the essential task of eradicating poverty as an indispensable requirement for sustainable development, in order to decrease the disparities in standards of living and better meet the needs of the majority of the people of the world.

Principle 6

The special situation and needs of developing countries, particularly the least developed and those most environmentally vulnerable, shall be given special priority. International actions in the field of environment and development should also address the interests and needs of all countries.

Principle 7

States shall cooperate in a spirit of global partnership to conserve, protect and restore the health and integrity of the Earth's ecosystem. In view of the different contributions to global environmental degradation, States have common but differentiated responsibilities. The developed countries acknowledge the responsibility that they bear in the international pursuit of sustainable development in view of the pressures their societies place on the global environment and of the technologies and financial resources they command.

Principle 8

To achieve sustainable development and a higher quality of life for all people, States should reduce and eliminate unsustainable patterns of production and consumption and promote appropriate demographic policies.

Principle 9

States should cooperate to strengthen endogenous capacity-building for sustainable development by improving scientific understanding through exchanges of scientific and technological knowledge, and by enhancing the development, adaptation, diffusion and transfer of technologies including new and innovative technologies.

Principle 10

Environmental issues are best handled with the participation of all concerned citizens, at the best relevant level. At the national level, each individual shall have appropriate access to information concerning the environment that is held by public authorities, including information on hazardous material and activities in their communities, and the opportunity to participate in decision-making processes. States shall facilitate and encourage public awareness and participation by making information widely available. Effective access to judicial and administrative proceedings, including redress and remedy, shall be provided.

Principle 11

States shall enact effective environmental legislation. Environmental standards, management objectives and priorities should reflect the environmental and developmental context to which they apply. Standards applied by some countries may be inappropriate and of unwarranted economic and social cost to other countries, in particular developing countries.

Principle 12

States should cooperate to promote a supportive and open international economic system that would lead to economic growth and sustainable development in all countries, to better address the problems of environmental degradation. Trade policy measures for environmental purposes should not constitute a means of arbitrary or unjustifiable discrimination or a disguised restriction on international trade. Unilateral actions to deal with environmental challenges outside the jurisdiction of the importing country should be avoided. Environmental measures addressing transboundary or global environmental problems should, as far as possible, be based on an international consensus.

Principle 13

States shall develop national law regarding liability and compensation for the victims of pollution and other environmental damage. States shall also cooperate in an expeditious and more determined manner to develop further international law regarding liability and compensation for adverse effects of environmental damage caused by activities within their jurisdiction of control of areas beyond their jurisdiction.

Principle 14

States should effectively cooperate to discourage or prevent the relocation and transfer to other States of any activities and substances that cause severe environmental degradation or are found to be harmful to human health.

Principle 15

In order to protect the environment, the precautionary approach shall be widely applied by States according to their capabilities. Where there are threats of serious or irreversible damage, lack of full scientific certainty shall not be used as a reason for postponing cost-effective measures to prevent environmental degradation.

Principle 16
National authorities should endeavour to promote the internationalization of environmental costs and the use of economic instruments, taking into account the approach that the polluter should, in principle, bear the cost of pollution, with due regard to the public interest and without distorting international trade and investment.

Principle 17
Environmental impact assessment, as a national instrument, shall be undertaken for proposed activities that are likely to have a significant adverse impact on the environment and are subject to a decision of a competent national authority.

Principle 18
States shall immediately notify other States of any natural diseases or other emergencies that are likely to produce sudden harmful effects on the environment of those States. Every effort shall be made by the international community to help States so afflicted.

Principle 19
States shall provide prior and timely notification and relevant information to potentially affected States on activities that may have a significant adverse transboundary environmental effect and shall consult with those States at an early stage and in good faith.

Principle 20
Women have a vital role in environmental management and development. Participation is therefore essential to achieve sustainable development.

Principle 21
The creativity, ideals and courage of the youth of the world should be mobilized to forge a global partnership in order to achieve sustainable development and ensure a better future for all.

Principle 22
Indigenous people and their communities and other local communities have a vital role in environmental management and development because of their knowledge and traditional practices. States should recognize and duly support

their identity, culture and interest and enable their effective participation in the achievement of sustainable development.

Principle 23

The environment and natural resources of people under oppression, domination and occupation shall be protected.

Principle 24

Warfare is inherently destructive of sustainable development. States shall therefore respect international law providing protection for the environment in times of armed conflict and cooperate in its future development, as necessary.

Principle 25

Peace, development and environmental protection are interdependent and indivisible.

Principle 26

States shall resolve all their environmental disputes peacefully and by appropriate means in accordance with the Charter of the United Nations.

Principle 27

States and people shall cooperate in good faith and in a spirit of partnership in the fulfilment of the principles embodied in this Declaration and in the further development of international law in the field of sustainable development.

United Nations Conference on Environment and Development, *Agenda 21*, Rio de Janeiro, 1992, p. 6-11.

Note

1   *Report of the United Nations Conference on the Human Environment,* Stockholm, 5-16 June 1972 (United Nations publication, sales No. E.73.II.A.14 and corrigendum), chap. I.

CHAPTER 31

Earth in the Balance

Forging a New Common Purpose

A. GORE

A Global Marshall Plan

Human civilization is now so complex and diverse, so sprawling and massive, that it is difficult to see how we can respond in a coordinated, collective way to the global environmental crisis. But circumstances are forcing just such a response; if we cannot embrace the preservation of the earth as our new organizing principle, the very survival of our civilization will be in doubt.

That much is clear. But how should we proceed. How can we create practical working relationships that bring together people who live in dramatically different circumstances? How can we focus the energies of a disparate group of nations into a sustained effort, lasting many years, that will translate the organizing principle into concrete changes—changes that will affect almost every aspect of our lives together on this planet?

We find it difficult to imagine a realistic basis for hope that the environment can be saved, not only because we still lack widespread agreement on the need for this task, but also because we have never worked together globally on any problem even approaching to this one in degree of difficulty. Even so, we must find a way to join this common cause, because the crisis we face is, in the final analysis, a global problem and can only be solved on a global basis. Merely addressing one dimension or another or trying to implement solutions in only one region of the world or another will, in the end, guarantee frustration, failure, and a weakening of the resolve needed to address the whole of the problem.

While this is true that there are no real precedents for the kind of global response now required, history does provide us with at least one powerful model of cooperative effort: the Marshall Plan. In a brilliant collaboration that was itself unprecedented, several relatively wealthy nations and several relatively poor nations—empowered by a common purpose—joined to reorganize an entire region of the world and change its way of life. The Marshall Plan shows how a large vision can be translated into effective action, and it is worth recalling why the plan was so successful.

Immediately after World War II, Europe was so completely devastated that the resumption of normal economic activity was inconceivable. Then, in the early spring of 1947, the Soviet Union rejected US proposals for aiding the recovery of German industry, convincing General George Marshall and President Harry Truman, among others, that the Soviets hoped to capitalize on the prevailing economic distress—not only in Germany but also in the rest of Europe. After much discussion and study, the United States launched the basis for the Marshall Plan, technically known as the European Recovery Program (ERP).

The commonly held view of the Marshall Plan is that it was a bold strategy for helping the nations of Western Europe rebuild and grow strong enough to fend off the spread of communism. That popular view is correct—as far as it goes. But the historians emphasize the strategic nature of the plan, with its emphasis on the structural causes of Europe's inability to lift itself out of its economic, political and social distress. The plan concentrated on fixing the bottlenecks—such as the damaged infrastructure, flooded coal mines, and senseless trade barriers—that were impeding the potential for growth in each nation's economy. ERP was sufficiently long-term that it could serve as an overall effort to produce fundamental structural reorientation, not just offer more emergency relief or another "development" program. It was consciously designed to change the dynamic of the systems to which it extended aid, thus facilitating the emergence of a healthy economic pattern. And it was brilliantly administered by Averell Harriman.

Historians also note the Marshall Plan's regional focus and its incentives to promote European integration and joint action. Indeed, from the very beginning the plan tried to facilitate the emergence of a larger political framework—unified Europe; to that end, it insisted that every action be coordinated with all the countries in the region. The recent creation of a unified European parliament and the dramatic steps toward a European political community to accompany the European Economic Community (EEC) have all come about in large part because of the groundwork of the Marshall Plan.

But when it was put in place, the idea of a unified Europe seemed even less likely than the collapse of the Berlin Wall did only a few years ago—and every bit as improbable as a unified global response to the environmental crisis seems today. Improbable or not, something like the Marshall Plan—a Global Marshall Plan, if you will—is now urgently needed. The scope and complexity of this plan will far exceed those of the original; what's required now is a plan that combines large-scale, long-term, carefully targeted financial aid to developing nations, massive efforts to design and then transfer to poor nations the technologies needed for sustained economic progress, a worldwide program to stabilize world population, and binding commitments by the industrial nations to accelerate their own transition to an environmentally responsible pattern of life.

But despite the fundamental differences between the late 1940s and today, the model of the Marshall Plan can be of great help as we begin to grapple with enormous challenge we now face. For example, a Global Marshall Plan must, like the original, focus on strategic goals and emphasize actions and programs that are likely to remove the bottlenecks presently inhibiting the healthy functioning of the global economy. The new global economy must be an inclusive system that does not leave entire regions behind—as our present system leaves out most of Africa and much of Latin America. In an inclusive economy, for instance, wealthy nations can no longer insist that Third World countries pay huge sums of interest in old debts even when the sacrifices necessary to pay them increase the pressure on their suffering populations so much that revolutionary tensions build uncontrollably. The Marshall Plan took the broadest possible view of Europe's problems and developed strategies to serve human needs and promote sustained economic progress; we must now do the same on a global scale.

But strategic thinking is useless without consensus, and here again the Marshall Plan is instructive. Historians remind us that it would have failed if the countries receiving assistance had not shared a common ideological outlook, or at least a common leaning towards a set of similar ideas and values. Postwar Europe's strong preference for democracy and capitalism made the regional integration of economies possible; likewise, the entire world is far closer to a consensus on basic political and economic principles than it was even a few short years ago, and as the philosophical victory of Western principles becomes increasingly apparent, a Global Marshall Plan will be increasingly feasible.

It is fair to say that in recent years most of the world has made three important choices: first, that democracy will be the preferred form of political organization on this planet; second, that modified free markets will be the preferred form of economic organization; and, third, that most individuals now feel themselves to be part of a truly global civilization—prematurely heralded many times in this century but now finally palpable in the minds and hearts of human beings throughout the earth. Even those nations that still officially oppose democracy and capitalism—such as China—seem to be slowly headed in our philosophical direction, at least in the thinking of younger generations not yet in power.

Another motivation for the Marshall Plan was a keen awareness of the dangerous vacuum created by the end of the Axis nations' totalitarian order and the potential for chaos in the absence of any positive momentum toward democracy and capitalism. Similarly, the resounding philosophical defeat of communism (in which the Marshall Plan itself played a significant role) has left an ideological vacuum that invites either a bold and visionary strategy to facilitate the emergence of democratic government and modified free markets throughout the world—in a truly global system—or growing chaos of the kind that is already too common from Cambodia to Colombia, Liberia to Lebanon, and Zaire to Azerbaijan.

The Marshall Plan, however, depended in part for its success on some special circumstances that prevailed in postwar Europe yet do not prevail in various parts of the world today. For example, the nations of Europe had developed advanced economies before World War II, and they retained a large number of skilled workers, raw materials, and the shared experience of modernity. They also shared a clear potential for regional cooperation—although it may be clearer in retrospect than it was at the time, when the prospect of warm relations between, say, Germany and England seemed remote.

In contrast, the diversity among nations involved in a Global Marshall Plan is simply fantastic, with all kinds of political entities representing radically different stages of economic and political development—and with the emergence of "post-national" entities, such as Kurdistan, the Balkans, Eritrea, and Kashmir. In fact, some people now define themselves in terms of an ecological criterion rather than a political subdivision. For example, "the Aral Sea region" defines people in parts of several Soviet republics who all suffer the regional ecological catastrophe of the Aral Sea. "Amazonia" is used by peoples of several nationalities in the world's largest rain forest, where national boundaries are often invisible and irrelevant.

The diversity of the world's nations and peoples vastly complicates the model used so successfully in Europe. Even so, another of the Marshall Plan's lessons can still be applied: within this diversity, the plans for catalyzing a transition to a sustainable society should be made with regional groupings in mind and with distinctive strategies for each region. Eastern Europe, for example, has a set of regional characteristics very different from those of the Sahel in sub-Saharian Africa, just as Central America faces challenges very different from those facing, say, the Southeast Asian archipelago.

Many of the impediments to progress lie in the industrial world. Indeed, one of the biggest obstacles to a Global Marshall Plan is the requirement that the advanced economies must undergo a profound transformation themselves. The Marshall Plan placed the burden of change and transition only on the recipient nations. The financing was borne entirely by the United States, which, to be sure, underwent a great deal of change during those same years, but not at the behest of a foreign power and not to discharge any sense of obligation imposed by an international agreement.

The new plan will require the wealthy nations to allocate money for transferring environmentally helpful technologies to the Third World and to help impoverished nations to achieve a stable population and a new pattern of sustainable economic progress. To work, however, any such effort will also require wealthy nations to make a transition themselves that will be in some ways more wrenching than that of the Third World, simply because powerful established patterns will be disrupted. Opposition to change is therefore strong, but this transition can and must occur—both in the developed and developing world.

And when it does, it will likely be within a framework of global agreements that obligate all nations to act in concert. To succeed, these agreements must be part of an overall design focused on devising a healthier and more balanced pattern in world civilization that integrates the Third World into the global economy. Just as important, the developed nations must be willing to lead by example; otherwise, the Third World is not likely to consider making the required changes—even in return for substantial assistance. Finally, just as the Marshall Plan scrupulously respected the sovereignty of each nation while requiring all of them to work together, this new plan must emphasize cooperation—in the different regions of the world and globally—while carefully respecting the integrity of individual nation-states.

This point is worth special emphasis. The mere mention of any plan that contemplates worldwide cooperation creates instant concern on the part of many—especially conservatives—who have long equated such language with the advocacy of some supranational authority, like a world government. Indeed, some who favor a common global effort tend to assume that a supranational authority of some sort is inevitable. But this notion is both politically impossible and practically unworkable. The political problem is obvious: the idea arouses so much opposition that further debate on the underlying goals comes to a halt—especially in the United States, where we are fiercely protective of our individual freedoms. The fear that our rights might be jeopardized by the delegation of even partial sovereignty to some global authority ensures that it's simply not going to happen. The practical problem can be illustrated with a question: What conceivable system of world governance would be able to compel individual nations to adopt environmentally sound policies? The administrative problems would be gargantuan, not least because the inefficiency of governance often seems to increase geometrically with the distance between the seat of power and the individuals affected by it; and given the chaotic state of some of the governments that would be subject to that global entity, any such institution would most likely have unintended side effects and complications that would interfere with the underlying goal. As Dorothy Parker once said about a book she didn't like, the idea of a world government "should not be tossed aside lightly; it should be thrown with great force".

But if world government is neither feasible nor desirable, how then can we establish a successful cooperative global effort to save the environment? There is only one answer: we must negotiate international agreements that establish global constraints on acceptable behavior but that are entered into voluntarily—albeit with the understanding that they will contain both incentives and legally valid penalties for noncompliance.

The world's most important supranational organization—the United Nations—does have a role to play, though I am skeptical about its ability to do very much. Specifically, to help monitor the evolution of a global agreement, the United Nations might consider the idea of establishing a Stewardship Council to deal with matters relating to the global environment—just as the Security Council now deals with matters of war and peace. Such a forum could be increasingly useful and even necessary as the full extent of the environmental crisis unfolds.

Similarly, it would be wise to establish a tradition of annual environmental summit meetings, similar to the annual economic summits of today, which only rarely find time to consider the environment. The preliminary discussions of a Global Marshall Plan would, in any event, have to take place at the highest level. And, unlike the economic summits, these discussions must involve heads of state from both the developed and developing world.

In any global agreement of the kind I am proposing, the single most difficult relationship is the one between wealthy and poor nations; there must be a careful balance between the burdens and obligations imposed on both groups of nations. If, for example, any single agreement has a greater impact on the poor nations, it may have to be balanced with a simultaneous agreement that has a greater impact on the wealthy nations. This approach is already developing naturally in some of the early discussions of global environmental problems. One instance is the implicit linkage between the negotiations to save the rain forests—which are found mostly in poor countries—and the negotiations to reduce greenhouse gas emissions—which is especially difficult for wealthy nations. If these negotiations are successful, the resulting agreements will become trade-offs for each other.

The design of a Global Marshall Plan must also recognize that many countries are in different stages of development, and each new agreement has to be sensitive to the gulf between the countries involved, not only in terms of their relative affluence but also their various stages of political, cultural, and economic development. This diversity is important both among those nations that would be on the receiving end of a global plan and among those expected to be on the giving end. Coordination and agreement among the donor countries might, for example, turn out to be the most difficult challenge. The two donor participants in the Marshall Plan, the United States and Great Britain, had established a remarkably close working relationship during the war, which was then used as a model for their postwar collaboration. Today, of course, the United States cannot conceivably be the principal financier for a global recovery program and cannot make the key decisions alone or with only one close ally. The financial resources must now come from Japan and Europe and from wealthy, oil-producing states.

The Western alliance has frequently been unwieldy and unproductive when large sums of money are at stake. Nevertheless, it has compiled an impressive record of military, economic, and political cooperation in the long struggle against communism, and the world may be able to draw upon that model just as the United States and Great Britain built upon their wartime cooperation in implementing the Marshall Plan. Ironically, the collapse of communism has deprived the alliance of its common enemy, but the potential freeing up of resources may create the ideal opportunity to choose a new grand purpose for working together.

Still, a number of serious obstacles face cooperation among even the great powers—the United States, Japan, and Europe—before a Global Marshall Plan could be considered. Japan, in spite of its enormous economic strength, has been reluctant to share the responsibility for world political leadership and thus far seems blind to the need for it to play such a role. And Europe wil be absorbed for many years in the intricacies of becoming a unified entity—a challenge further complicated by the entreaties of the suddenly free nations in Eastern Europe that now want to join the EEC.

As a result, the responsibility for taking the initiative, for innovating, catalizing, and leading such an effort, falls disproportionately on the United States. Yet in the early 1990s our instinct for world leadership often seems not nearly so bold as it was in the late 1940s. The bitter experience of the Vietnam War is partly responsible, and the sheer weariness of carrying the burden of world leadership has taken a toll. Furthermore, we are not nearly as dominant in the world economy as we were then, and that necessarily has implications for our willingness to shoulder large burdens. And our budget deficits are now so large as to stifle our willingness to consider even the most urgent of tasks. Charles Maier points out that the annual US expenditures for the Marshall Plan between 1948 and 1951 were close to 2 per cent of our GNP. A similar percentage today would be almost $100 billion a year (compared to our total nonmilitary foreign aid budget of about $15 billion a year).

Yet the Marshall Plan enjoyed strong bipartisan support in Congress. There was little doubt then that government intervention, far from harming the free enterprise system in Europe, was the most effective way to foster its healthy operation. But our present leaders seem to fear almost any form of intervention. Indeed, the deepest source of their reluctance to provide leadership in creating an effective environmental strategy seems to be their fear that if we do step forward, we will inevitably be forced to lead by example and actively pursue changes that might interfere with their preferred brand of laissez-faire, nonassertive economic policy.

Nor do our leaders seem willing to look as far into the future as did Truman and Marshall. In that heady postwar period, one of Marshall's former colleagues, General Omar Bradley, said, "it is time we steered by the stars, not by the lights of each passing ship". This certainly seems to be another time when that kind of navigation is needed, yet too many of those who are responsible for our future appear to be distracted by such "lights of passing ships" as overnight public opinion polls.

In any effort to conceive of a plan to heal the global environment, the essence of realism is recognizing that public attitudes are still changing—and that proposals which are today considered too bold to be politically feasible will soon be derided as woefully inadequate to the task at hand. Yet while public acceptance of the magnitude of the threat is indeed curving upward—and will eventually rise almost vertically as awareness of the awful truth suddenly makes the search for remedies an all-consuming passion—it is just as important to recognize that at the present time, we are still in a period when the curve is just starting to bend. Ironically, at this stage, the maximum that is politically feasible still falls short of the minimum that is truly effective. And to make matters worse, the curve of political feasibility in advanced countries, where the immediate threats to well-being and survival often make saving the environment seem to be an unaffordable luxury.

It seems to make sense, therefore, to put in place a policy framework that will be ready to accommodate the worldwide demands for action when the magnitude of the threat becomes clear. And it is also essential to offer strong measures that are politically feasible now—even before the expected large shift in public opinion about the global environment—and that can be quickly scaled up as awareness of the crisis grows and even stronger action becomes possible.

With the original Marshall Plan serving as both a model and an inspiration, we can now begin to chart a course of action. The world's effort to save the environment must be organized around strategic goals that simultaneously represent the most important changes and allow us to recognize, measure, and asses our progress toward making those changes. Each goal must be supported by a set of policies that will enable world civilization to reach it as quickly, efficiently, and justly possible.

In my view, five strategic goals must direct and inform our effort to save the global environment. Let me outline each of them briefly before considering each in depth.

The first strategic goal should be *the stabilizing of world population*, with policies designed to create in every nation of the world the conditions necessary for the so-called demographic transition—the historic and well-documented change from a dynamic equilibrium of high birth rates and death rates to a stable equilibrium of low birth rates and death rates. This change has taken place in most of the industrial nations (which have low rates of infant mortality and high rates of literacy and education) and in virtually non of the developing nations (where the reverse is true).

The second strategic goal should be *the rapid creation and development of environmentally appropriate technologies*—especially in the fields of energy, transportation, agriculture, building construction, and manufacturing—capable of accommodating sustainable economic progress without the concurrent degradation of the environment. These new technologies must then be quickly transferred to all nations—especially those in the Third World, which should be allowed to pay for them by discharging the various obligations they incur as participants in the Global Marshall Plan.

The third strategic goal should be *a comprehensive and ubiquitous change in the economic "rules of the road" by which we measure the impact of our decisions on the environment*. We must establish—by global agreement—a system of economic accounting that assigns appropriate values to the ecological consequences of both routine choices in the marketplace by individuals and companies and larger, macroeconomic choices by nations.

The fourth strategic goal should be *the negotiation and approval of a new generation of international agreements* that will embody the regulatory frameworks, specific prohibitions, enforcement mechanisms, cooperative planning, sharing arrangements, incentives, penalties, and mutual obligations necessary to make the overall plan a success. These agreements must be especially sensitive to the vast differences of capability and need between developed and undeveloped nations.

The fifth strategic goal should be *the establishment of a cooperative plan for educating the world's citizens about our global environment*—first by the establishment of a comprehensive program for researching and monitoring the changes now under way in the environment in a manner that involves the people of all nations, especially students; and, second, through a massive effort to disseminate information about local, regional, and strategic threats to the environment. The ultimate goal of this effort would be to foster new patterns of thinking about the relationship of civilization to the global environment.

Each of these goals is closely related to all of the others, and all should be pursued simultaneously within the larger framework of the Global Marshall Plan. Finally, the plan should have as its more general, integrating goal *the establishment, especially in the developing world—of the social and political conditions most conducive to the emergence of sustainable societies*—such as social justice (including equitable patterns of land ownership); a commitment to human rights; adequate nutrition, health care, and shelter; high literacy rates; and greater political freedom, participation, and accountability. Of course, all specific policies should be chosen as part of serving the central organizing principle of saving the global environment.

Gore, A., *Earth in the Balance; Forging a New Common Purpose*, Earthscan Publications Ltd, London, 1992, p. 295-307.

CHAPTER 32

Changing Course

A Global Business Perspective on Development and the Environment

S. SCHMIDHEINY

The Business of Sustainable Development

"With greater freedom for the market comes greater responsibility".
  Gro Harlem Brundtland
  Prime Minister, Norway

Running a company requires daily assessments of opportunities, risks, and trends. Corporate leaders who ignore economic, political, or social changes will lead their companies toward failure. So too will those who overreact to change and perceived risk.

Many global trends offer hope. Life expectancy, health care, and education have all improved dramatically in the second half of this century. World food production has stayed well ahead of population growth. Average per capita incomes have increased by the highest sustained rates ever. No shortage of raw materials looms in the foreseeable future. Given the right technology, the planet's soils can supply more than the basic food needs of much larger populations.

But neither business nor any other leaders can afford to see only the positive, especially when these optimistic signs are based on averages that mask alarming departures from the norm. Several other linked global trends demand any thinking person's attention. Each is replete with scientific uncertainty. To overreact to any of these would be dangerous, but to ignore any of them would be irresponsible.

First, the human population is growing extremely rapidly, according to the most optimistic estimates, an already crowded planet is likely to have to support twice as many people next century. Environmental ills have varying causes, but all are made worse by the pressure of human numbers.

Second, the last few decades have witnessed an accelerating consumption of natural resources—consumption that is often inefficient and ill planned. Resources that biologists call renewable are not being given time to renew. The bottom line is that the human species is living more off the planet's capital and less off its interest. This is bad business.

Third, both population growth and the wasteful consumption of the resources play a role in the accelerating degradation of many parts of the environment. Productive areas are hardest to hit. Agriculturally fruitful drylands are turning into desert; forests into poor pastures; freshwater wetlands into salty, dead soils, rich coral reefs into lifeless stretches of ocean.

Fourth, as ecosystems are degraded, the biological diversity and genetic resources they contain are lost. Many environmental trends are reversible; this loss is permanent.

Fifth, this overuse and misuse of resources is accompanied by the pollution of atmosphere, water, and soil—often with substances that persist for long periods. With a growing number of sources and forms of pollution, this process also appears to be accelerating. The most complex and potentially serious of these threats is a change in climate and in the stability of air circulation systems.

There are also alarming trends apparent in patterns of "development"; these, too, begin with population projections. More than 90 per cent of population growth takes place in the developing world—that is, in poorer countries.[1] This means that when the present world population of more than 5 billion doubles next century, there will be an extra 4.5 billion people in nations where today it is hardest to secure jobs, food, safe homes, education and health care.

Already, population growth means that the number of people who belong to the underclass—those unable to secure such basics of life as adequate food, shelter, clothing, health care, and education—is rising yearly in much of the developing world. In the mid-1980s more than a billion people on the planet, almost a third of the population of the developing world, were trying to survive on an income equivalent to about $1 per day.[2]

Poverty, rapid population growth, and the deterioration of natural resources often occur in the same regions, creating a huge imbalance between the quarter of the planet living in rich, industrial nations and the three quarters residing in developing nations. The national income of Japan's 120 million people is about to overtake the combined incomes of the 3.8 billion people in the developing world.[3] The industrial nations have generally cut their aid to the developing world. And because of debt servicing and repayment and reduced foreign investment, total capital flows were reversed in the second half of the 1980s, with money flowing from poor to rich.

These two sets of alarming trends—environment and development—cannot be separated. Economic growth in most of the developing world will depend for some time on agricultural production, so a reduction in the productivity in ecosystems tends to mean declining farm production and loss of revenue.

The resulting global structural challenge was concisely summed up by

Maurice Strong, Secretary General of the UN Conference on Environment and Development in Brazil in June 1992: "The gross imbalances that have been created by concentration of economic growth in the industrial countries and population growth in developing countries is at the centre of the current dilemma. Redressing these imbalances will be the key to the future security of our planet—in environmental and economic as well as traditional security terms. This will require fundamental changes both in our economic behavior and our international relations."[4]

Clearly action is required. But which actions, and when, given the huge uncertainties involved? This is the sort of issue that business copes with daily. Corporate leaders are used to examining uncertain, negative trends, making decisions, and then taking action, adjusting, and incurring costs to prevent damage. Insurance is just one example. There are costs involved, but these are costs the rational are willing to bear and costs the responsible do not regret, even if things turn out not to have been as bad as they once seemed. We can hope for the best, but the "precautionary principle" remains the best practice in business as well as in other aspects of life.

This principle was agreed to at the World Industry Conference on Environmental Management in 1984 and at the 1989 Paris Summit of the leaders of the seven richest industrial nations (the G7). It was strengthened in the Ministerial Declaration of the 1990 UN Economic Commission for Europe meeting in Bergen: "In order to achieve sustainable development, policies must be based on the precautionary principle. Environmental measures must anticipate, prevent and attack the causes of environmental degradation. Where there are threats of serious or irreversible damage, lack of full scientific certainty should not be used as a reason for postponing measures to prevent environmental degradation."[5]

Yet risk and uncertainty are usually accompanied by new opportunities, and business has long been adept at seizing such chances. For example, using energy more efficiently, and thus reducing global and local pollution, decreases costs and increases competitiveness. Many of what politicians and economists call "no regrets policies"—actions that make sense no matter what the real threat of global warming—are from a business point of view opportunities and good investments; examples including providing energy efficiency and developing new energy sources, new drought-resistant crops, and new resource management techniques.

There will be many opportunities for business. Having researched various aspects of competitive advantage among nations, Harvard Business School Professor Michael Porter reported: "I found that the nations with the most rigorous [environmental standards] requirements often lead in exports of affected products . . . The strongest proof that environmental protection does not hamper competitiveness is the economic performance of nations with the strictest laws." He mentions the successes of Japan and Germany, as well as that of the United States in sectors actually subject to the greatest environmental costs: chemicals, plastics, and paints.[6]

Viewing environmental threats from a business perspective can help guide both governments and companies toward plausible policies that offer protection from disaster while making the best of the challenges.

This book is about the steps required of business, and of the governments that set the frameworks for industry, to ensure that humans and all other species continue to occupy a safe and bountiful planet.

Sustainable Development

During the first great wave of environmental concern in the late 1960s and early 1970s, most of the problems seemed local: the products of individual pipes and smokestacks. The answers appeared to lie in regulating these pollution sources.

When the environment reemerged on the political agenda in the 1980s, the main concerns had become international: acid rain, depletion of the ozone layer, and global warming. Analysts sought causes not in pipes and stacks but in the nature of human activities. One report after another concluded that much of what we do, many of our attempts to make "progress", are simply unsustainable. We cannot continue in our present methods of using energy, managing forests, farming, protecting plant and animal species, managing urban growth, and producing industrial goods. We certainly cannot continue to reproduce our own species at the present rate.

"This is the moment of truth for Western Europe and the industrialized world. Will we prove to be strong enough and refrain from part of our own consumption in order to secure a peaceful and democratic development in Eastern Europe? Will we be able to give hope to all the poor, who for so long have been oppressed by an inhuman system and denied economic development as well as an acceptable environment?"

Percy Barnevik
Chief Executive Officer
ABB Asea Brown Boveri Ltd.

Energy provides a striking example of present unsustainability. Most energy today is produced from fossil fuels: coal, oil, and gas. In the mid 1980s, the world was burning the equivalent of 10 billion metric tons of coal per year, with people in industrial nations using much more than those of the developing world. At these rates, by 2025 the expected global population of more than 8 billion would be using the equivalent of 14 billion metric tons of coal. But if all the world used energy at industrial country levels, by 2025 the equivalent of 55 billion metric tons would be burned. Present levels of fossil fuel use may be warming the globe; a more than fivefold increase is unthinkable. Fossil fuels must be used more efficiently while alternatives are being developed if economic development is to be achieved without radically changing the global climate.[7]

Given such widespread evidence of unsustainability, it is not surprising that the concept of "sustainable development" has come to dominate the environment/development debate. In 1987, the World Commission on Environment and Development, appointed three years earlier by the UN General Assembly and headed by Norwegian Prime Minister Grow Harlem Brundtland, made sustainable development the theme of its entire report, *Our Common Future*. It defined the concept simply as a form of development or progress that "meets the needs of the present without comprising the ability of future generations to meet their own needs."[8]

The phrase itself can be misleading, as the word development might suggest that it is a chore for "developing" nations only. But development is more than growth, or quantitative change. It is primarily a change in quality. More than a decade ago, the influential World Conservation Strategy, compiled by the United Nations and organizations representing governments and private bodies, defined development as "the modification of the biosphere and the application of human, financial living and non-living resources to satisfy human needs and improve the quality of human life.[9]

Thus all nations are, or would wish to be, developing. And sustainable development will require the greatest changes in the wealthiest nations, which consume the most resources, release the most pollution, and have the greatest capacity to make the necessary changes. These nations must also respond to the criticism from many leaders in the poor parts of the world that industrial countries risk reversing the relationship between production and the satisfaction of needs. They charge that increase production in wealthy nations no longer serves primarily to satisfy needs; rather, the creation of needs serves to increase production.

The idea that much of what humanity does in name of progress is unsustainable and must be changed has gained rapid acceptance. In 1987, the UN General Assembly passed a resolution adopting the World Commission's report as a guide for future UN operations, and commending it to governments. Since then, many governments have tried to bend their policies to its recommendations. The July 1989 G7 summit called for "the early adoption, worldwide, of policies based on sustainable development".[10]

Business has also taken up the challenge, at the international, national and sectoral levels. The International Chamber of Commerce drafted a "Business Charter for Sustainable Development", which was launched in April 1991 at the Second World Industry Conference on Environmental Management. The Charter, endorsed by 600 firms worldwide by early 1992, encourages companies to "commit themselves to improving their environmental performance in accordance with these [the Charter's] 16 Principles, to having in place management practices to effect such improvement, to measuring their progress, and to reporting this progress as appropriate internally and externally".[11]

The senior business group in Japan, the Keidanren, adopted an Environmental Charter in 1991 that sets out codes of behavior toward the environment.[12] Malaysia has established a corporate environmental policy that calls on companies "to give benefit to society; this entails . . . that any adverse effects on the environment are reduced to a practicable minimum".[13] India's Confederation of Engineering Industry has also urged an "Environment Code for Industry" upon its members.[14]

Chemical industry associations in several countries have agreed to a Responsible Care program to promote continuous improvement in environmental health and safety. Begun in Canada and taken up in the United States, Australia, and many European countries, the scheme encourages associations to draft codes of conduct in many areas of operations, and it recommends that large companies help smaller ones with environmental and safety improvements.[15]

Sustainable development will obviously require more than pollution prevention and tinkering with environmental regulations. Given that ordinary people—consumers, business people, farmers—are the real day-to-day environmental decision makers, it requires political and economic systems based on the effective participation of all members of society in decision making. It requires that environmental considerations become a part of the decision-making process of all government agencies, all business enterprises, and in fact all people. It requires levels to and enforcing treaties to manage global commons such as the atmosphere and oceans. It requires, beyond immediate environmental concerns, an end to the "arms culture" as a method of achieving security, and new definitions of security that include environmental threats.

Recently the nations of the world seem to have begun to move, albeit slowly, in these directions. Environmental concern has gradually begun to infuse all areas of decision making. Democracy has become a more prevalent form of government throughout the developing world, Eastern Europe, and the former Soviet Union. The Montreal protocol on the ozone layer and an emerging treaty on the atmosphere suggest that nations may be able to cooperate along the harder paths toward a cleaner global environment. Definitions of security are changing, and the end of the cold war may free resources for work on environmental security.

Will these changes last, and are they happening fast enough?

The Growth Controversy

Perhaps the most controversial conclusion of the World Commission was that sustainable development requires rapid economic growth. This assumption is based on the reality that growing populations and poor populations require goods and services to meet essential needs and that "meeting essential needs depends in part on achieving full growth potential, and sustainable development clearly requires economic growth in places where such needs are not being met. Elsewhere, it can be consistent with economic growth, provided the content of growth reflects the broad principles of sustainability and non-exploitation of others."[16]

There are critics who argue that the limits to economic growth have already been reached and there must be "zero growth" from now on. Some of them do not explain how zero growth will meet the needs of a planet with more than 10 billion people. But other who feel that the environmental limits to growth have been reached, such as Robert Goodland and Herman Daly of the World Bank's Environment Department, argue that "development by the rich must be used to free resources . . . for growth and development so urgently needed by the poor. Large-scale transfers to the poorer countries also will be required".[17]

The World Commission itself noted that growth alone is not enough, as high levels of productivity and widespread poverty can coexist and can endanger the environment. So sustainable development requires societies to meet human needs both by increasing environmentally sustainable production and "by ensuring equitable opportunities for all".[18]

The Commissioners argued that growth was limited at present by both the nature of technologies and the nature of social organization. For example, humanity will eventually be using much more energy than today, but an increasing part will have to come from sources other than fossil fuels. And if societies remain organized so that many people remain impoverished despite sustained economic growth, then this poverty will both degrade environmental resources and eventually act as a brake on growth.

The Business Challenge

The requirement for clean, equitable economic growth remains the biggest single difficulty within the larger challenge of sustainable development.

Proving that such growth is possible is certainly the greatest test for business and industry, which must devise strategies to maximize added value while minimizing resource and energy use. Given the large technological and productive capacity of business, any progress toward sustainable development requires its active leadership.

Open, prospering markets are a powerful force for creating equity of opportunity among nations and people. Yet for there to be equal opportunity, there must first be opportunity itself. Open, competitive markets create the most opportunities for the most people. It is often the nations where markets most closely approach the ideal of "free", open, and competitive that have the least poverty and the greatest opportunity to escape from that poverty.

The World Commission listed as the first prerequisite for sustainable development a political system in which people can effectively participate in decision making. But freedom to participate in political decisions and freedom to participate in markets are inseparable over the long run. The citizens of Central Europe, having achieved political freedom, are now building market freedom. The Asian nations that achieved thriving market economies under authoritarian regimes are now moving toward more democratic governments.

Yet no market can be called "free" in which the decisions of a few can cause misuse of resources and pollution that threaten the present and future of the many. Today, for instance, the earth's atmosphere is providing the valuable service of acting as a dump for pollutants; those enjoying this service rarely pay a reasonable price for it.

"Eco-efficiency"

The present limits to growth are not so much those imposed by resources, such as oil and other minerals, as was argued by the 1972 Club of Rome report *The Limits to Growth*.[19] In many cases they arise more from a scarcity of "sinks", or systems that can safely absorb wastes. The atmosphere, many bodies of water, and large areas of soil are reaching their own absorptive limits as regards wastes of all kinds.

"We believe a business cannot continue to exist without the trust and respect of society for its environmental performance."
  Shinroku Morohashi
  President Mitsubishi Corporation

Business has begun to respond to this truth. It is moving from a position of limiting pollution and cleaning up waste to comply with government regulations toward one of avoiding pollution and waste both in the interest of corporate citizenship and of being more efficient and competitive. The economies of the industrial countries have grown while the resources and energy needed to produce each unit of growth have declined. Chemical companies in industrial nations have doubled output since 1970 while more than halving energy consumption per unit of production.[20]

Industry is moving toward "demanufacturing" and "remanufacturing"—that is, recycling the material in their products and thus limiting the use of raw materials and of energy to convert those raw materials. (see chapter 7.) That this is technically feasible is encouraging; that it can be done profitably is more encouraging. It is the more competitive and successful companies that are the forefront of what we call "eco-efficiency".

But eco-efficiency is not achieved by technological change alone. It is achieved by profound changes in the goals and assumptions that drive corporate activities, and change in the daily practices and tools used to reach them. This means a break with business-as-usual mentalities and conventional wisdom that sidelines environmental and human concerns.

A growing number of leading companies are adopting and publicly committing themselves to sustainable development strategies. They are expanding their concepts of who has a stake in their operations beyond employees and stockholders to include neighbors, public interest groups (including environmental organizations), customers, suppliers, governments, and the general public. They are communicating more openly with these new stakeholders. They are coming to realize that "the degree to which a company is viewed as being a positive or negative participant in solving sustainability issues will determine, to a very great degree, their long-term business viability", in the words of Ben Woodhouse, director of Global Environmental Issues at Dow Chemical.

The Challenge of Time

As the World Commission noted, sustainable development requires forms of progress that meet the needs of the present without compromising the ability of future generations to meet their own needs. In the late twentieth century, we are failing in the first clause of that definition by not meeting the basic needs of more than 1 billion people. We have not even begun to come to grips with the second clause: the needs of future generations. Some argue that we have no responsibility for the future, as we cannot know its needs. This is partly true. But it takes no great leap of reason to assume that our offspring will require breathable air, drinkable water, productive soils and oceans, a predictable climate, and abundant plant and animal species on the planet they will share.

Yet it is a hard thing to demand of political leaders, especially those who rely on the votes of the living to achieve and remain in high office, that they ask those alive today to bear costs for the sake of those not yet born, and not yet voting. It is equally hard to ask anyone in business, providing goods and services to the living, to change their ways for the sake of those not yet born, and not yet acting in the marketplace. The painful truth is that the present is a relatively comfortable place for those who have reached positions of mainstream political or business leadership.

This is the crux of the problem of sustainable development, and perhaps the main reason why there has been great acceptance of it in principle, but less concrete actions to put it into practice: many of those with the power to effect the necessary changes have the least motivation to alter the status quo that gave them the power.

When politicians, industrialists, and environmentalists run out of practical advice, they often take refuge in appeals for a new vision, new values, a new commitment, and a new ethic. Such calls often ring hollow and rhetorical. But given that sustainable development requires a practical concern for the needs of people in the future, then it does ultimately require a new shared vision and a collective ethic based on equality of opportunity not only among people and nations, but also between this generation and those to come. Sustainability will require new technology, new approaches to trade to spread the technology and the goods necessary for survival, and new ways of meeting needs through markets. Business leadership will be required, and expected, in all these areas.

However, sustainable development will ultimately be achieved only through cooperation among people and all their various organizations, including businesses and governments. And leaders elected to decision-making and executive offices retain a fundamental obligation to inform and educate their constituencies about the urgent necessity and the reasons for changing course.

We believe that the best aspects of the human propensity to buy, sell, and produce can be an engine of change. Business has helped to create much of what is valuable in the world today. It will play its part in ensuring the planet's future.

Shaping the Future

The inevitable process of change towards sustainable forms of development will determine the future course of human civilization and shape our life-styles and thereby the way we do business. Yet many business leaders have so far been relatively passive in dealing with these issues.

Perhaps this ultimately has to do with the way we each react to the dimension of time. Those who have little interest in the future of nature and humanity will have little regard for the sustainability of their actions and will not be concerned to grapple with and understand the challenges facing us all. Those who do care about society and its progress have learned to understand that business never operates in a vacuum. It interacts at many levels with society, and society is now entering a period of rapid and fundamental changes.

Business has developed remarkable skills in market intelligence to spot and to certain extent predict changing demand patterns. It must also construct a system of "social intelligence" to spot, understand, and interpret signals of change in development patterns. Those who are quickest to receive and act on such signals will have a great advantage over competitors who react only when changes in society become apparent in the form of changed consumer habits.

The environmental challenge has grown from local pollution to global threats and choices. The business challenge has likewise grown—from relatively simple technical fixes and additional costs to a corporatewide collection of threats, choices, and opportunities that are of central importance in separating tomorrow's winners from tomorrow's losers. Corporate leaders must take into account when designing strategic plans of business and deciding the priorities of their own work.

Sustainable development is also about redefining the rules of the economic game in order to move from a situation of wasteful consumption and pollution to one of conversation, and from one of privilege and protectionism to one of fair and equitable chances open to all. Business leaders will want to participate in devising the rules of the new game, striving to make them simple, practical, and efficient.

No one can reasonably doubt that fundamental change is needed. This fact offers us two basic options: we can resist as long as possible, or we can join those shaping the future. *Changing Course* is an invitation and a challenge to business leaders to choose the more promising and more rewarding option of participation.

> Schmidheiny, S., *Changing Course; A Global Business Perspective on Development and the Environment*, MIT Press, Cambridge, 1992, p. 1-13.

Notes and References

1 UN Environmental Programme, *Environmental Data Report 1989/90*, Oxford: Basil Blackwell Ltd., 1989.

2 World Bank, *World Development Report 1990*, Washington DC, 1990.

3 Havelock Brewster, "Third World's Prospects in the World Economy in the 1990s", *Third World Economics* July 16-31, 1991.

4 Maurice Strong, Speech to UNEP-UK Committee, London, April 1991.

5 International Chambers of Commerce (ICC), *WICEM II; Conference Report and Background Papers*, Paris 1991.

6 Michael Porter, "Green Competitiveness", *New York Times*, June 5, 1991.

7 World Commission on Environment and Development (WCED), *Our Common Future*, Oxford University Press, 1987.

8 Ibid.

9 International Union for Conservation of Nature and Natural Resources et al., *World Conservation Strategy*, Gland, Switzerland, 1980.

10 "Economic Declaration", Summit of the Arch, Paris, July 16, 1989.

11 ICC, *WICEM II*.

12 Keidanren, "Keidanren Global Environmental Charter", Tokyo, April 21, 1991.

13 Malaysian Environmental Quality Council, "Malaysia's Corporate Environmental Policy", 1991.

14 Confederation of Engineering Industry, "Environment Code for Industry", New Delhi, 1991.

15 Chemical Manufacturers Association, "Responsible Care Brochure", Washington DC, 1991.

16 Ibid.

17 Robert Goodland et al. (eds.), *Environmentally Sustainable Economic Development: Building on Brundtland*, Paris, UNESCO, 1991.

18 WCED, *Our Common Future*.

19 Donella Meadows et al., *The Limits to Growth*, New York, Universe Books, 1972.

20 Organisation for Economic Co-operation and Development, *The State of the Environment*, Paris, 1991.

CHAPTER 33

The Environment; Towards a Sustainable Future

DUTCH COMMITTEE FOR LONG-TERM ENVIRONMENTAL POLICY

Our approach and our basic assumptions

Thinking about the long-term environmental future is complicated by the fact that the long-term can be interpreted in several ways. For metereologists it is very difficult and often impossible to give a reliable prediction of tomorrow's weather. Tomorrow is already too long a term for prediction. For them it makes no sense to talk about a period of 40 to 50 years. On the other hand we know that dynamics in ecosystems and the development of evolutionary processes may take centuries and sometimes even more than hundreds of centuries. Then a period of 40 to 50 years is too short a term to conclude something relevant. Paleontologists use thousands of years as a frame of reference for their description of the evolutionary process. A period of 40 to 50 years is no more than a meaningless episode in their perception of time. We as committee have to avoid the Scylla of the metereologist and the Charybdis of the paleontologist.

Our approach

From the offset we have used a well-defined scheme of thinking to guide us in committing research activities. As stated, this scheme is based on some assumptions.

1 The first is that in our present western society the first signs of a sustainable future should be recognizable: hence it is worthwhile to look for *signs of hope*.

2 The second assumption is that in order to promote a sustainable future, it is necessary that some basic *transformations* take place. For these processes we have used the following terms: "socialization", "actor-ization" and "dialogue-ization" concerning environmental problems.

3 The third is that the transformations mentioned take place within the different levels of our daily life (local, regional, continental, global). Due to internationalization of our existence we emphasize the importance of the process of *internationalization* of environmental policy.

4 The fourth assumption is that a sustainable future implies radical changes in the *basic conditions and institutions* of society: biosphere, population, economy, education, (wo)men's emancipation, political order, science, technology and physical planning.

We will now briefly comment on these four assumptions of our study.

Signs of hope

CLTM is convinced that mankind is entering a new stage in its development, in which a reformulation takes place of the relation between society and environment. There are signs of hope that can be seen as the first indications of a new age. Governments at different levels, industry, agriculture, consumer organizations, nongovernmental organizations, schools and universities, mass media, artists, youth organizations, etc., are mobilizing for a sustainable future. We can see shifts in the environmental consciousness of people, a growing awareness of the importance of environmental problems, an expanding offering and willingness to act for the environment, the growth of environmentally sound behaviour, attention paid to environmental problems in the mass media, the development of ecologically sound products, an agriculture based on ecological principles, the use of sustainable materials, the exclusion of environmentally hazardous substances, the expansion of environmental laws and regulations, the Rio conference, the university courses on environmental problems, and much more. What is going on is a process of a social mobilization against the deterioration of the environment and a restructuring of society in the direction of an ecological society (Nelissen, 1990). Without doubt, the number and diversity of signs of despair nowadays are obviously larger than the number and diversity of signs of hope. The recognizable signs of hope do not mean that we have already reached the objective of a sustainable future. Nevertheless, the social mobilization to stop the deterioration of the environment is on its way, at least in a number of countries. This process must not be seen as a small success that has little to do with the threats the earth is facing, but as a first sign of a starting serious attack of these threats.

Transformations

The elaboration of a sustainable future asks for important transformations in society.

1 In the first place the "*socialization*" of a policy which promotes sustainability is necessary. In this respect the concept of *environmental marketing* could help to reduce the social dilemmas which prevent people to behave in line with their pro-environmental attitudes.

2 Secondly we need an "*actor-ization*" of environmental policy objectives, resulting in a shared responsibility for the environment by all actors in society. This can be achieved by building and maintaining *policy networks* in which all actors in the environmental field positively participate.

3 Thirdly the *"dialogue-ization"* of environmental policy objectives between science and policy should be developed. This can be promoted by a process of *multilogue*, which means the simultaneous exchange of experiences by all actors in the environmental area in pursuit of a greater understanding of environmental problems and the ways to solve them.

The international context

CLTM is convinced that achieving the objective of a sustainable future depends highly on a concerted international effort. Environmental problems have become more and more international. We are witnessing a globalization of environmental problems and policies to attack them. EC, UN, OECD and other international institutions are becoming important actors in developing and implementing environmental policy. The Rio conference has emphasized again the strong relationship between environment and development. Agenda 21 can be seen as an important guide for future activities concerning environmental problems and development. However, environmental policy is still young and lacks the maturity needed to solve international, often global, problems, which are usually of a long-term nature. Adequate policies, instruments and institutions must be developed. The fact that the Rio declaration in its principles has stressed the interdependence and indivisibility of environment, development and peace is a complicating factor of considerable importance.

Radical changes in the basic institutions of society

To reach a sustainable future radical changes in the basic institutions of society are necessary. But what are these basic institutions? In which direction do the radical changes have to take place? What about the diversity between societies, especially the differences between developed and developing countries? We are convinced that the diversity between societies is so large that it is perhaps irresponsible to talk about "the" society. Nevertheless, we do talk about society in general, albeit fully aware of its diversity. Differentiation between societies will affect the grade in which these radical changes are needed, but not the principle. In one society there may be a tendency towards radical changes; they may even already occur. In other societies the radical changes in the institutions still have to start. It is clear to our committee that those radical changes have to be promoted which work in favour of a sustainable future. Although the concept "sustainable future" is multi-interpretable, we already know that anti-pollution, reduction of every consumption and closing of ecological cycles are important elements of a strategy aimed at reaching a sustainable future.

Which basic institutions need radical changes, and of what nature should these changes be? We have selected and analysed the following:
a Sustainable living requires the integrity of the *biosphere*: the basic ecological processes and biogeochemical cycles, which maintain the functions of the envi-

ronment for man and which are presently severely disrupted by man's activities, are of major importance.

b Population size is an important factor for the future of our planet. In developing countries population growth has a different quality with respect to the environment as compared with more developed countries, because environmental pressure is decisive and an individual from more developed countries is a comparatively larger burden on the environment than one from less developed countries.

c International economic cooperation to obtain sustainability is crucial for the environment. New trade relationships between countries, especially between developed and developing countries, are needed.

d Environment-conscious *education* only works when it is carried out by parents and adults who believe in what they say, and who show a credible attitude themselves.

e From the point of view of *(wo)men's emancipation*, a redistribution of "caring" tasks (maintenance, self-provision, care providing) between men and women is necessary.

f We are in need of a *political order* that integrates democracy and ecology. Especially international political institutions have to play an important role in developing a justice system in which ecology has its place.

g Science which focuses on sustainability should try to gain an intermediate position between "alternative" and "regular" science. There is a need for constructivistic realism.

h Together with culture and structure, *technology*, formerly seen as the primary originator of environmental problems, must find its new challenge in providing for a sustainable future.

i All these basic institutions together have their impact on the *spatial picture* of society. We are in need of physical planning inspired by sustainability. Earlier in this book, a creative vision of the Netherlands was presented.

Are we moving towards a sustainable future?

Although many studies have been published on the essence of sustainable development, a well-defined and operational definition does not exist. Nevertheless, the Brundtland definition is considered useful: sustainable development is a development that ensures the needs of the present generation without compromising the ability of future generations to meet their own needs. Essentially, sustainable development implies limits, not absolute limits but limitations imposed by the present state of technology and social organization on environmental resources and by the ability of the biosphere to absorb the effects of human activities. As the Brundtland report states, sustainable development is not a fixed state of harmony, but rather a process of change in which the ex-

ploitation of resources, the direction of investments, the orientation of technological development, and institutional changes are made consistent with future
as well as present needs.

Sustainable future

We are using the concept of a sustainable future. In essence there is no great
difference between this concept and the Brundtland report's concept of sustainable development. Maybe the sustainable future concept has the connota-
tion of being a little bit static. According to the committee one should see the
future in terms of process and development. The future is always disappearing
the moment we reach it; it produces its own new future. In this sense a sustainable future is a basic principle; a guiding idea concerning a desirable future; a
notion to describe the will of existing people to take care for the future of new
generations. The concept of sustainable future has ethic aspects; it emphasizes
the need for responsibility of the present generation for the world of future
generations.

Are there signs of hope for a sustainable future?

Are we leaving the polluting society and on the threshold of a new, ecological
society? Some years ago the answer to these questions would certainly have
been negative. The ongoing environmental degradation, the continuing exploitation of natural resources, and the expansion of environment-unfriendly attitudes and behaviour were "proof" of a development that was in essence unsustainable. Have changes occurred in the last few years? Do we see the first
symptoms of changing attitudes and behaviour in connection with the environment? Are environmental policies developed? Are we on our way to deal with
environmental problems in a more effective way than we did before? In our
first study, some years ago, we were fully convinced that the environmental
situation was dramatic and that radical changes were needed to save the world.
We then already recognized the first element of radical changes in the way in
which society deals with the environmental problems. Now, the basis of all
studies developed within the context of our second period, we as CLTM have
the opinion that radical changes are really taking place in every sector of society. These radical changes are—for a great part—in favour of a sustainable
future. One can no longer hold the idea that our society is still the same, when
we see such a large number of changes taking place in the basic institutions
oriented towards a sustainable future. Of course, we must not neglect the overwhelming number of ways in which our planet is affected by depletion of the
ozone layer, the greenhouse effect, acidification, eutrophication, reduction of
biodiversity, waste disposal, etc. When we look at the actual state of our planet,
it seems even dangerous to say that we are on the way to a sustainable future.
Nevertheless, we must not be blind to all the signs of hope that already exist.

When we look at them in an isolated way, we could consider these signs as unimportant and misleading, but when we look at them as being interrelated, as signs of the growing responsibility of humanity for its own destiny, then we have to say that we are entering a new stage in human history; a stage characterized by more human concern for its own ecological roots and dependencies.

Problems of first and second evidence

One of the great contemporary French philosophers Edgar Morin uses the concepts "problems of first evidence" and "problems of second evidence" in his latest book *Terre-Patrie* (1993). The problems of first evidence are the problems with which the actual world society is confronted: the confusion of the world economic order, the enormous growth of the world population, the expansion of the world ecological crisis and the growing shift between welfare development in the North and in the South. Morin emphasizes the growth of antagonisms in our world: the antagonism between nations, religion, political systems, cultures, welfare, etc. We are fully uncertain about the future. He uses the phrase "universal crisis of the future"; that is the problem of second evidence. We have to become aware that we are now facing problems of second evidence. Morin states: "Ainsi on ne saurait détacher un problème numéro un qui subordonnerait tous les autres; il n'y a pas un seul problème vital, mais plusieurs problèmes vitaux, et c'est cette inter-solitarité complexe des problèmes, antagonismes, crises, processus incontrôlés, crise générale de la planète, qui constitue le problème vital numéro un." (Morin, 1993, p. 110).

Complete awareness

As we have already said, the world is becoming aware of this general crisis of our planet, we are looking at world problems as being interrelated, and what is even more important, we are looking at problem solutions as being interrelated. The world is in search of a sustainable future, a world in which economic, demographic, ecological and developmental problems and solutions are interpreted in an interrelated way. The above-mentioned Edgar Morin speaks of "la grande confluence": the complementary awareness. We are becoming aware of the unity of the earth, of the unity and diversity of the biosphere, of the unity and diversity of mankind, of the anthropo-bio-physical status of our existence, of our simple "Dasein" without knowing why, of the planetary era, of our common future on this planet. These interrelated and complementary awarenesses produce the basis for a new hope related to our planet and our future. They are the first steps in a direction of a global consciousness concerning the destiny of mankind.

The need for further action

Many people could argue that we are too positive, even too optimistic concerning the state of the world and the presented solutions to the problems. If we face the impression that we only have to wait for a sustainable future, we have to correct ourselves. We have to work hard to reach a sustainable future. All existing policies, all the work done by government, industry, nongovernmental organizations and individuals, are not enough to reach sustainability; they can only be considered as the first important steps. We have to accelerate the transformation process; to elaborate the policy strategies and instruments in favour of a sustainable future. Our achievements dare not lead to misleading satisfaction, but form a stimulating atmosphere for further action. The social mobilization for the environment during the last years will only be successful if we continue our environmental actions. If not, all will have been in vain.

Collective enterprise

It is important to re-emphasize that actions from all kinds of actors are needed. It is sometimes said that government has to play a crucial role in developing environmental policy. Of course, the role of government at the different levels is necessary and important, but governmental actions will only be successful if they are accepted and applied by private institutions and by the citizens. We have to see environmental policy from a multi-actor perspective. If there is one thing that has become clear during the last years, it is that only a combined action of all the actors in society will lead to success. We have to see environmental policy as a collective enterprise.

International cooperation

A collective enterprise is not only necessary at a local, regional and national level, but also internationally. It has often been said: environmental problems are not stopped by national boundaries; in essence they are global. This implies that environmental measures have to be taken at the international level. Here also a multi-actor perspective is needed. The way in which societies deal with problems is actually on a network basis. The concept of policy network has been introduced to describe the new way in which societies are tackling problems. One of the implications of this concept is that existing policy networks have to focus on environmental problems, and to refresh the way of tackling environmental problems. International governmental institutions have to cooperate with nongovernmental ones. Often one has the impression that actors are working separately and even antagonistically, but a sustainable future demands network cooperation between all actors in society. This will need new strategies and procedures. Rio 1992 has emphasized the importance of international cooperation, not only between national governments, but also, and particularly, between governments and societal groups.

Shared responsibility

Working on a sustainable future is not only a question of collective enterprise; it has also a normative and ethic component in the sense that there exists a shared responsibility. As formulated in the *Fifth European Action Programme for the Environment: Towards a sustainable future*, all the parties involved are responsible for the future of mankind and the planet. Hence the responsibility does not rest only with government at the different levels, but with all actors: government, industry, agriculture, transport, consumers, citizens. Shared responsibility should not mean that everyone is waiting for the other to take responsibility; it means that everybody who has the opportunity to contribute to a sustainable future, must not hesitate or fail to do so.

Towards an evolving green strategy

All studies published so far have not given a solution for the problem of the so-called second evidence. We can see the different suggestions for improving the planet as elements of a "jigsaw"; albeit with the complication that "the ultimate picture" of the jigsaw is unknown. We are making suggestions for another world, without knowing what the better world will be, or without being absolutely sure that the suggestion will positively contribute to a sustainable future. This means that it is difficult to design a "greenprint" for the future. Processes have been set in motion and people are influencing them continuously. There is no end state of mankind, there is no programmed scheme inherent in mankind. History, and therefore the future, depends on millions, if not billions of actions of people all over the world, and each action has a relatively unpredictable influence on the future. The future is not a fully planned end product of an omnipotent entity; it is the result of actions of a multitude of people, each person is trying to realize his own dream. But as we know, the total product of all those individual dreams can result in a collective nightmare.

Government as an important actor

This statement has severe consequences for the strategy that policy-makers have to develop to reach a desirable state of the environment. We must be aware that some actors are more influential than others, and that for example government can play a strategic role in producing a society based on ecological principles. Government is not only *an* actor, it is a very important one in influencing the direction of history. Other races are also important. We use the term "collective enterprise" to underline the importance of collective action.

Evolving green strategies

What is the desirable strategy for the collective enterprise, the shared responsibility, in favour of the environment? We would like to call this strategy "*the evolving green strategy"*. What do we mean by this concept and what is it all about? We have entered a new stage in the history of mankind. Societies are aware of the environmental problems and have at least developed schemes and programmes to handle them. Some are in the stage of implementation of environmental measures. In a lot of countries and also at the international level, the environmental issue ranks high on the political agenda, in spite of the well-known barriers that had to be overcome. There is no government, industry, agriculture or traffic institution that is fully unaware of environmental implications of its own actions or the actions of other parties. Governments at all levels, industries all over the world, agriculture in different climate zones, airline companies and public transport organizations are all facing environmental problems and developing ways in which pollution can be reduced or stopped. The results are not yet fully satisfactory, but can be seen as signs of hope for a better future regarding the environment. In this sense it is important to state that we need an evolving green strategy. Green policy must not be seen as an incidental initiative to reduce a local problem. It must be interpreted as a continuous way of looking at society from an ecological point of view and as a continuous package of measures in favour of the environment. Green strategy must be a cornerstone in the policy of all actors in society. This means that the first successes in the direction of an ecological society have to be followed by new ones. We have to be aware that an ecological society is necessary and vital for the future of mankind, and in fact of our planet as such.

Dutch Committee for Long-Term Environmental Policy, *The Environment; Towards a Sustainable Future*, Kluwer Academic Publishers, Dordrecht, 1994, p. 39-47.

Conclusions and Perspective

In the previous sections an outline was given of the developments in the field of environmental studies on the basis of fragments of classical texts. All considered, its history spans one to two centuries, in which studies by Malthus and Darwin are taken as the starting points. The history, although the order of texts might indicate otherwise, has not really developed in a systematic fashion. It has been one in which an evolution has been discovered only in hindsight. In science, numerous studies are published, but many are never referred to and disappear overnight. Some 'survive', which can be likened to a Darwinian process of 'survival of the fittest'. They remain a source of inspiration to modern scientists. They are a part of the 'collective memory' of science.

Now that we have finished our tour, we will draw some conclusions and provide a perspective on the discipline of environmental studies. In the first place, some remarks will be made on the development, the actual state, and the future of the discipline of environmental studies. Secondly, some comments will be given on the results of environmental studies. Thirdly, we want to sketch the future of the discipline within the context of the ecologization of society.

1 Development, current state and future of environmental studies

The codification of knowledge in the field of the environment has only recently begun to take shape. One could say it has only commenced in this decade. At present, we are witnessing the convergence of different points of view as they start to take on a transparent unity. Environmental studies has matured. It has developed its own body of knowledge and frame of reference. Methods and practices have been developed and have their own history.

Head and tail
It can be stated that the field of environmental studies has a 'Head and Tail'. Therefore, introductory books in this field always have a very recognizable structure. They systematically focus on the evolution of life on earth, the different environments in which the life of plants, animals and humans are found, the relation between biotic and abiotic elements in ecosystems, the structure

and functioning of ecosystems, the dynamics of ecosystems, the damage done to ecosystems by human activity, the great diversity in environmental problems, the attempts to tackle environmental problems by means of focused policy, the role of government, business, and environmentalists in this field, etc. Issues of population growth, reduction of natural resources, pollution of oceans, soil, surface water, drinking water and air, the influence of pesticides, the depletion of the ozone layer, the greenhouse effect, the acidification of soil and ground water, the extinction of species, waste dispersal, the spread of environmentally dangerous substances, among others, are the subjects that pass in review in the textbooks on environmental studies. Sometimes, special attention is given to the environment in particular parts in the world: in Europe, the United States, the Antarctic, or the Mediterranean Sea. In short, the body of knowledge within the field of environmental studies has been largely codified and has a recognizable structure.

Discipline, multidiscipline, interdiscipline, applied science?

People are less than unanimous about the question whether or not environmental studies should be seen as a discipline, a multidiscipline, an interdiscipline, or a skill. To a certain extent, this discussion is merely an academic one, and not really edifying. It is an issue of whether or not environmental studies has its own material and formal object of study. Should environmental studies be seen as a selection of insights that focus on the environment as they exist in disciplines and, therefore, be seen as a multidiscipline? Another view is that environmental studies is really more than the sum of insights of different disciplines and that the field of studies should be characterized by the mutual influence of different scientific insights and should subsequently have an interdisciplinary character. There is no consensus. In our view, this has to do with the developmental phase of the field. It can be said that many disciplines have been created out of a combination of previously existing sciences and, at one point or another, had a multi- and interdisciplinary character. As a cumulative result of the endeavors in the field, an independent domain of scientific knowledge has come into existence and, therefore, one can speak of a discipline. In our view, a developmental process is involved. There is good reason for claiming that environmental studies is on the way to becoming an independent discipline, if that is not already the case.

Another issue is whether environmental studies should be seen as a pure or an applied science. An applied science implies that the scientific insights gained are applied to practical issues that require a solution. We try not to be too dogmatic about this issue. At present, however, science for the sake of science is more fiction than reality. Much of the scientific knowledge currently gained is directly related to practical applications. Given the social function of science, there is not much room for science to exist solely for the sake of discovery. Moreover, it is very difficult to precisely demarcate the line between pure science and applied science. It is probably best to leave this dicussion in the realm of informal debate and reserve our energies for the study of the environment itself.

Environmental discipline and environmental specialities

Related to and yet distinct from the aforementioned discussion is the issue of environmental studies in the future. How will environmental studies develop? There are at least two options: a further development into a single coordinated environmental discipline, or a development into different environmental disciplines such as environmental economics, environmental biology, environmental chemistry, etc. In practice, developments in both directions are taking place. In different countries, however, there are different configurations. The Netherlands in recent years, for example, is leaning towards developing the field of environmental studies through its environmental disciplines. Both paths have their pros and cons. It remains to be seen which choice is wisest. Science develops relatively autonomously and, despite preferences one might have, it will keep on developing. It would undoubtedly be best if environmental studies developed in both directions.

Nature, social, and policy sciences

Environmental studies has historically been dominated by professionals from the natural sciences. That is not surprising, as environmental issues mainly concern important natural scientific aspects. The field of environmental studies includes insights from biology, physics, and chemistry. Yet, the hegemony of the natural sciences has been broken in recent years. Environmental issues are no longer strictly natural science problems. They are caused by individuals and societies, which means that knowledge of social, behavioral and policy aspects have gained in significance. It is striking to see how many social-scientific studies on environmental issues have been published in recent years. Old contrasts, in particular between natural and social sciences, have lost their sharp edges. In the academic world, more understanding has grown towards different approaches and there is a greater willingness to apply them side-by-side. Nowadays, it is no longer a matter of or-or, but one of and-and.

Specialization

Although environmental studies has been described in overall terms, it should be realized that by developing the field of environmental studies, an ongoing specialization has come to life. There are people who are specialized in special areas in the field. They know about greenhouse gases or have great expertise in the field of chemical pesticides. They are specialists in the field of tropical rain-forests and experts on water pollution, oceanic pollution, ozone issues, and the meltdown of ice caps. In short, the whole field is fragmented into expertise do-mains, which have been expanding into so many different areas that it is hard to even provide an overview. This ongoing specialization has advantages and disadvantages. The advantages lie in the specific, detailed knowledge that can be gained. But there are also some drawbacks. Specialists have a tendency to become isolated, which makes it harder for them to communicate with other professionals in the field.

2   Results of environmental studies

The number of people involved in the field of environmental studies has in-creased considerably. Nowadays, there are a great many scientists who study natural and environmental problems. There are scientists with roles in educa-tion. Professors at universities give courses and provide for education in the field of the environment. The number of people holding research positions has also increased considerably. In many countries, environmental research in-stitutes have been established. They study sections and aspects of environmen-tal problems. There are also many government officials at the international, national, regional, and local level with tasks in the field of issuing environmen-tal permits and environmental law and rule enforcement. Non-governmental organizations also have environmental experts. Until recently, there were hardly any qualified environmental experts to fill the job openings available. That situation has changed due to an influx of students and researchers into the field who are now becoming available for the labour market. There is still some debate about the fine tuning between education and professional prac-tice. Some say that the studies are too academic and that they have little con-sideration for the real-life issues which the graduates have to deal with. None-theless, it seems to put more stress on basic insights in academic studies than on practical tricks. In practice, one learns fast; for a basic orientation, one needs to invest years of study.

Environmentalist movement in retreat?

Some of the intellectuals in the field of environmental studies are active in the environmentalist movement. This movement possesses a great deal of internal diversity: there are in truth many movements. We can briefly characterize these movements as follows. There is a group of 'Reformers', people within the environmentalist movement who have chosen the path of deliberation and negotiation with the establishment. Secondly, there is the group of 'Revolutionaries', a group who hardly believes in deliberation and negotiation; they use focused action as a way of making sure that the interests of the environment come first at all times. The third movement is that of the 'Alternatives'. This group chooses to live an alternative way of life. They shield their world from as many influences of consumer society as they possibly can. They live their life with few material means and close to nature. In some countries, the ecologists have come to embody political movements. They have established political parties with such names as Ecolo, Grünen, the Green, Amis de Terre, Agalev, etc. They participate in legislative elections at the local, regional, national, and European level. The environmental movement is currently experiencing a crisis of identity, in particular the green parties. The number of listed members is stagnating and in some cases decreasing. Popular support for the green parties is fluctuating. In a number of countries, including Germany, France, and Belgium, the number of seats in representational organs has decreased. The green parties are adrift. What should their new strategy for the future be? Back to basics and with as few political concessions as possible? Or try to become a more broadly-oriented movement and look for ways of joining the current social-economic debate. There is some reason to believe that the future of the green movement and of the green parties will be less favourable. After all, part of their heritage has already been anchored in society and has been embraced by other political parties. The question raised here is if declining interest in the environmentalist movement can be seen as an advantage. It could indicate that green ideas have become commonplace, not that they have been pushed to the background. The ultimate goal of a one-issue movement should be to become superfluous because its goals have been accomplished.

More knowledge, fewer problems?

This raises the question whether, given the increase of knowledge, the increasing number of experts in the field of environmental studies, and the ongoing mobilization for the environment, the environment has actually profited. In part it is a case of the law of the conservation of misery. The increasing number of experts has not led to a substantial decrease in the number of environmental problems. Most important environmental problems still exist and some of them, e.g., the greenhouse effect, have only gotten worse. On the other hand, one could say that as a result of the increasing attention given to environmental problems and the many experts focusing on environmental issues, an excessive growth of the problem was avoided. If increasing the number of experts resulted in a decrease of problems, one would think that the number of diseases would have decreased, given the ever-increasing number of medical professionals, and that economic crisis would cease to exist, given the number of economists.

Do we have sufficient knowledge?

There is also the question of whether we have enough knowledge concerning environmental issues and if our insights are correct. This concern comes to mind when we discuss the problem of global warming. Is it a result of the increased emission of greenhouse gases, such as CO_2, or are other factors important? Some scientists say that there is not enough conclusive evidence to support the claim that the rising temperatures are a direct result of greenhouse gases. They say it might be due to natural fluctuations and that there is no reason for panic. Many scientists are of the opinion that the hot summers of recent years are not coincidental and are evidence of climatic change. There is uncertainty about other issues as well. A number of these can be seen as decision-making issues. Daily, the question arises whether or not a given product is environmentally friendly. Judging this issue is dependent on the factors that are included in the judgement and how these factors are weighed. The life cycle analyses that are often used are frequently not conclusive, and the result is that government, business, and citizens are left with uncertainty.

3   Ecologization of society

Because of the ever-increasing attention given to environmental problems and the tendency to include the environment factor in our everyday activities, it could be said that an ecologization of society has taken place. Increasingly, people are taking the environment into consideration. The environment is becoming a factor that cannot be ignored. The economy is becoming increasingly permeated with environmental thinking. Transport and traffic are being viewed in terms of their environmental implications, as are tourism, recreation, agriculture, and industry. Tax reforms are being defined from an environmental perspective. People are being prompted by environmental considerations. In short, one could speak of an ecologization of society. The environment is, in this sense, no longer a sector of social life, but a facet that needs to be taken into consideration in many different social domains.

La grande confluence

In his book *Terre—Patrie* (1993), the French philosopher Edgar Morin uses the concepts of 'problems of first evidence' and 'problems of second evidence.' The problems of first evidence are the problems with which the global society is confronted: the confusion of the world economic order, the enormous growth of the world population, the expansion of the world ecological crisis, and the growing shift between welfare development in the North and in the South. Morin emphasizes the growth of antagonisms in our world: the antagonism between nations, religions, political systems, cultures, welfare groups, etc. We are filled with uncertainty about the future. He uses the phrase 'universal crisis of the future'; this refers to the problem of second evidence. We have to become aware that we are now facing problems of second evidence. There is no longer one vital problem, but several vital problems. The interwoven character of all the vital problems is problem number one. The world is becoming aware of this general crisis of our planet and is in search of a sustainable future, one in which demographic, social, cultural, economic, ecological, and developmental problems will be interpreted in an interrelated fashion. Morin speaks of 'la grande confluence', the complementary awarenesses. We are becoming aware of the unity of the earth, of the unity and diversity of the biosphere, of the unity and diversity of mankind, of the anthropobiophysical status of our existence, of our simple 'Dasein' without knowing why, of the planetary era, and of our common future on this planet. These interrelated and complementary awarenesses produce the basis for a new hope for our planet and future, a hope for an ecologization of society.

Breaking through isolation

Environmental studies has now moved out of its isolated position. Until recently, environmental issues were viewed as a more or less independent domain, with little relation to other domains. Nowadays, the environment is consciously related to other actual issues. Environmental problems are increasingly being linked to issues of war and peace, development and underdevelopment, and safety and crime. The efforts of the Brundtland Committee have led environmental problems to be linked to North-South issues. These issues are not separate ones, they are interrelated in essential ways. The causes of and solutions to environmental problems cannot be disconnected from issues of over- and underdevelopment. Desertification is closely linked to the exploitation of the natural environment in areas where there is underdevelopment. The depletion of natural resources has much to do with the exploitation and use of these sources by underdeveloped countries. Issues of war and peace are also related to environmental problems. Conflicts between states or peoples often have their roots in environmental issues. In many cases, countries are dependent on the waterways of other countries. People find themselves battling over claims to lands with important natural resources. Internal security is closely linked to the environment. Criminal organizations often succeed in dodging the rules enforced in any given legal system. The consequence might be that forms of environmental crime arise that the national government cannot adequately control.

Crisis science?

Environmental studies could be viewed as a crisis science, a science that was created as a result of the environmental crisis. The question is whether the science will disappear with the disappearance of the environmental problems. Generally, one could say that there is little need for keeping a science alive that has no social function. This is not, however, the case with this field of studies. One could even say that the crisis has only recently become apparent and much needs to be done to find a solution. In that sense, it is expected that environmental studies still has a long future ahead of it, though the science will have to adapt to changing conditions. Environmental studies as it existed in the sixties was much different from the field today. Science is not a static entity. It develops in dialogue with its object of study, as well as with the developments in the material and formal object of study. From that perspective, the end of the history of environmental studies is far from being reached; in a sense, it has only just begun. Therefore, the *'Classics in Environmental Studies'* should not be seen as the swansong of environmental studies, but an inaugural note. We hope that the study of these classical texts may contribute to the awareness that we stand on each other's shoulders in the world of science, and that there are clearly discernible lines in history, even if there is no rigid agenda and there is no opportunity for exact prediction. Therefore, we must be guided by the German saying *'Wir lassen uns überraschen'* and continue to revel in our surprise and amazement.

References

Albert, M., 1993. *Capitalisme contre Capitalisme*, Paris.

Beck, U., 1992. *Risk Society; Towards a New Modernity*, London.

CLTM, 1990. *Het Milieu; Denkbeelden voor de 21ste Eeuw*, Zeist.

CLTM, 1994. *The Environment; Towards a Sustainable Future*, Dordrecht.

Morin, E., 1993. *Terre—Patrie*, Paris.

Naisbitt, J. and P. Aburdene, 1990. *Megatrends 2000; Ten New Directions for the 1990s*, New York.

Ponting, C., 1991. *A Green History of the World*, London.

Toffler, A., 1990. *The Power Shift*, New York.

About the Authors

Nico Nelissen graduated in sociology at Tilburg University. He is a professor of environmental studies at Tilburg University (the Netherlands), as well as a professor of public administration at the Catholic University of Nijmegen (the Netherlands). His doctoral dissertation was entitled 'Social ecology' (1970). He has written several books on environmental consciousness and environmentally sound behaviour and on environmental movements and environmental policy. He was a member of the National Council for Environmental Problems and was the president of the Dutch Committee for Long-Term Environmental Policy. His scientific work focusses on the quality of the natural and the built environment. In this context, he has also written several books on urban renewal, monument care, urban planning, and architecture.

Leon Klinkers (M.Sc., M.A.) holds degrees in biology, political science, and business administration from the Catholic University of Nijmegen. He worked as an environmental management consultant for Det Norkse Veritas (DNV) and advised companies on how to set up and implement environmental management systems and how to integrate E.M.S. in other management systems. He has a position at Tilburg University as an academic tutor in the European Association of Environmental Management Education Program (E.A.E.M.E.). His Ph.D. dissertation, forthcoming, is focused on: The role of internal communication as a critical success factor in implementing an environmental management system.

Jan van der Straaten studied economics at the Erasmus University in Rotterdam and defended his Ph.D. dissertation, entitled *Acid Rain, Economic Theory and Dutch Policies*, at Tilburg University. He is currently affiliated with the European Centre for Nature Conservation and the Department of Leisure Studies at Tilburg University. He is doing research on the relationship between economic theories and environmental policies as well as European rural development, paying special attention to the relationship between tourism, nature, and agriculture. He is a member of the editorial board of *Milieu, Business Strategy and the Environment*, and *Environmental Politics*; in addition he is Vice-President of the *European Society for Ecological Economics*.